NG
'ifornia,

10-25-90

HANDBOOK OF POLYMER COATINGS FOR ELECTRONICS

MATERIALS SCIENCE AND PROCESS TECHNOLOGY SERIES

Editors

Rointan F. Bunshah, University of California, Los Angeles *(Materials Science and Process Technology)*

Gary E. McGuire, Microelectronics Center of North Carolina *(Electronic Materials and Process Technology)*

DEPOSITION TECHNOLOGIES FOR FILMS AND COATINGS: by *Rointan F. Bunshah et al*

CHEMICAL VAPOR DEPOSITION IN MICROELECTRONICS: by *Arthur Sherman*

SEMICONDUCTOR MATERIALS AND PROCESS TECHNOLOGY HANDBOOK: edited by *Gary E. McGuire*

SOL-GEL TECHNOLOGY FOR THIN FILMS, FIBERS, PREFORMS, ELECTRONICS AND SPECIALTY SHAPES: edited by *Lisa A. Klein*

HYBRID MICROCIRCUIT TECHNOLOGY HANDBOOK: by *James J. Licari* and *Leonard R. Enlow*

HANDBOOK OF THIN FILM DEPOSITION PROCESSES AND TECHNIQUES: edited by *Klaus K. Schuegraf*

IONIZED-CLUSTER BEAM DEPOSITION AND EPITAXY: by *Toshinori Takagi*

DIFFUSION PHENOMENA IN THIN FILMS AND MICROELECTRONIC MATERIALS: edited by *Devendra Gupta* and *Paul S. Ho*

SHOCK WAVES FOR INDUSTRIAL APPLICATIONS: edited by *Lawrence E. Murr*

HANDBOOK OF CONTAMINATION CONTROL IN MICROELECTRONICS: edited by *Donald L. Tolliver*

HANDBOOK OF ION BEAM PROCESSING TECHNOLOGY: edited by *Jerome J. Cuomo, Stephen M. Rossnagel,* and *Harold R. Kaufman*

FRICTION AND WEAR TRANSITIONS OF MATERIALS: by *Peter J. Blau*

CHARACTERIZATION OF SEMICONDUCTOR MATERIALS—Volume 1: edited by *Gary E. McGuire*

SPECIAL MELTING AND PROCESSING TECHNOLOGIES: edited by *G.K. Bhat*

HANDBOOK OF PLASMA PROCESSING TECHNOLOGY: edited by *Stephen M. Rossnagel, Jerome J. Cuomo,* and *William D. Westwood*

FIBER REINFORCED CERAMIC COMPOSITES: edited by *K.S. Mazdiyasni*

HANDBOOK OF SEMICONDUCTOR SILICON TECHNOLOGY: edited by *William C. O'Mara, Robert B. Herring,* and *Lee P. Hunt*

HANDBOOK OF POLYMER COATINGS FOR ELECTRONICS: by *James J. Licari* and *Laura A. Hughes*

Related Titles

ADHESIVES TECHNOLOGY HANDBOOK: by *Arthur H. Landrock*

HANDBOOK OF THERMOSET PLASTICS: edited by *Sidney H. Goodman*

SURFACE PREPARATION TECHNIQUES FOR ADHESIVE BONDING: by *Raymond F. Wegman*

HANDBOOK OF POLYMER COATINGS FOR ELECTRONICS

Chemistry, Technology and Applications

Second Edition

by

James J. Licari

Microelectronic Circuits Division
Hughes Aircraft
Newport Beach, California

Laura A. Hughes

Autonetics Electronic Systems
Rockwell International
Anaheim, California

np

NOYES PUBLICATIONS
Park Ridge, New Jersey, U.S.A.

Library of Congress Catalog Card Number: 89-70994
ISBN 0-8155-1235-X
Printed in the United States

Published in the United States of America by
Noyes Publications
Mill Road, Park Ridge, New Jersey 07656

10 9 8 7 6 5 4 3 2 1

Library of Congress Cataloging-in-Publication Data

Licari, James J., 1930-
Handbook of polymer coatings for electronics : chemistry, technology, and applications / by James J. Licari, Laura A. Hughes, -- 2nd ed.
p. cm.
Includes bibliographical references.
ISBN 0-8155-1235-X :
1. Electronic packaging--Materials. 2. Polymers. I. Hughes, Laura. II. Title
TK7870.L486 1990
621.381'046--dc20 89-70994
CIP

Foreword

Electronic materials development and manufacturing has become a complex and varied science that encompasses many technologies. Electrical/electronic design of a function or system is only the start of a process that must be accomplished to achieve a successful end product. My experience with microelectronic research and manufacturing has provided an appreciation for reference sources with the details and accuracy given in this book.

Selecting a material for printed circuit boards, adhesive bonding of components in hybrid and surface mount circuits or other applications requires a careful balance of properties for successful products. The data on materials such as parylene, photoresists, polymer encapsulants, and many others expertly link the structural, electrical, physical and chemical properties that allow designers to choose the best materials.

The first edition of this book was published as a text of plastic coatings for electronics. However, the subject matter was much more inclusive than the title implied and comprised a broad range of data for electronic packaging and assembly technology. That edition was a constant and accurate source of information in the NASA microelectronics laboratory. This new edition expands and further broadens the scope and will be a valuable and practical reference to materials scientists, engineers, and chemists working in the field of electronic packaging.

Dr. James Licari has been a professional associate and personal friend for more than twenty years. His experience in his company as manager in the chemical materials and process laboratory and the microelectronic development department make him eminently qualified to blend the technologies of electronics and chemistry. The excellent data and professional presentation of the subject matter reflects Dr. Licari's and Dr. Hughes' comprehensive knowledge and experience.

This book will be a valuable and reliable source of information to the electronics, chemical, and materials science industries for current and future use.

NASA/Marshall Space Flight Center
Huntsville, Alabama
March 1990

Salvadore V. Caruso

Preface

For almost two decades and several printings, including a Japanese translation, *Plastic Coatings for Electronics* remained the primary book for this subject. However, since its first publication in 1970, many advancements and new applications of organic coatings have materialized, closely following the rapid advancements in microelectronics. Thus a revision and complete updating of the book were deemed necessary.

Much new material and data have been added: obsolete product designations have been deleted and current designations introduced. To achieve this, the material was coordinated with hundreds of manufacturers.

New advancements in dielectric coatings for high-density multilayer interconnect substrates for the new wave of wafer-scale integrated circuits are covered for the first time. Further, sections have been included to address particle-immobilizing coatings for use inside hermetically sealed hybrid microcircuits; "glob-top" adhesives for the selective protection of chip devices; polymer thick films used as conductors, dielectrics, and resistors; and coatings used as particle getters inside hybrid circuit packages.

While changes have been made to bring the book current with the latest electronic applications, fundamentals of the chemistry of coatings and basic material properties have been retained since these have not changed over the years. The discussion of some of the more popular polymer coatings (polyimides, silicones, and polyxylylenes) however has been enhanced to include photosensitive, solvent-soluble, and high purity variations.

This book should therefore benefit a wide cross-section of personnel working in the field of electronic circuits including design engineers, materials and process chemists, and quality assurance and management personnel.

NOTICE

To the best of the Publisher's knowledge the information contained in this publication is accurate; however, the Publisher assumes no responsibility nor liability for errors or any consequences arising from the use of the information contained herein. Final determination of the suitability of any information, procedure, or product for use contemplated by any user, and the manner of that use, is the sole responsibility of the user.

The book is intended for informational purposes only. The reader is warned that caution must always be exercised when dealing with chemicals, products, or procedures which might be considered hazardous. Expert advice should be obtained at all times when implementation is being considered.

Mention of trade names or commercial products does not constitute endorsement or recommendation for use by the Publisher.

Contents

1. **CHEMISTRY AND PROPERTIES OF COATINGS**—*Polyurethanes—Epoxies—Phenoxies—Silicones* 1
Polyurethanes 1
The Chemistry of Polyurethanes 2
Classifications and Properties 5
One-Component Blocked Polyurethanes (ASTM Type 3) 7
Two-Component Prepolymer: Polyol (ASTM Type 5) 8
Two-Component Catalyst-Cured Systems (ASTM Type 4) 10
One-Component Oil-Modified Systems: Uralkyds (ASTM Type 1) 12
One-Component Moisture-Cured Systems (ASTM Type 2) 13
One-Component Nonreactive (ASTM Type 6) 14
Epoxies 15
The Chemistry of Epoxies 15
Bisphenol A Resins 16
Novolac Resins 18
Other Epoxy Types 19
Curing Agents and the Polymerization Process 21
Catalytic Polymerization 21
Catalytic Cure 22
Polymerization with Amine Hardeners 23
Polymerization with Acid and Anhydride Hardeners 26
Polymerization with Polyamide Resins 27
Latent Curing Agents 28
Classification of Epoxy Coatings 29
Nonesterified Solution Coatings 29
Esterified Solution Coatings 29
100-Percent-Solids Coatings 30
Properties of Epoxies 30

Adhesion 30
Water and Salt-Spray Resistance 31
Electrical Properties 31
Thermal Stability 33
Properties of 100-Percent-Solids Coatings 36
Properties of Esterified Epoxies 36
Phenoxies 37
Silicones 37
The Chemistry of Silicones 38
Silicone Alkyds 41
Halogenated Silicones 42
Silicone Properties 42
RTV (Room Temperature Vulcanizing) Silicones 48
References 53

2. CHEMISTRY AND PROPERTIES OF COATINGS—*Polyimides—Fluorocarbons—Polyxylylenes* 55
Polyimides 55
The Chemistry and Structure of Polyimides 55
Condensation-Type Polyimides 56
Addition-Type Polyimides 57
Solvent-Soluble Polyimides 58
Photosensitive Polyimides 61
Properties of Polyimides 61
Thermal Stability 61
Solvent and Chemical Resistance 64
Moisture Resistance 65
Electrical Properties 65
Fluorocarbons 65
The Chemistry of Fluorocarbons 65
Properties of Fluorocarbons 67
Electrical Properties 67
Thermal Properties 68
Mechanical Properties 68
Chemical and Solvent Resistance 69
Fluorocarbon Dispersion Coatings 72
Polyxylylenes 73
Chemistry 73
Properties of Polyxylylenes 75
Electrical Properties 76
Other Properties 78
Electronic Applications 80
References 83

3. CHEMISTRY AND PROPERTIES OF COATINGS—*Polyesters—Polyvinyls—Polystyrenes—Acrylics—Diallylphthalates—Polyamides—Phenolics—Polysulfides* 85
Polyesters 85

The Chemistry of Unsaturated Polyesters . . . 85
The Chemistry of Alkyd Polyesters . . . 86
Properties of Alkyd Coatings . . . 88
Polyvinyls . . . 88
Properties of Polyvinyls . . . 90
Polyvinyl Formals . . . 91
Polyvinyl Alcohols and Acetates . . . 91
Polyvinyl Chlorides and Fluorides . . . 92
Polystyrenes . . . 93
Acrylics . . . 94
Diallylphthalate and Other Allylic Polymers . . . 95
Polyamides . . . 96
Properties of Polyamides . . . 97
Phenolics . . . 97
Properties of Phenolics . . . 99
Polysulfides . . . 100
Properties of Polysulfides . . . 101
References . . . 101

4. FUNCTIONS OF COATINGS . . . 103
Electrical Properties . . . 103
Resistance and Resistivity . . . 104
The Effect of Variables on Resistivity . . . 105
Surface Resistivity . . . 109
Conductance and Conductivity . . . 111
Metal-Filled Polymers . . . 111
Molecularly Conductive Polymers . . . 112
Capacitance . . . 113
Dielectric Constant . . . 114
Dissipation Factor and Power Factor . . . 117
Variation of Dielectric Constant and Dissipation Factor with Temperature . . . 118
Variation of Dielectric Constant and Dissipation Factor with Cure . . . 119
Dielectric Strength and Breakdown Voltage . . . 122
Arc Resistance . . . 124
Environmental Protection . . . 125
Protection Against Moisture . . . 125
Water Absorption . . . 125
Permeability . . . 126
Humidity Testing . . . 128
Protection Against Corrosion . . . 129
Magnesium . . . 130
Tin-Lead Solder . . . 133
Aluminum . . . 133
Magnesium-Lithium Alloys . . . 133
Wash Primers . . . 134
Other Primers . . . 134

Mechanical Protection 135
Abrasion Resistance 135
Friction Resistance 136
Hardness 136
Protection Against Electromagnetic and Radiofrequency Interference 139
Protection Against Electrostatic Discharge 139
Microbial Protection 140
Thermal Conductivity 142
Test Methods 145
The Guarded-Hot-Plate Method 145
Heat-Flow Meter Method 146
References 147

5. MANUFACTURING TECHNOLOGY 150
Cleaning 150
The Nature and Effects of Contaminants 150
Prevention of Contamination 154
Removal of Contaminants 154
Solvents and Solutions Used for Cleaning 154
Solvent Purity 155
Cleaning Methods 156
Verification of Surface Cleanliness 161
Application Methods 162
Spray Coating 162
Pressurized Liquid 162
Airless Spray 162
Hot-Melt Flame Spraying 165
Other Spray Equipment 165
Electrostatic Spray Coating 165
Dip Coating 166
Fluidized-Bed Coating 168
Sprayed-Powder Process 172
Electrostatic Fluidized-Bed Process 172
Vacuum Impregnation 172
Screen Printing 173
Spin Coating 173
Flow and Curtain Coating 175
Roller Coating 175
Brushing 175
Automated Coating Processes 175
Spray Coating 175
Dip Coating 176
Curing and Polymerization of Solid Films 177
Heat Curing 177
Step Curing 178
Equipment 178
Photocuring 178

Photoinitiation . . . 178
Free-Radical Photopolymerization . . . 179
Cationic Photopolymerization . . . 179
Properties . . . 179
Equipment . . . 182
Thin-Film Deposition . . . 182
Spin Coating . . . 184
Vapor Deposition . . . 184
Ultraviolet-Light Polymerization . . . 186
Electron-Bombardment Polymerization . . . 188
Glow-Discharge/Plasma Polymerization . . . 189
Gamma-Radiation Curing . . . 190
General Properties of Thin-Film Polymer Coatings . . . 191
References . . . 192

6. CIRCUIT COATINGS . . . 196
Circuit-Board Coatings . . . 196
Types of Coatings . . . 196
Reliability . . . 200
Failure Due to Moisture . . . 203
Coating Thickness and Coverage . . . 206
Reworkability of Coated Assemblies . . . 209
Multilayer-Circuit Insulation . . . 210
Resist Coatings . . . 211
Photoresists . . . 211
Negative Resists . . . 212
Positive Resists . . . 215
Dry Film Resists . . . 215
Screened Resists . . . 215
Other Resists and Maskants . . . 216
Coatings for Thin- and Thick-Film Circuits . . . 217
Polymer Thick Films . . . 218
PTF Conductors . . . 218
PTF Resistors . . . 219
PTF Dielectrics . . . 220
Conformal Coatings for Thick-Film Circuits . . . 220
Particle Immobilizing Coatings . . . 221
Polymeric Dielectrics for High-Density, High-Speed Circuits . . . 221
References . . . 224

7. COATINGS FOR SPACE ELECTRONICS . . . 226
Outgassing . . . 226
Requirements and Tests for Spacecraft Materials . . . 229
Toxicity of Outgassing Products . . . 231
Sterilization . . . 232
Sterilization Requirements . . . 232
Materials Problems . . . 232
Thermal-Sterilization Effects . . . 233

Ethylene Oxide Effects . . . 234
Space and Nuclear Radiation . . . 236
The Space-Radiation Environment . . . 236
The Nuclear-Radiation Environment . . . 237
Effects of Radiation on Polymeric Materials . . . 237
Permanent Effects . . . 237
Transient Effects . . . 238
Side Effects . . . 238
Radiation Effects on Plasticizers and Fillers . . . 242
Correlation of Organic Structures with Radiation Resistance . . . 242
Relative Radiation Resistance of Coatings . . . 244
Ultraviolet Irradiation . . . 244
Combined Space-Environment Effects . . . 246
References . . . 248

8. WIRE AND COIL COATINGS . . . 250
Magnet-Wire Classifications . . . 250
Wire-Coating Types . . . 252
Plain-Enamel or Oleoresinous Coatings . . . 253
Polyvinyl Formal and Modified-Polyvinyl Formal Coatings . . . 259
Polyurethane Coatings . . . 260
Polyester Varnishes and Enamels . . . 260
Polyimide Coatings . . . 261
Impregnating Varnishes . . . 262
Application and Winding Methods . . . 264
Testing Wire Coatings . . . 265
Film Thickness . . . 265
Adhesion . . . 265
Flexibility . . . 265
Heat Shock . . . 265
Elongation . . . 265
Scrape Resistance . . . 265
Springback . . . 266
Thermoplastic Flow . . . 266
Dielectric Strength . . . 266
Continuity . . . 267
Solvent Resistance and Completeness of Cure . . . 267
Solderability . . . 268
Electrical Overload . . . 268
Bond Strength of Self-Bonding Wire . . . 268
Alcohol Tack . . . 268
Thermal Rating and Thermal Stability . . . 268
Differential Scanning Calorimetry (DSC) . . . 269
Thermogravimetric Analysis (TGA) . . . 271
Thermomechanical Analysis (TMA) . . . 271
Effects of Radiation . . . 271
Stripping of Wire Coatings . . . 273
References . . . 276

9. COATINGS FOR ELECTRONIC COMPONENTS AND DEVICES 278
Functions of Plastic Packaging Materials for Discrete Devices 279
Plastic Packaging Materials 280
Glob-Top Coatings. 288
Thin-Film Coatings 288
Fabrication Aids 288
Photosensitive Polyimides 290
Interlayer Insulation and Passivation 291
Functions of Coating Materials for Hybrid Microcircuits 293
Particle Immobilizing Coatings 293
Parylene Coatings 293
Solvent-Soluble Coatings. 293
Particle Getters 293
Application Methods 294
Transfer Molding. 295
Casting, Dipping, Dropping, and Spinning Processes 298
Fluidized-Bed Coating 300
Parameters Affecting Electronic Devices 300
Stresses and Stress Relief 300
Relieving Stress by Means of Barrier Coatings (Glob-Top) 300
Mechanical Protection with Rigid Encapsulants. 300
Design Considerations in Minimizing Stresses 301
Stress Testing 304
Purity 304
Ionic Impurities 305
Other Impurities 307
Optimum Resin-to-Hardener Weight Ratios 308
Tests for Plastic Purity 308
Moisture Resistance 310
Water Absorption 311
Moisture-Vapor Transmission. 313
The Pressure-Cooker Test 314
The Dye-Penetration Test 314
Corrosivity. 314
Corrosion Tests. 316
Adhesion 318
Tests for Adhesion. 319
Commercial and Military Uses of Plastic Packages 319
References 320

10. SPECIFICATIONS 324
Material Specifications 324
Process Specifications. 325
Electrical-Performance Specifications 348
Commercial, Military, and Federal Specifications. 348
Test Methods for Organic Coatings 348
References 362

APPENDIX 363
Table A-1: Abbreviations and Symbols Used in Text 363
Table A-2: Thickness Units and Equivalents Commonly Used for Coatings 365
Table A-3: Thermal-Conductivity Units and Equivalents 365
Table A-4: Thermal-Conductivity Data 366
Table A-5: Coefficients of Linear Thermal Expansion 368
Table A-6: Water-Absorption Data 370
Table A-7: Moisture-Vapor-Transmission Rates of Plastic Coatings and Films 371
Table A-8: Sward Hardness Values for Coatings 372
Table A-9: Abrasion-Resistance Values for Organic Coatings 373
Table A-10: Dielectric Strength 374
Table A-11: Arc Resistance 376
Table A-12: Volume Resistivities 377
Table A-13: Dielectric Constants of Polymer Coatings (at 25°C) 378
Table A-14: Dielectric Constants of Materials Other Than Polymers 380
Table A-15: Dielectric Constants of Inorganic Insulators 380
Table A-16: Dissipation Factors (at 25°C) 381

INDEX 382

1

Chemistry and Properties of Coatings

Polyurethanes—Epoxies—Phenoxies—Silicones

An understanding of the chemistry and molecular structure of the polymers on which coatings are based is extremely important in determining the properties and predicting the behavior of plastic coatings. Materials and design engineers can no longer afford to use an entirely empirical or "shotgun" approach in screening and selecting materials. With the increasing complexity and sensitivity of the electronic equipment being produced, the engineer must, more than ever, have a basic understanding of the materials he is working with. The general chemistry and properties of the materials considered to be most widely used and specifically applicable as coatings are described in the following sections. Further details on their chemistry and synthesis may be found in the numerous books and articles cited throughout this chapter.

POLYURETHANES

Polyurethanes constitute one of the most important classes of synthetic coatings used for high-reliability electrical applications. When properly formulated and cured, they combine numerous desirable properties; among these are toughness, flexibility, high resistance to abrasion, excellent electrical insulating properties, and good moisture and chemical resistance. Polyurethanes are outstanding chiefly for their high abrasion resistance and superior toughness. A comparison of these two properties for polyurethanes, alkyds, and other coating types is given in Tables 1-1 and 1-2.

Almost all the early work and, in fact, most of the elucidation of the basic chemistry of the polyurethanes were performed in Germany around 1937. Considerable work was also done in the United States by Du Pont around 1940, resulting in a series of patents on the reaction products of polyisocyanates with various

Table 1-1: Toughness (Energy of Rupture) for Clear Films[2]

Coating type	*Energy of rupture, in.-lb/in.*^{3}
Polyurethane, two-component system	4,000–8,000
Polyurethane, one-component system	1,000–2,000
Alkyd	200– 600

Table 1-2: Abrasion Resistance[2–5]

Coating Type	Taber Wear Index (mg/1,000 cycles)
Polyurethane, Type 1*	43–155
Polyurethane, Type 2	28–33
Polyurethane, Type 5	60
Alkyd	147
Vinyl	85--106
Amine-catalyzed epoxy varnish	38
Epoxy polyamide	95
Phenolic varnish	91–172
Epoxy ester	116–121
Chlorinated rubber	220

*Types defined under Classifications and Properties.

glycols, alkyd resins, polyamides, and polyesters. However, it was not until after World War II, when much of the practical technology developed in Germany was transferred to the United States, that commercialization and wide use of polyurethanes were effected.[1] The German technology was based on a series of addition polymers formed from polyesters called *desmophens* and polyisocyanates called *desmodurs.* These polymers soon became popular and widely used as coatings because of their flexibility (which could be varied widely, depending on composition), toughness, good electrical properties, and excellent moisture, abrasion, chemical, and corrosion resistance.

The Chemistry of Polyurethanes

The parent compounds for all polyurethane coatings are the di- or polyisocyanates, and the one most widely used is tolylene diisocyanate (TDI). It consists of 2,4-tolylene diisocyanate or various mixtures of the 2,4- and 2,6-isomers:

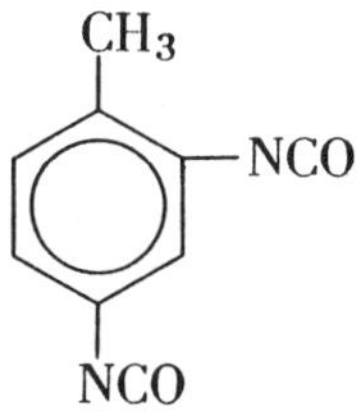

2,4–Tolylene diisocyanate

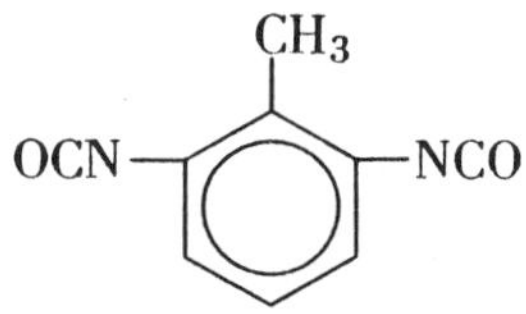

2,6–Tolylene diisocyanate

Another commonly used diisocyanate is methylene bis(4-phenylisocyanate), also referred to as *p,p'*-diphenylmethane diisocyanate (MDI):

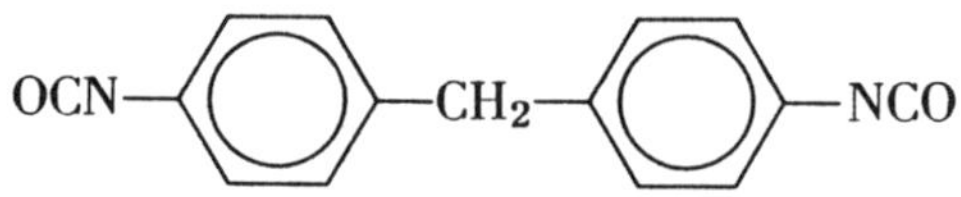

p,p'–Diphenylmethane diisocyanate

These basic isocyanates, however, are not employed as such in commerical formulations because of their high vapor pressures and the highly toxic nature of free isocyanate (NCO) groups. In order to lower their vapor pressures and render them safe, they are modified by co-reacting all or a large number of the isocyanate groups. As an example, prepolymers are formed by reaction with trimethylolpropane. This compound combines with three of the six available isocyanate groups, yielding an adduct which is safe to work with and which can serve as one component of a two-component polyruethane system. In this adduct, three NCO groups are still available for polymerization with a dihydroxyl compound (diol) or with other resins containing several active hydroxy groups:

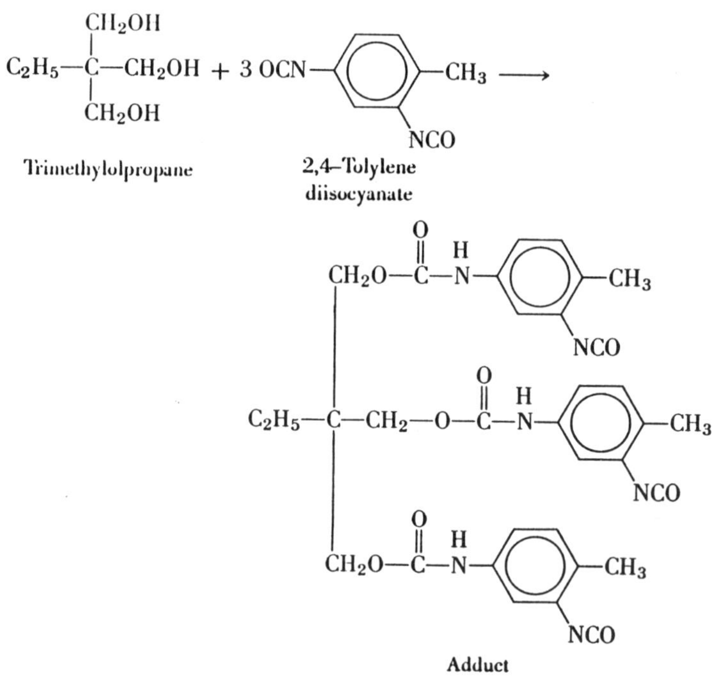

Preparation of low-vapor-pressure Urethane Adducts

Some of the basic and modified isocyanates are described in Table 1-3.

Table 1-3: Typical Isocyanates Used in Polyurethane Coating Formulations (Available from Mobay Corporation)

Trade name	Chemical type	Specific Gravity (15.5C)	% Solids	% NCO
Mondur[a] TDS	2,4-TDI	1.22	100	48
Mondur TD80	80/20 mixture of 2,4- and 2,6-TDI isomers	1.22	100	48
Mondur TD	65/35 mixture of 2,4- and 2,6-TDI isomers	1.22	100	48
Desmodur[b] H	Hexamethylene diisocyanate (HDI)	1.04	100	50
Desmodur W	Methylene Bis-(4-Cylco-hexyl-isocyanate)	1.07	100	31.8
Mondur M	Diphenyl methane di-isocyanate (solid)	1.19	100	33.6
Desmodur N-75	Biuret of HDI	1.06	75	16.5
Desmodur N-3390	Trimer of HDI	1.13	90	19.5
Mondur CB-75	TDI adduct, 75% nonvol-atiles in ethyl acetate	1.19	75	13
Mondur CB-60	TDI adduct, 60% nonvol-atiles in PM acetate	1.12	60	10.0-11.6
Mondur XP-743	MDI polyisocyanate-polyether prepolymer	1.16	100	8.0
Mondur MRS	MDI adduct, 50% monomeric 50% polymeric- increased 2,4'-monomer	1.24	100	31.5
Desmodur T.P. LS-2800	Blocked aliphatic polyiso-cyanate (MEKO)	1.06	75	11.3
Desmodur AP Stabil	Phenol-blocked TDI adduct	1.3	---	12.1
Desmodur T.P. LS-2759	HDI trimer and proprietary blocking agent	1.09	75	8.9
Desmodur CT Stabil	Phenol-blocked TDI cyclic trimer	1.3	89	14.0

[a] Registered trademark of Mobay Corporation
[b] Registered trademark of Bayer AG, Germany

The reaction common to all polyurethane chemistry is the addition of a hydroxyl group to an isocyanate group, resulting in a *urethane* linkage:

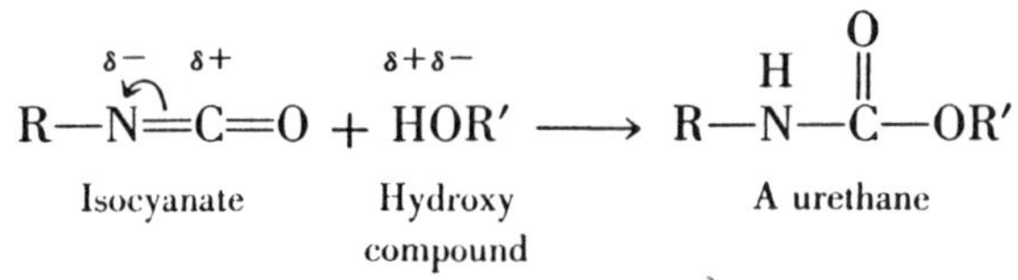

The Basic Urethane Reaction*

Besides the hydroxyls, other groups, having active hydrogens (replaceable hydrogen atoms), such as the amino (—NH_2) or carboxyl (—COOH) groups, will react with the isocyanate group. In these cases, however, the linkages formed are not called urethanes, but *ureas* or *amides,* respectively. Furthermore, by combining polyfunctional reactants, that is, compounds containing two or more isocyanate groups, with compounds containing two or more active hydrogens, high-molecular-weight polyurethane polymers may be formed:

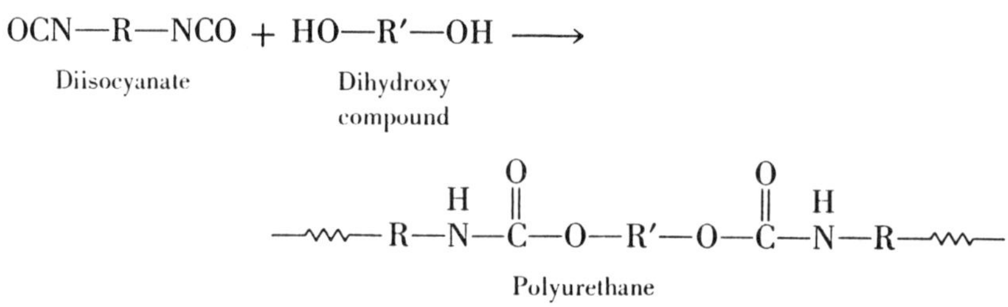

The Basic Polyurethane Reaction

A large variety of di- and polyhydroxy resins may be used in coating formulations. Hydroxyl-terminated polyesters, polyesters, and other commonly used polyol compounds are listed in Table 1-4.

Classifications and Properties

There are essentially six polyurethane classifications. Four of these are one-component systems and two are two-component systems. These types have been established and defined by the American Society for Testing and Materials (ASTM) sec. 06.02 D16 as follows.

> *Type 1: One component prereacted.* These are the urethane-oil or uralkyd types in which polyisocyanates have been reacted with a polyhydric alcohol ester of a vegetable fatty acid. They cure by oxidation at ambient or slightly higher temperature in 5 min to 1 hr.

---

*In these structures and in others throughout this book, the symbol R will be used to designate an organic group, either aliphatic (straight chain) or aromatic (ring structure). This is a general method of indicating that a series of compounds are possible in which the R group varies in structure.[6]

Table 1-4: Typical Polyols Used in Polyurethane Coating Formulations

Trade name	Manufacturer	Chemical type	% Solids	Hydroxyl no.	Coating applications
Desmophen 800	Mobay	Polyester	100	290	Coatings where a high order of chemical resistance combined with hardness.
Multron R-12A	Mobay	Moderately branched polyester	100	150-175	General purpose coatings with good abrasion resistance.
Multron R-18	Mobay	Polyester	100	57-63	Flexible coatings.
Desmophen 1300	Mobay	Polyester	100	132	Surface coatings.
Multron R-351-65	Mobay	Polyester	100	55	Surface coatings.
Desmophen 650A	Mobay	Polyester	100	261	Coatings with good chemical resistance and weather resistance.
Multron R-221	Mobay	Polyester	100	148	Coatings with good outdoor weathering and flexibility.
Kelpol 835-50M	Spencer Kellogg	Modified Soya/Castor oils	50	---	Brush-applied fast drying coatings.
Kelpol 890-60X	Spencer Kellogg	Modified Soya/Castor/ Tung oils	60	---	Coatings with hard finish.
Kelpol 3710-X-80	Spencer Kellogg	Modified Soya oil	80	---	General purpose enamels.
Polyol 4294	Spencer Kellogg	Polyester	100	275-300	Coatings with high hardness, chemical and solvent resistance.
Polyol 4295	Spencer Kellogg	Polyester	100	175-195	Flexible coatings with good impact resistance
Polyol 4357	Spencer Kellogg	Product-finishing polyol	50	90-110	Fast drying for production line finishing.
Polyol 8100	Spencer Kellogg	Product-finishing polyol	80	150-172	Fast drying, higher solids coatings.
Polyol 4573	Spencer Kellogg	Acrylic-polyol	60	93-119	Durable surface coatings.

Type 2: One-component moisture-cured. These are resins having free reactive isocyanate groups which can cross-link and harden with ambient moisture. They are slower curing than Type 1, requiring 1 to 12 hr.

Type 3: One-component heat-cured. These are the latent phenol-blocked isocyanates which at elevated temperatures will release the blocking agent and regenerate the active isocyanate.

Type 4: Two-component catalyst-cured. These systems consist of a prepolymer or adduct having free isocyanate groups as one component and a catalyst as a second component. Small amounts of the catalyst will cross-link the prepolymer. Catalysts may consist of monomeric polyols or polyamines. The pot life after the two components are mixed is generally quite short and varies inversely with the amount of catalyst used.

Type 5: Two-component polyol-cured. In this classification, one component is a prepolymer or adduct having free isocyanate groups, and the other is a resin having reactive hydrogen atoms such as hydroxyl-terminated polyesters or polyols such as castor oil.

Type 6: One-component nonreactive. These are polymerized urethanes that are thinned with a solvent. The systems are characterized by the absence of any significant quantity of free isocyanates or other functional groups. Such coatings convert to solid films primarily by solvent evaporation.

The order of the following discussion of these types reflects their relative importance and use in the electronics and electrical industries.

One-component Blocked Polyurethanes (ASTM Type 3). The one-component systems most widely used for electrical applications, especially for wire coatings, are those with blocked isocyanate groups. To extend pot life and provide more desirable one-component systems, the active isocyanate groups are rendered latent by reacting them with other organic compounds having active hydrogen atoms. In this condition, referred to as a *blocked isocyanate,* the prepolymer is nonreactive and has an indefinite shelf life. When it is heated the blocking compound splits off, and the isocyanate group which is liberated may then polymerize with suitable hardener components. Phenol is the most commonly used blocking agent, but other agents include cresol, methyl ethyl ketoxime, butanol, and caprolactam.

The chemical mechanism for blocking and unblocking is as follows:

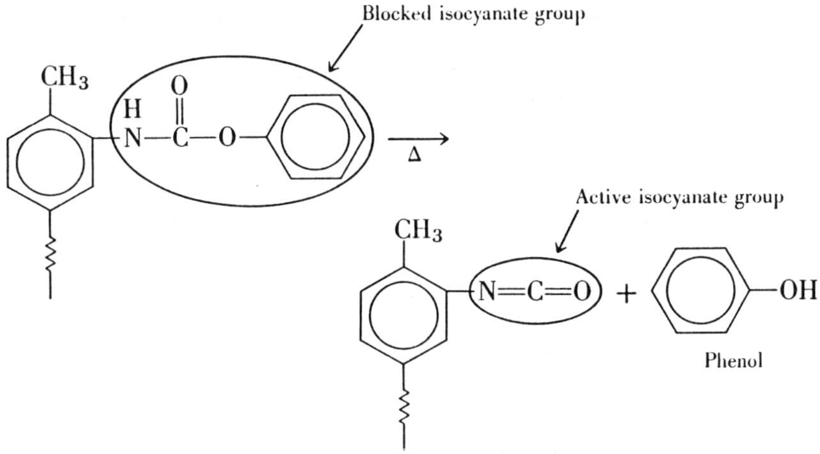

Mechanism for Unblocking Type 3 Polyurethane Resin

The temperature required to unblock the phenol adduct is about 160°C. Such high processing temperatures limit the use of the Type 3 polyurethanes; they cannot be used for temperature-sensitive electronic components and hardware. However, they are ideal for wire and coil insulation and for coating other metal or ceramic objects prior to the attachment of temperature-sensitive parts. Other adducts with lower unblocking temperatures are available, such as the malonic ester adduct, which unblocks at 130°C.

A commonly used blocked isocyanate for wire coatings and magnet-wire enamels is Mondur HCB, used in combination with Multron R-12A hardener. A typical one-component bake-type coating formulation is given in Table 1-5.

Two-component Prepolymer: Polyol (ASTM Type 5). Two component Type 5 polyurethanes are widely used for corrosion-protective and high-reliability electrical insulating coatings. Component A is usually a tolylene diisocyanate or a polyisocyanate prepolymer, also known as an adduct (see Table 1-3 for representative examples). Component B is a hydroxyl containing resin. The most commonly used B components, are hydroxyl-terminated polyesters, polyethers, polyols, castor oil, and some epoxy resins (these last react through the free hydroxyl groups along the polymer chain). The polyethers are based on polypropylene oxide and are termed *polypropylene glycols.* A large variety of these glycols, differing in molecular weight and viscosity, are available.

Both the polyethers and polyesters range from long linear structures, as in the generalized formula for a hydroxyl-terminated polyether,

$$\underset{\substack{| \\ CH_3}}{HOCH}CH_2OCH_2\underset{\substack{| \\ CH_3}}{CH}—O—\sim\sim—CH_2\underset{\substack{| \\ CH_3}}{CH}—O—CH_2\underset{\substack{| \\ CH_3}}{CH}OH$$

to highly branched structures. By proper selection, the properties of the coating may be varied from highly flexible to very hard or brittle. As a rule, the more linear the structure, the more flexible the resulting film; the more branched the structure, the harder and more chemically resistant the coatings will be. Accordingly, the

Table 1-5: One-Component Bake-Type Polyurethane Formulation[7]

Components	Weight	Volume (gal)	Weight Solids	Volume Solids	Supplier
Desmophen 670–80	267.88	29.12	214.31	21.83	Mobay
Lactimon	1.51	0.02	1.51	0.20	Byk-Mallinkrodt
Bentone 34*	30.18	3.51	2.19	0.34	NL Industries
Bayferrox AC-5303	95.85	2.21	95.85	2.21	Mobay
Bayferrox AC-5047	83.94	1.92	83.94	1.92	Mobay
Ti-Pure R-960	29.91	0.85	29.91	0.85	DuPont
Carbon Black FR Lak	29.87	2.15	29.87	2.15	Mobay/Harman
Modaflow**	10.69	1.31	10.69	1.31	Monsanto
Byk 300	3.21	0.40	1.60	0.18	Byk Mallinkrodt
PMA	145.98	18.25	0.00	0.00	–
Solvesso 150	72.98	9.85	0.00	0.00	–
T-12***	37.77	5.09	3.78	0.88	Air Products
Desmodur BL-3175	222.50	25.14	166.87	17.49	Mobay
Total	1032.27	100.00	640.52	48.68	

Weight Solids:	62.05%	Weight/Gallon	10.32
Volume Solids:	49.02%	NCO:OH Ratio	1.10
VOC	4.52	Viscosity @ 77°F #2 Zahn Cup	28 seconds
		Cure Condition:	PMT = 460°F for 60 seconds or 280°F for 30 minutes

*15% gel in Xylene
**40% in PMA
***10% in PMA

Properties

MEK double rubs	>100
Flexibility	
Reverse impact	160 in-lb
Hardness	2H
60° Gloss	50

highly branched Multrons R-2 and R-4 will yield very durable corrosion-resistant and chemical-resistant coatings, whereas Multrons R-16, R-12, and R-22, which have no branching or only a very slight degree of branching, yield general-purpose rubbery coatings. Almost all the polyethers and polyesters are hygroscopic and capable of absorbing as much as 2 percent by weight of moisture under high-humidity conditions. With proper containment and storage, however, moisture absorption may be kept to a minimum. The isocyanate component A is also sensitive to moisture. Water reacts with the free isocyanate groups, releasing carbon dioxide gas and forming an amine compound. This compound, which is also reactive, combines with more free isocyanate groups to give substituted urea polymers. Also, the carbon dioxide formed can cause bubbles in the coatings. In summary, then, moisture should be excluded because it competes with component B for the free isocyanate groups.

Formulations consisting of either Mondur CB-60 or Mondur CB-75, and any one of a number of Multron polyester resins are widely used for surface coatings.

Castor oil is also a very good choice for curing isocyanates and gives excellent insulating and protective coatings. However, oil-based coatings show strong yellowing and are therefore used primarily as primer coatings.

Typical physical properties of Type 5 coatings are given in Table 1-6, and typical electrical properties are presented in Table 1-7.

Table 1-6: Properties of ASTM Type 5 Polyurethanes[8]

Property	Conathane EN-2521	Conathane EN-2522
Cure schedule		
25 C	7 days	7 days
60 C	6 hours	6 hours
100 C	4 hours	4 hours
Hardness, Shore D	65	45
Tensile strength, psi	1600	2500
Elongation, %	40	80
Linear Shrinkage, %	0.71	0.91
Water absorbtion, %		
24 hours	0.19	0.22
7 days	0.55	0.56
Operating temp range	-55 to 130 C	-55 to 130 C

Courtesy of Conap, Inc.

Table 1-7: Electrical Properties of ASTM Type 5 Polyurethanes[9]

Electrical property	Conathane CE-1155	Conathane CE-1165
Dielectric strength, volts/mil	3000	3000
Dielectric constant (10^6 hz)	3.42	3.10
Dissipation factor (10^6 hz)	0.016	0.020
Insulation Resistance, ohm-cm	2.5×10^{13}	2.8×10^{13}

Courtesy of Conap, Inc.

Two-component Catalyst-cured Systems (ASTM Type 4). Type 4 systems consist of an isocyanate resin as one component and a catalyst as the second component. The urethane reaction is base-catalyzed, and hence small amounts (0.5 to 4 percent) of basic (alkaline) materials may be used to initiate polymerization. Among the commonly used catalysts are the tertiary amines such as tri-

ethylamine and others shown in Figure 1-1. Curves showing the relative polymerization rate of urethanes with various catalysts are given in Figure 1-2. It is apparent that cobalt naphthenate is the most active catalyst of the group tested.

Some of these catalysts may also be used to accelerate the cure of other two-component polyurethane systems (Type 5). Although this may be desirable from a manufacturing standpoint, precautions should be taken that the addition of such catalysts does not alter essential engineering properties of the coating. Combinations of two different catalysts have been found to be more active than either catalyst used alone. This enhancement achieved by combining two catalysts, called *synergistic action,* has been reported for combinations of tin catalysts and tertiary amines.[10]

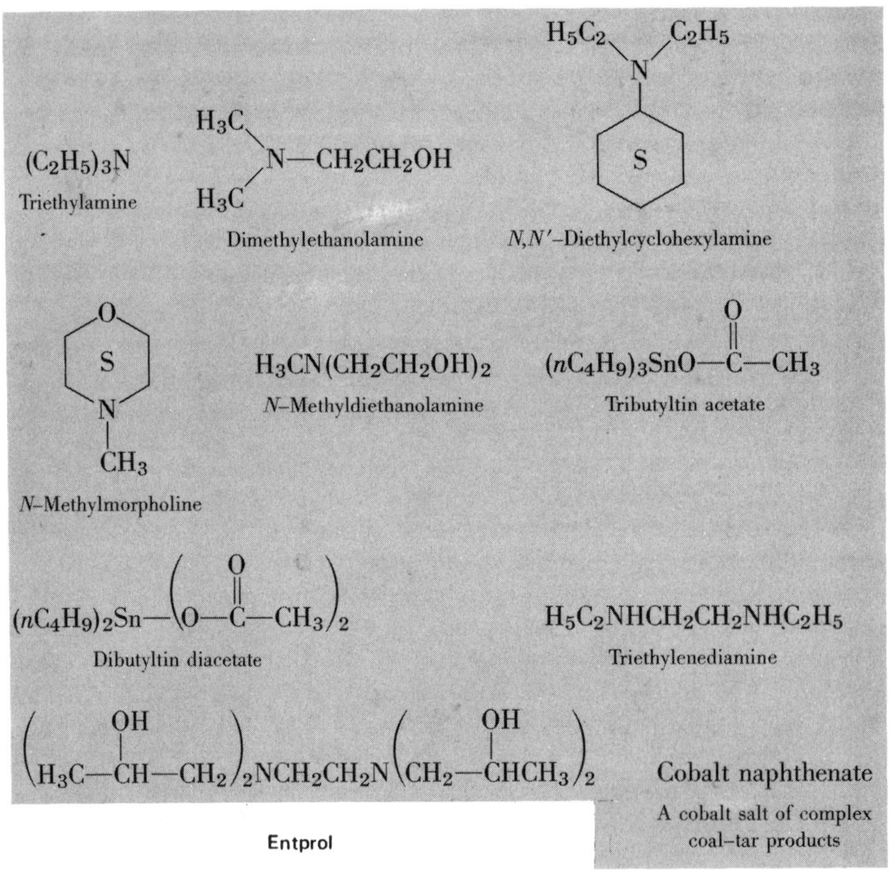

Figure 1-1: Typical polyurethane catalysts and their structures.

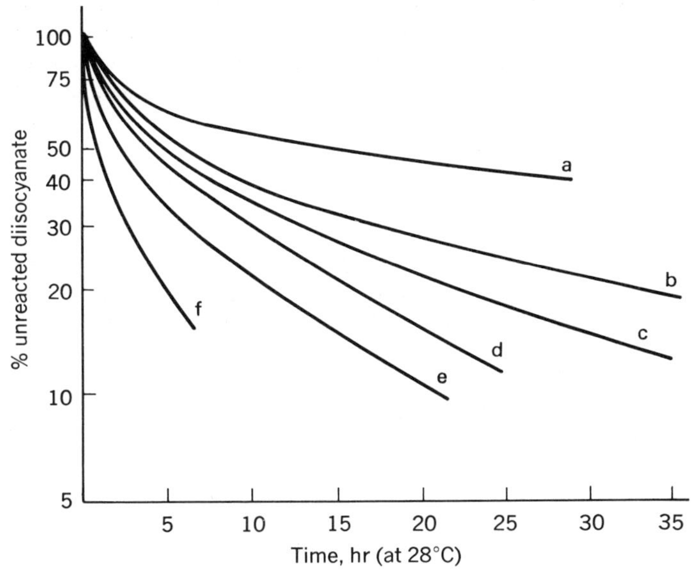

Figure 1-2: Polymerization rates of Nacconate 80-diethylene glycol adipate polyester with various catalysts: (a) no catalyst, (b) N-methyl morpholine, (c) N,N'-dimethylethanolamine, (d) N,N'-dimethylcyclohexylamine, (e) triethylamine, (f) cobalt naphthenate.

One-component Oil-modified Systems: Uralkyds (ASTM Type 1). One-component polyurethane systems are commercially available. They are based on urethanes containing as part of their structure modified drying oils such as linseed oil,[11] and are referred to as *uralkyds, urethane oils,* or *oil-modified urethanes.* In their manufacture the oil is first partially deesterified to give a mixed hydroxy ester. The free hydroxyl group of this ester can then react with an diisocyanate to give the prepolymer resin:

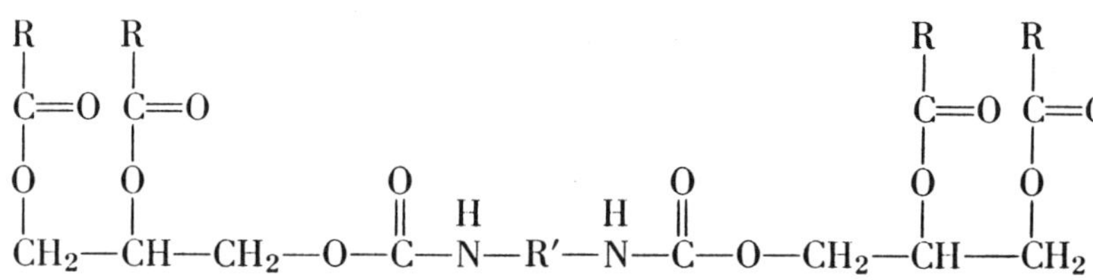

Uralkyd–Prepolymer Generalized Structure

where *R* represents a fatty acid portion containing double bonds, such as: $—(CH_2)_7—CH=CH—CH_2—CH=CH—CH_2CH=CH—CH_2CH_3$

It should be noted that this resin cannot polymerize further through free isocyanate groups because none are available. The polymer does, however, contain

many double bonds (ethylenic groups) in the fatty acid portion (R) of the oil and therefore cures by air oxidation. Free radicals are formed from the active methylene groups (—CH_2—) adjacent to the double bonds, and several units may then combine to form a high molecular weight polymer. The addition of a drying catalyst such as cobalt or manganese naphthenate will accelerate the air curing. Tack-free times of 2 to 5 min at room temperature may be achieved by the addition of such driers. More specific information on the mechanism for this type of cure may be found in a number of references.[12-15] The main feature of uralkyd resin coatings is their lower cost in comparison to other polyurethane types. This is a result of the lower cost of the naturally occurring oils used in their synthesis. Like the alkyds, they exhibit very good water and alkali resistance, but they are marginal in color and gloss retention. They are not as extensively employed for electrical applications as the blocked polymers or the two-component systems, since they do not have the high overall performance characteristics. A typical uralkyd formulation is shown in Table 1-8. A rapid air-drying formulation which becomes tack-free in 2 to 5 min is shown in Table 1-9.

Table 1-8: Uralkyd Formulation

Ingredient	*Proportion, pbw*
Linseed oil (alkali refined)	325.7
Glycerol	46.9
Calcium naphthenate (5% metal)	0.9
Mondur TD-80	126.5
Xylene	500.0
Total	1,000.0

Table 1-9: Rapid Air-Drying Uralkyd Formulation

Ingredient	*Proportion, pbw*
Spencer Kellogg Spenkel F78-50T	340
Nitrocellulose, 18 centipoises, 30% solution	100
Manganese naphthenate, 6%	1
Nuodex Exkin #2	0.4
Toluene	100
Butyl acetate	260
Total	801.4

One-component Moisture-cured Systems (ASTM Type 2). The moisture-cured prepolymers are obtained by reacting hydroxyl-containing materials such as castor oil, hydroxyl-terminated polyesters, or hydroxyl-terminated polyethers with an excess of tolylene diisocyanate. Prepolymer adducts having free isocyanate groups are thus obtained.

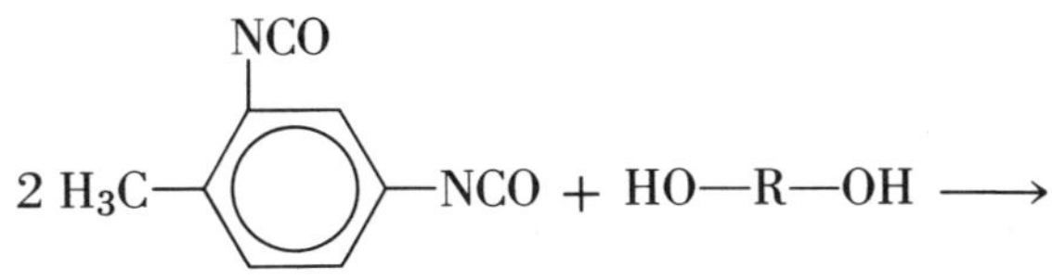

Tolylene diisocyanate

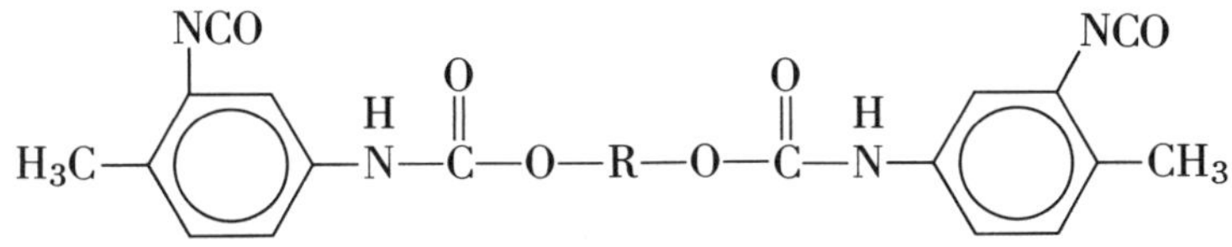

Prepolymer adduct

Thin films of these prepolymers will cure on exposure to air moisture. The curing mechanism involves the cross-linking of molecules through the formation of urea linkages and may be depicted as follows:

$$\text{—R—N=C=O} + \text{HOH} \longrightarrow \left[\text{—R—}\overset{\text{H}}{\text{N}}\text{—}\overset{\overset{\text{O}}{\|}}{\text{C}}\text{—OH}\right] \xrightarrow{-CO_2} \text{—RNH}_2$$

A transient species   Amine

$$\text{—RNH}_2 + \text{O=C=N—R—} \longrightarrow \text{—RNH—}\overset{\overset{\text{O}}{\|}}{\text{C}}\text{—NHR—}$$

Moisture-curing Mechanism for Polyurethanes

Moisture-cured polyurethane coatings are among the toughest and most abrasion-resistant coatings available (see Tables 1-10 and 1-11). They command the widest use as floor, wood, and appliance coatings. Their use in electronics is limited because the time required for a complete cure depends on the relative humidity and may therefore vary considerably from day to day and place to place. It would thus be difficult to adhere strictly to production specifications; films 1 to 3 mils thick might require 45 min to 12 hr to harden, depending on the formulation and the relative humidity. It is also difficult to prepare good coatings by baking, again because the humidity in a curing oven is very low and the coating may remain in an uncured or semicured condition.

One-component Nonreactive (ASTM Type 6). These polymers are prereacted and thinned with solvents such as alcohols, N-methyl-2-pyrrolidone, toluene, or methyl ethyl ketone. They may also be purchased as anionic colloidal urethane elastomer dispersions in a water/solvent mixture. These systems are primarily used as impregnants for fabrics, leather and paper and as primers and top coats for other plastics. Type 6 urethanes are not generally used for coating electronics.

Table 1-10: Properties of a Typical Moisture-Cured Coating[16]

Property	*Rating*
Nonvolatile content, %	45
Solvent	Xylene
Viscosity (bubble)	A^-
NCO-OH ratio	1.65
TDI, %	42.8
Sward hardness value	48
Impact	>160
Elongation, %	> 14
Drying time, hr	
Set to touch	0.4
Tack-free	6.0
Hard dried	6.0

Table 1-11: Abrasion Resistance of Polyurethane Coatings (mg/1,000 cycles)[17]

Isocyanate curing agent	Resin				
	Mondur TD-80	E-244	E-262	E-268	Mondur CB-60
Modified alkyd (uralkyd)	72				
Polyether prepolymer, moisture-cured	31	29	33		
Multron R-12	...	...	...	35	38

EPOXIES

Epoxies comprise another important class of polymers widely used for protective and electrical-grade coatings. Initial experimental work was done in the early 1930s, but production for commercial applications did not start until 1950. Epoxy resins were developed and promoted initially by Ciba and the Shell Chemical Company. Today there are numerous producers. Since 1950 new uses for epoxies have developed at a very rapid pace, and production has increased each year.

The Chemistry of Epoxies

The chemistry of epoxies is based primarily on the high reactivity of the strained three-membered epoxy (also called oxirane) ring. With acidic or alkaline compounds or compounds containing active hydrogen atoms, the epoxy ring opens up and a polymerization reaction takes place, resulting in useful epoxy coatings, adhesives, encapsulants, and laminates.

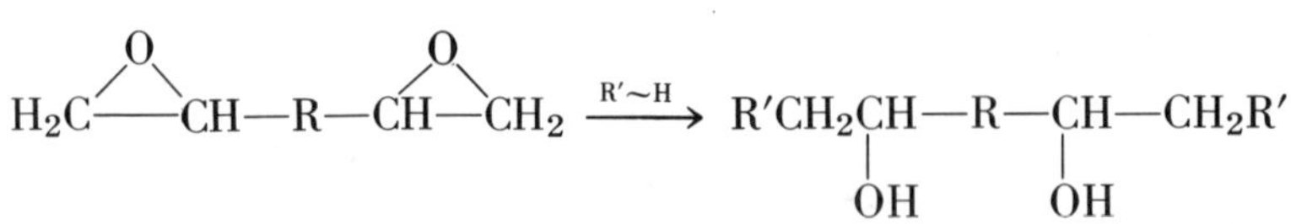

A diepoxide

Table 1-12: Typical Polyurethane Coatings and Manufacturers

Trade Name	Manufacturer or Distributor
Type 1 (one-component uralkyd)	
Spenkel F48, F77, F78 series	Spencer Kellogg
Uralane 7525 and 8267	Furane, Ciba Geigy
Type 2 (one-component moisture-cured)	
Conathane CE-1163, -1164, and -1166	Conap
Spenkel M21-40X, M21-Z-40, M37-A6X-42, and M86-A6X-50	Spencer Kellogg
Eccocoat 933-63	Emerson & Cuming
PR-1568	Products Research
Humiseal 1A20	Columbia Chase
Type 3 (one-component blocked polyurethane)	
Desmodur AP and CT, Mondur S and SH	Mobay
Type 4 (two-component catalyst-cured)	
Humiseal 1A27, 1A33 (plus catalyst)	Columbia Chase
Type 5 (two-component prepolymer polyol)	
Conathane CE-1155 and -1165	Conap
Spenkel P49-A6-60, P81-A6-75, and P82-K4-75	Spencer Kellogg
Eccocoat LN1461-96	Emerson & Cuming
Uralane 5750-A/B and 7754-A/B	Furane, Ciba Geigy
Scotchcast 237, 216, 250, and 226	3M
PR-1538, -1546, -1564, -1570	Products Research
PC15, PC18, PC28, and PC29	Hysol
Humiseal 2A64	Columbia Chase
Solithane 113	Morton Thiokol
Type 6 (one-component nonreactive)	
Spensol L51, L53, L91-40K	Spencer-Kellogg

Besides the epoxy group, the pendant hydroxyl groups of an epoxy resin are also reactive, and further cross-linking or lengthening of the polymer chain can occur through these groups.

Bisphenol A Resins. The synthesis of di- and polyfunctional epoxy resins has already been well described in the literature[18-20] and will not be treated in any detail here. The most commonly used resins are those based on bisphenol A. When bisphenol A reacts with epichlorohydrin, the simplest epoxy resin, called the *diglycidyl ether of bisphenol A* (DGEBA), is formed (Figure 1-3).

A series of epoxy resins based on bisphenol A can be prepared and are commercially available. The generalized formula for this series is

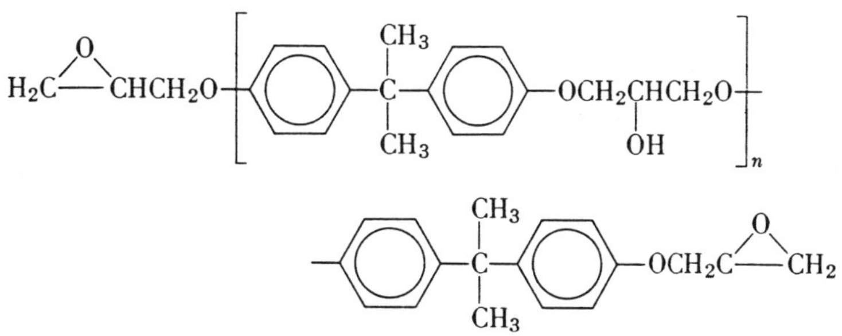

Generalized Structure for Bisphenol A Epoxy Resins

In this structure, *n*, the number of repeating units, may range from 0 to 24, and the resins will be liquids at low *n* values of less than 1 and solids of increasing melting points as the *n* value and molecular weight increase. These resins therefore differ in their degree of polymerization, molecular weight, and viscosity. They also differ in their epoxy equivalents, defined as the weight of resin which contains one gram equivalent of epoxy.

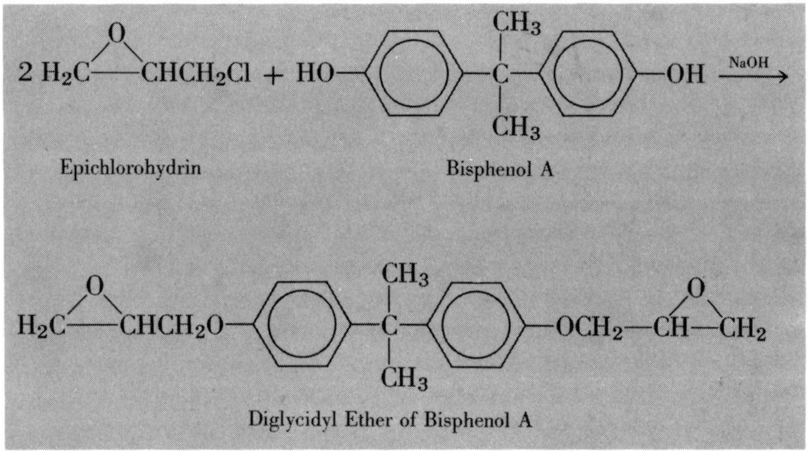

Figure 1-3: Preparation of diglycidylether of bisphenol A epoxy resin.

There are many manufacturers of bisphenol A-type epoxy resins, each using a different designation for essentially the same resin. These designations and some of their properties are given in Table 1-13.

Table 1-13: Equivalent Epoxy Resins (Bisphenol A Types)

Average molecular weight	Viscosity at 25 C (centipoises)	Epoxy equiv.	Shell Epon	Ciba-Geigy Araldite	Interez Epi-Rez	Dow D.E.R.
340-350	6,500-10,000	175-200	826	GY6004	509	330
350-400	10,000-16,000	185-200	828	GY6010	510	331
340-400	500-700	180-250	813,815	506	5071	337,324
	5,000-15,000	175-210	---	GY6005	---	383
	>15,000	175-200	---	---	---	317
900	solid	450-550	1001F	GT7071	520-C	661
1,100	solid	550-650	1002F	GT7072	522-C	662
1,400	solid	875-1000	1004F	6084	530-C	664
2,900	solid	1600-2000	1007F	6097	540-C	667
3,875	solid	2400-4000	1009F	---	---	668

Numerous bisphenol A analogs have been synthesized and investigated; the syntheses of 84 bisphenol compounds and their corresponding epoxy resins have been reported.[21] Of these, only a few, such as the halogenated bisphenol A types, are of any commercial importance. Halogenated bisphenol A epoxy resins are useful because of their inherent flame-retardant properties. Examples are the chlorinated and brominated bisphenol A derivatives:

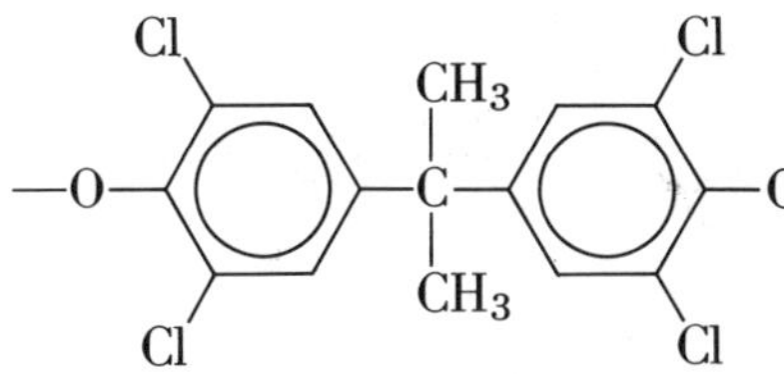

Tetrachlorobisphenol A group

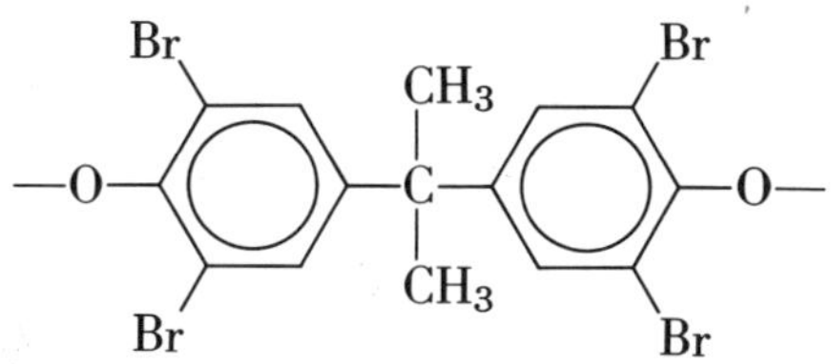

Tetrabromobisphenol A group

These and other epoxy resins are listed in Table 1-14.

Novolac Resins. Besides bisphenol A, other phenolic and polyhydroxy aromatic compounds react with epichlorohydrin to yield epoxy resins having a variety

Table 1-14: Additional Epoxy Resin Types and Properties

Trade name	Manufacturer	Chemical type	Epoxy equiv.	Specific Gravity	Viscosity at 20-25 C centipoises
EPN 1138	Ciba-Geigy	Epoxy Phenol Novolac	176-181	---	semisolid
EPN 1139	Ciba-Geigy	Epoxy Phenol Novolac	172-179	---	70,000-80,000
Epi-Rez 5018	Interez	Aliphatic monogylcidyl ether	290	---	15
Epi-Rez 5044	Interez	Aliphatic triglycidyl ether	165	---	250
Epi-Rez 5022	Interez	1,4-butanediol digycidyl ether	140	---	20
Araldite PT 810	Ciba-Geigy	Triglycidyl isocyanurate	101-111	1.30	solid
ERL 4206	Union Carbide	Vinylcyclohexene dioxide	70	1.09	13
ERL 4221	Union Carbide	3,4-epoxycyclohexylmethyl-3,4-epoxycyclohexane carboxylate	131-143	1.18	350-450
ERL 4299	Union Carbide	Cycloaliphatic epoxy	205	1.15	650
Araldite CY 179	Ciba-Geigy	Alicyclic diepoxy carboxylate	131-143	---	350-450
Araldite 502	Ciba-Geigy	Dibutyl phthalate modified bisphenol epichlorohydrin	233-250	---	2000-2200
Epi-Rez 5011	Interez	*o*-Cresol glycidyl ether	190	---	solid
Araldite 507	Ciba-Geigy	Cresyl glycidyl ether-modified bisphenol	185-193	---	500-700
Epoxy Resin 0163	Ciba-Geigy	Tetrafunctional phenol	200-260	1.26-1.28	solid
Epoxy Resin 0500	Ciba-Geigy	*p*-Aminophenol-epichlorohydrin	110	1.22	3000
Araldite CY 8043	Ciba-Geigy	Brominated epoxy resin	---	---	3000-4500

of properties. A very widely used resin is that derived from novolac, a phenolformaldehyde resin. Reaction of the hydroxyl groups of this resin with epichlorohydrin yields an epoxy-novolac resin[22] (Figure 1-4). These resins are characterized by their high epoxy functionality, resulting in coatings with high cross-linking density and high temperature stability.

Other Epoxy Types. Epoxy resins may also be prepared by reacting unsaturated compounds (compounds containing double bonds) with oxidizing agents such as peroxides or peracids.[21] Both linear and cyclic unsaturated compounds may be prepared in this way (Figure 1-5). The peracetic acid method is being used commercially, and epoxy resins prepared in this way have become known in the trade as *peracetic acid resins.*

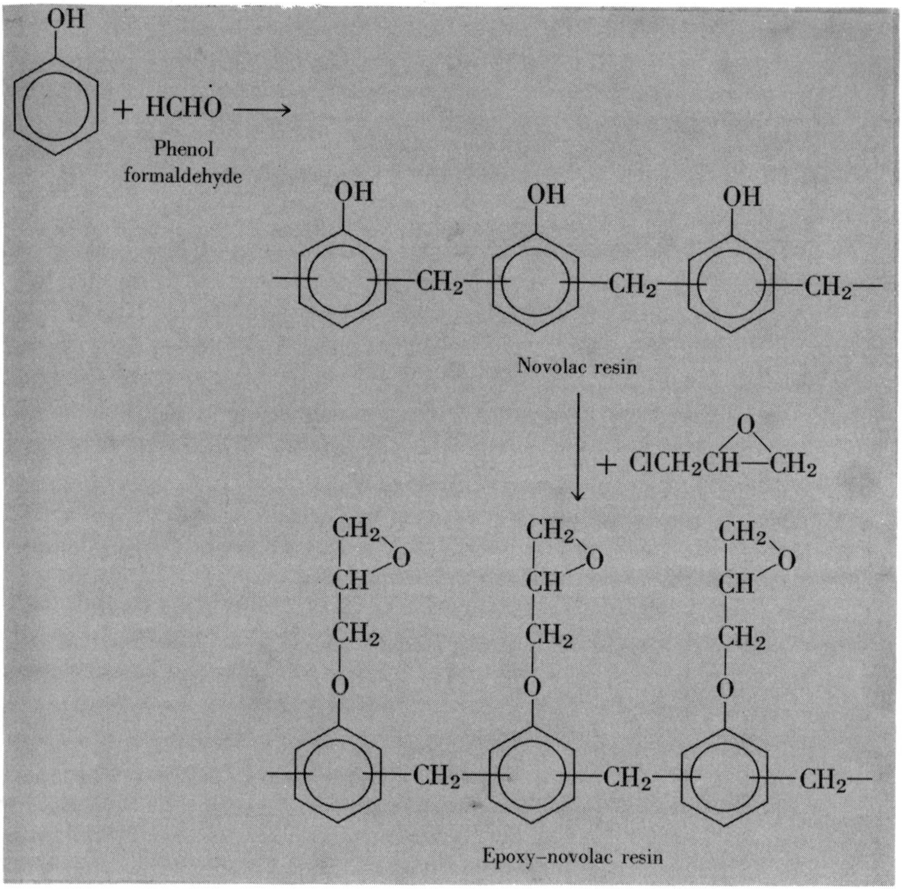

Figure 1-4: Formation of novolac and epoxy-novolac resins.

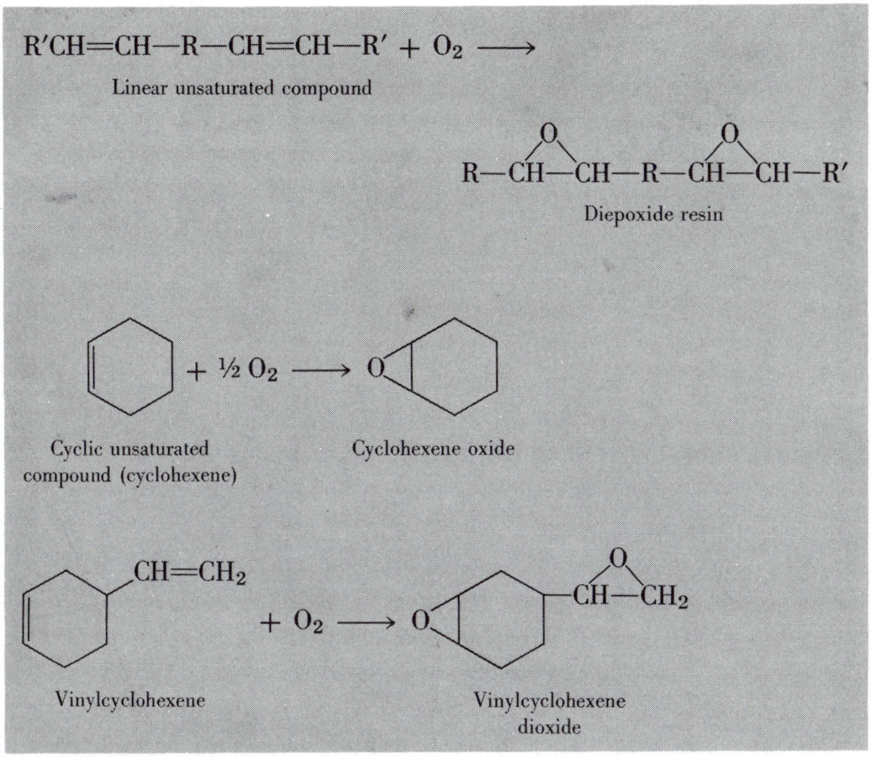

Figure 1-5: Preparation of epoxy resins by epoxidation of unsaturated compounds.

Epoxidized polyolefin resins[22-25] possess a molecular structure rather different from that of the bisphenol A series. The one thing they have in common is the epoxy ring, but even in this respect there is a difference in the position of the rings. In the epoxidized polyolefins the rings are present along the chain as well as at the end of the chain, whereas in the bisphenol A series the epoxy rings are always at the end of the chain.

Epoxy polyolefin resins may be cured with any of the hardeners or catalysts commonly used to cure the bisphenol resin types, and also with free-radical initiators because of the many double bonds along the chain. Hence dicumyl peroxide and other peroxide catalysts may be used.

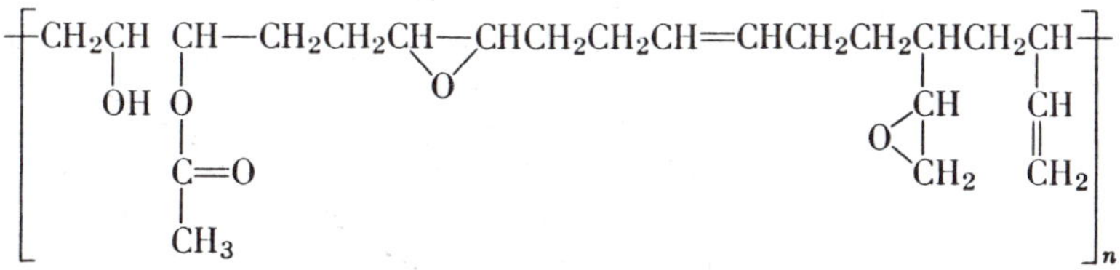

Structure of Oxiron 2000 epoxy resin (trade name of FMC Corporation)

The peracetic acid resins have the advantage over bisphenol epoxy resins in possessing lower densities. Uncured resins are about 20 percent lighter in weight, a property which is desirable for some airborne electronic applications. These resins also have better high-temperature performance, providing a smaller weight loss. In other respects, such as electrical properties and chemical and solvent resistance, they are comparable to the bisphenol resins.

Curing Agents and the Polymerization Process

With the addition of suitable curing agents, epoxy resins polymerize to highly cross-linked three-dimensional solids by either an addition or a catalytic mechanism. According to the addition reaction, the curing agent, also known as a *hardener,* combines chemically into the polymer molecule and serves as a bridge for cross-linking a number of resin molecules. This type of polymerization is sometimes referred to as *heteropolymerization.* The catalytic reactions simply promote the self-polymerization of the resin, referred to as *homopolymerization.* A very large number and variety of both catalysts and hardeners are available, and as in the case of the resins, an almost infinite number are possible synthetically. The hardening agent or catalyst selected governs the polymerization reaction rate and dictates the resulting properties of the cured polymer coating. The hardening agent will also determine the curing schedule (time and temperature necessary to achieve optimum properties) and whether the mixture needs to be cured at room temperature or elevated temperature to complete the reaction.

The choice of a suitable hardening agent for a specific application therefore becomes a most important factor in determining the success of the system. The advantages and limitations of various types of curing agents are summarized in Table 1-15. It should be pointed out that with proper precautions and suitable equipment many of the limitations listed may be avoided.

Catalytic Polymerization. A number of Lewis acid and Lewis base compounds will initiate epoxy polymerization to yield high molecular weight polyethers (Table 1-16). *Lewis acids* are compounds, such as boron trichloride, boron trifluoride, or aluminum chloride, which easily accept an electron pair from another compound to form a complex. Conversely, *Lewis bases* are compounds, such as the tertiary amines, which will easily donate a pair of electrons.

Most tertiary amines require slightly elevated temperatures when used for curing the low molecular weight liquid glycidyl ether resins. However, room-temperature cures are possible with the high molecular weight resins which contain a larger number of hydroxyl groups. Tertiary-amine catalysts are used in quantities of 5 to 15 parts phr (parts per hundred resin), whereas Lewis acid or amine complexes are used in amounts of 2 to 4 phr. Both free Lewis acids and bases present some difficulties in handling. They may be highly corrosive or irritating gases, liquids which may cause skin irritation, or liquids of high vapor pressure which may be noxious or toxic. Also, they are very reactive and have extremely short pot lives. Boron trifluoride, for example, is one of the most reactive catalysts known for epoxy curing. In fact, its extreme reactivity makes it almost impossible to handle in a practical situation. These highly reactive catalysts are therefore sold and used as latent complexes, or are modified in such a way that their vapor pressures and toxicities are reduced and their pot lives extended. Hence boron trifluoride is employed as an amine complex, such as aniline boron trifluoride or triethylamine

Table 1-15: Epoxy Curing Agents—General Comparison

General Type	Advantages	Limitations
Polyamines (primary secondary, and tertiary amines)	Rapid cures. Cure at room temperature Excellent wettability and adhesion.	Short pot life (20–30 min). Require careful and accurate weighing and mixing. High exotherms may be difficult to control. May result in slightly colored films. May have noxious odor or be irritating to skin.
Polyamides	Result in flexible coatings with high vibration, impact, and shock resistance. Are nonirritating. Coatings have high degree of adhesion; surface preparation and cleanliness are not as critical to adhesion as with other curing agents. Pot lives are longer than for amine-cured epoxies. Peak exotherm temperatures are lower than for amine-cured epoxies.	Electrical properties not as good as with other curing agents. Moisture absorption and permeability are higher than for polyamines.
Anhydrides	Long pot life at room temperature. Good thermal resistance (to 260°C). High heat-distortion temperatures. Low peak exotherm temperature. Coatings do not discolor on aging.	Require tertiary amine or other catalyst to accelerate cure. Require higher temperatures (80°–260°C) and longer time (up to 24 hr) to effect cure.
Boron trifluoride complexes	Very rapid cures	High exotherm. Some have short pot lives. Some are hygroscopic.

boron trifluoride. These complexes dissociate gradually at room temperature or more rapidly with the application of heat or other forms of energy, liberating the active catalysts. Catalytic curing agents have not been as popular in coating formulations as the polyamines and other additive hardeners.

Catalytic Cure. The mechanism for curing epoxies with tertiary-amine catalysts is as follows:

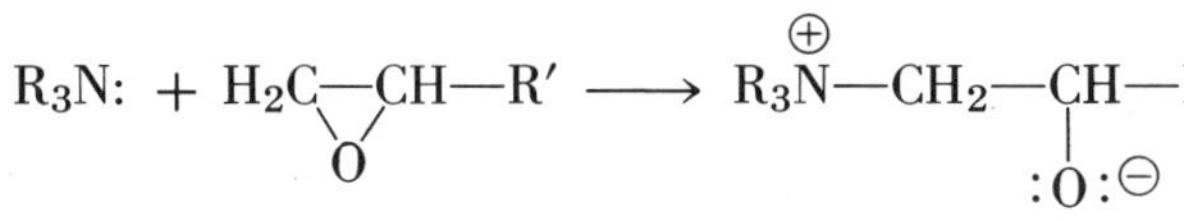

The tertiary amine first opens up an epoxy ring, forming a quaternary base and leaving an oxygen atom with a pair of unsatisfied electrons. The negatively charged oxygen atom then proceeds to open up another epoxy ring, producing another negatively charged oxygen; this in turn opens still another epoxy ring, and so on. In this manner, long chain polyethers are formed:

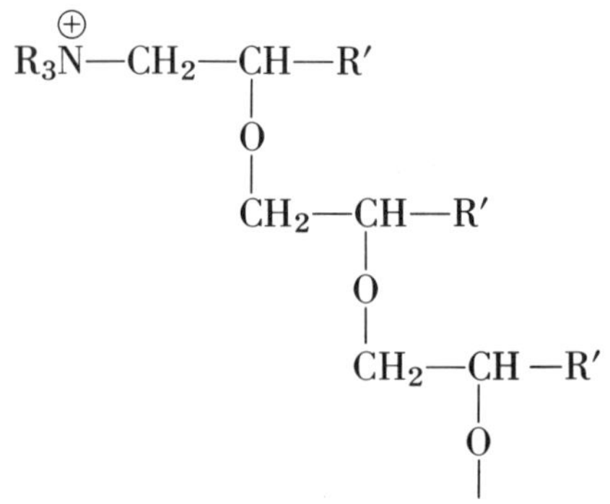

Table 1-16: Lewis Catalysts Used for Curing Epoxies

Lewis catalyst	*Typical sources*
Base catalysts	
Triethylamine	BASF, Union Carbide, Pennwalt, Ashland, Escambia
Benzyldimethylamine	BASF
a-Methylbenzylamine	Aldrich, Hexcel
2-(Dimethylaminomethyl)-phenol (DMP-10)	Rohm & Haas
2,4,6,Tris(dimethylaminomethyl)-phenol (DMP-30)	Rohm & Haas
Acid catalysts	
Boron trifluoride	Air Products, Airco, Matheson, Union Carbide
Boron trichloride	Air Products, Airco, Matheson, Union Carbide
Aluminum chloride	BASF, Ashland, Ethyl, Witco
Zinc chloride	BASF, Du Pont, Cerac, Ashland
Ferric chloride	Cerac, Du Pont, Aldrich, Baker, BASF
Stannic chloride	Witco, Cerac, Baker, M & T Chem.
Diethylether-boron trifluoride complex	BASF
Boron trifluoride etherate	BASF

Polymerization with Amine Hardeners. Amine compounds are very popular curing agents for epoxy coatings, providing rapid cures at room temperature in 1 to 2 hr or in shorter periods of time at higher temperatures. Polyamines are used in concentrations of 4 to 20 phr. The resultant coatings have excellent chemical and solvent resistance, electrical properties, and thermal and vacuum stability. In coatings which are postcured or cured at elevated temperatures, these properties are even better. Numerous amine compounds may be used but fall into 12 generic structural types (Table 1-17).

Specific examples of commercially available amine curing agents and their sources are given in Table 1-18. As with the tertiary amines, the primary and secondary amines may cause skin irritation, possess a noxious odor, or be corrosive and fuming in air. They are therefore sold as modified variations under many proprietary names. They may be used as eutectic mixtures, adducts with low molecular weight epoxy compounds, or complexes with boron trifluoride. In all

Table 1-17: Classification of Amine Hardeners and Catalysts

Type	Aliphatic	Aromatic	Alicyclic	Heterocyclic*
Primary......	R—NH_2	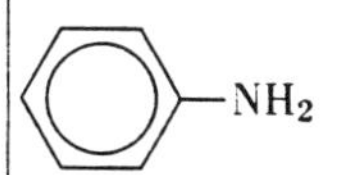	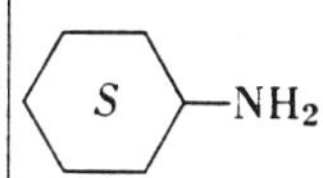	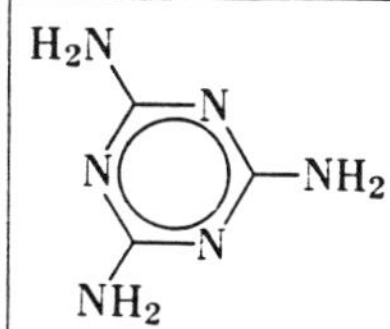
Secondary...	H R—N—R	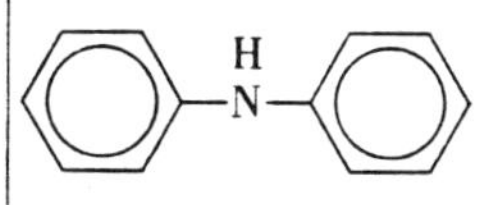	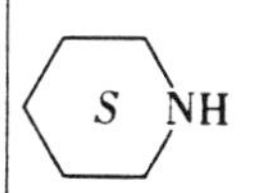	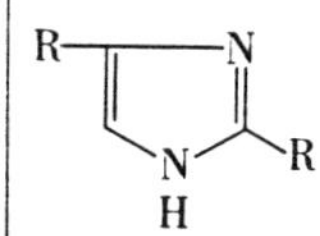
Tertiary......	R—N—R | R	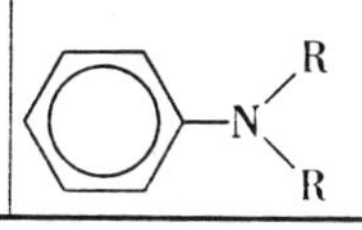	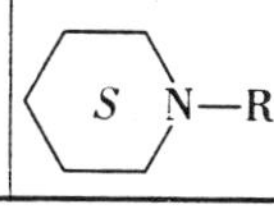	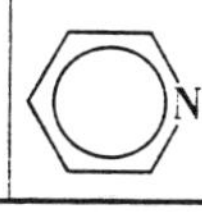

* Specific examples are given; other permutations involving combinations of primary, secondary, and tertiary amines are possible.

Table 1-18: Commonly Used Amine Curing Agents for Epoxies

Chemical Type	Commercial Designation	Typical Source
Aliphatic polyamines		
Diethylenetriamine	DTA or DETA	Ashland, BASF, Dow, Union Carbide
Triethylenetetramine	TETA	Air Products, BASF
Tetraethylenepentamine	TEPA	Dow
Cycloaliphatic amine	XU HY 265	Ciba-Geigy
Polyamine salt	Curing Agent C-111, C-112, U; XU HY 302	Shell, Ciba-Geigy
Modified aromatic amine	Curing Agent Y, Z; XU HY 350	Shell, Ciba-Geigy
Aromatic polyamine		
m-Phenylenediamine	MPD	Du Pont
Diaminodiphenyl sulfone	DDS; HT 976	Polychemical, Ciba-Geigy
Methylenedianiline	MDA	BASF, Ciba-Geigy
Cycloaliphatic polyamine	XU HY 355	Ciba-Geigy
Ketimine	Curing Agent H-1, -2, -3	Shell

cases, vapor pressures are reduced and the possibility of skin irritation is decreased. The use of modified amines is also beneficial in controlling and extending the pot life of a formulation.

Primary and secondary amines of both the aliphatic (straight-chain) and aromatic (benzenoid-ring) types react readily with epoxy resins in concentrations of 10 to 12 phr. An example of an epoxy-cure mechanism by means of ethylenediamine is given in Figure 1-6. Other amines react similarly to give high molecular weight thermosetting epoxy coatings.

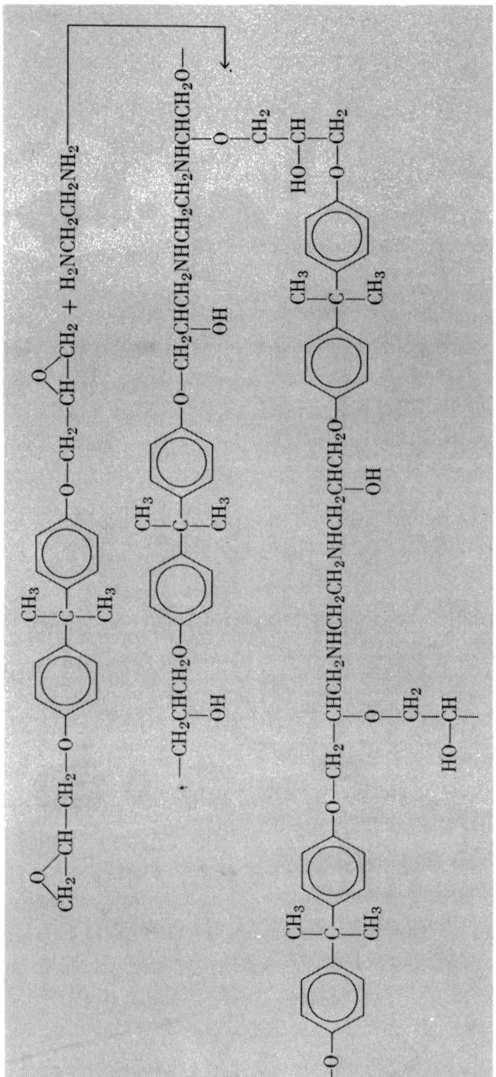

Figure 1-6: Mechanism of epoxy cure with ethylenediamine.

Polymerization with Acid and Anhydride Hardeners. Organic dicarboxylic acids and anhydrides derived from them are also widely used as epoxy-curing agents. They will open up an epoxy ring and become incorporated into the structure as an ester (Figure 1-7). Oxalic acid (HOOC-COOH), the simplest dicarboxylic acid, is used in a number of coating formulations. Formulations have a pot life of several days and may be cured at 125 to 175°C for 1 hr or longer. Other dicarboxylic acids of the series HOOC—$[CH_2]_n$—COOH, where n may be 1 to 8, may be used, but they have not become popular for solution coatings because of

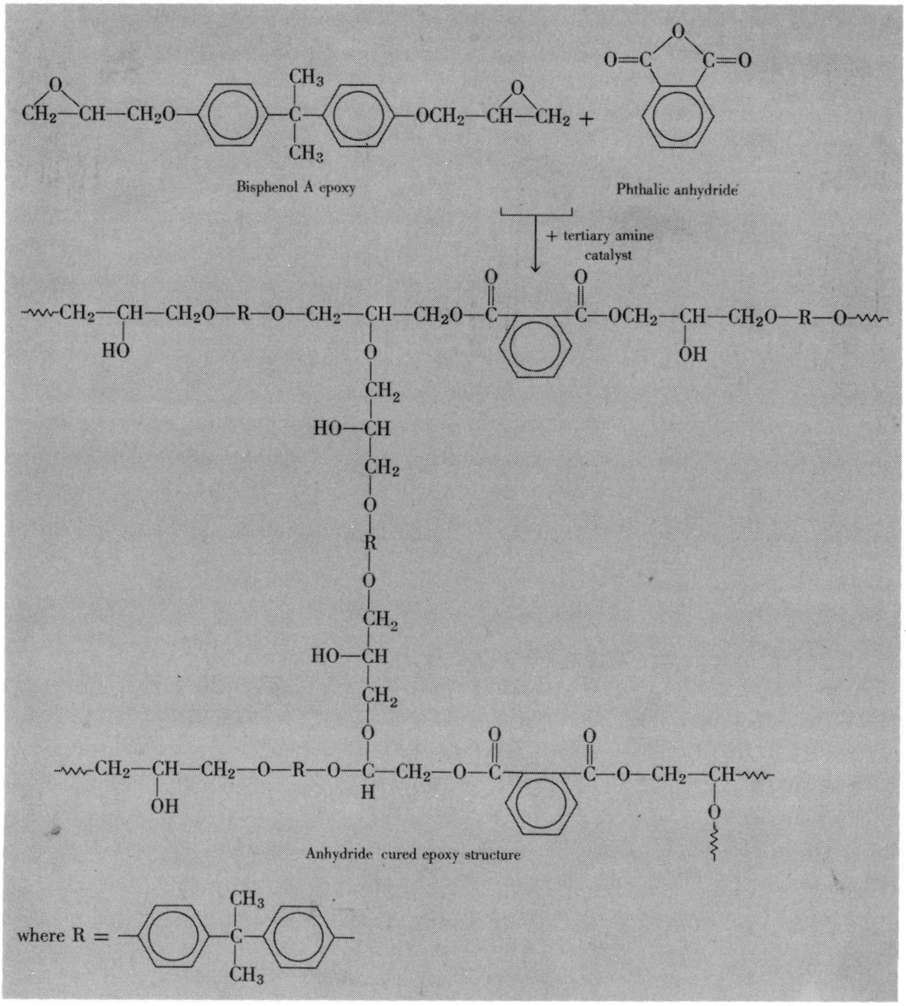

Figure 1-7: Mechanism of epoxy cure with anhydrides.

their insolubility, their difficulty in mixing with resins, and their tendency to foam the resin.[23] Anhydrides derived from the dicarboxylic acids are much more commonly used. Compared with amine-cured systems, the anhydride-cured systems are noted for their better thermal resistance, higher distortion temperatures (40 to 60°C higher), and improved electrical properties. Low dielectric constants, for example, 2.8 to 3, may be achieved. Anhydride curing agents are used in concentrations of 30 to 140 phr. They usually require accelerators or catalysts, without which they react sluggishly with epoxies, especially with the liquid types. With small amounts of a tertiary-amine catalyst (0.1 to 5phr), epoxy compositions can be formulated which are stable up to 12 hr, maintaining a low viscosity at room temperature, but which cure rapidly at 120°C.[24] If the epoxy resin contains more hydroxy groups than epoxy groups, as in the case of the solid bisphenol A resins, the higher hydroxy functionality is effective in initiating the reaction without the need for a tertiary amine catalyst. Some commonly used anhydrides and their manufacturers are given in Table 1-19.

Table 1-19: Commonly Used Anhydride Curing Agents for Epoxies

Chemical Type	Commercial Designation	Typical Source
Aliphatic		
Dodecenylsuccinic anhydride	DDSA	Anhydrides & Chemicals, Dixie
Aromatic		
Phthalic anhydride	PA	BASF, Exxon, Monsanto, Olin
Trimellitic anhydride	TMA	Amoco
Nadic methyl anhydride	NMA	Buffalo Color
Alicyclic		
Hexahydrophthalic anhydride	HHPA	Buffalo Color, Airchem
Chlorendic anhydride	CA	Velsicol, Hooker

Chlorinated and brominated anhydrides are used in special flame-retardant formulations. One of the most commonly used is chlorendic anhydride,

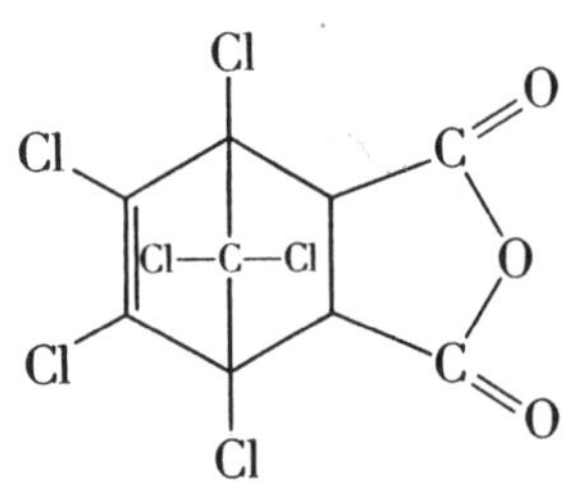

Chlorendic anhydride

This is a very reactive hardener which does not require a catalyst to accelerate the cure, as do other anhydrides. A discussion of other anhydrides is given elsewhere.[19]

Polymerization with Polyamide Resins. Polyamides, used as epoxy-curing agents, are resinous polymers based on vegetable oils and should not be confused with the nylon-type polyamides, which have an entirely different molecular struc-

ture.[25] The epoxy-curing polyamides consist of condensation polymers of dimerized or trimerized vegetable oils or of unsaturated fatty acids with polyamines. As epoxy-curing agents, they should be treated as another type of amino-containing compound. It is these amino groups, both primary and secondary types, in the polyamide structure which are epoxy reactive, and not the amide groups. The hydrogen of the amide group is not sufficiently acid or labile to be reactive. The following structure, based on dimerized linoleic acid and diethylenetriamine (DETA), is typical of the polyamide resins:

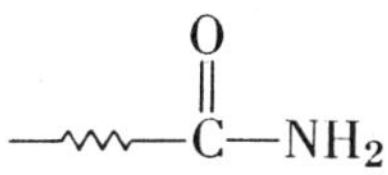

Amide group

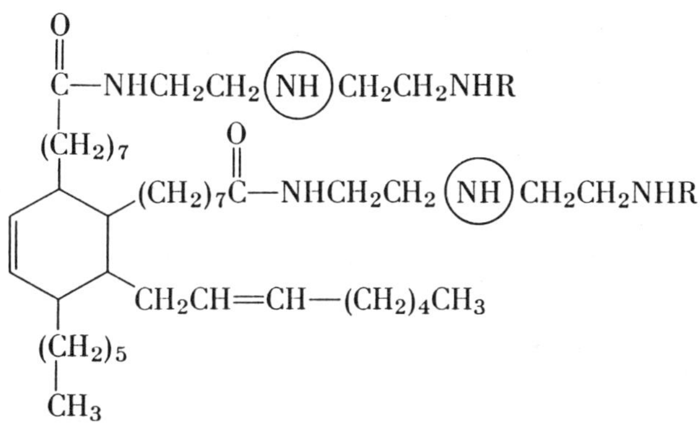

Linoleic acid—diethylenetriamine polyamide

The free secondary-amine groups (circled) provide the reactive sites for cross-linking epoxy-resin chains. The *R* group can represent another linoleic dimer portion or a hydrogen atom.

Latent Curing Agents. Some compounds similar in structure to the polyamides, but unreactive with epoxies at room temperature, may decompose into active amines at elevated temperatures. A notable example is dicyandiamide ("dicy"), a solid, latent-type curing agent which represents a special type of amide (an amidine):

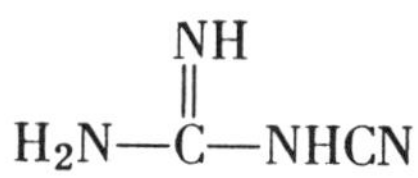

Dicyandiamide

Dicy is used extensively in solution coating formulations for impregnating glass fiber and fabricating epoxy prepreg used in laminates, for adhesive bonding of electrostatic shields or heat sinks to circuit boards, and for bonding layers of circuit laminates together to form multilayer circuit boards. Although it is reported that dicy decomposes at about 150°C into active amine curing agents, the exact curing mechanism appears to be more complex and has been studied by only a few investigators.[26]

A typical solution coating for epoxy NEMA grade G-10 prepreg fabrication consists of Epon 1001F, dicy, and benzyldimethylamine. Tertiary amines such as benzyldimethylamine or quaternary ammonium salts such as benzyltrimethylammonium chloride accelerate an otherwise sluggish reaction.[27]

Classification of Epoxy Coatings

Epoxy coatings may be classified into three main categories: nonesterified solution types, esterified solution types, and 100-percent-solids types.

Nonesterified Solution Coatings. These are two-component epoxy types in which the curing agent may be an amine, polyamide, acid anhydride, or other active hydrogen-containing compound. Many solution coatings are prepared by dissolving solid, high molecular weight resins such as Epons 1001F to 1009F (see Table 1-13) in suitable organic solvents. These solutions may be further modified with resins such as urea formaldehyde, phenol formaldehyde, or melamine formaldehyde. Solution coatings may be cured at room temperature or at elevated temperatures, depending on the reactivity and amount of curing agent used. In heat curing, care should be taken to permit removal of solvents by allowing an air-drying period or a step-by-step cure. The entrapment of volatiles or solvents often presents serious problems in solution coatings. The incorporation of certain pigments and fillers has been found helpful in releasing these solvents during the cure cycle.

Esterified Solution Coatings. Esterified solution coatings are those in which the free hydroxyl groups along the polymer chain have been prereacted with organic acids. Thus groups of unsaturated-fatty-acid esters may be grafted onto the epoxy backbone chain according to several generalized equations (Figure 1-8).

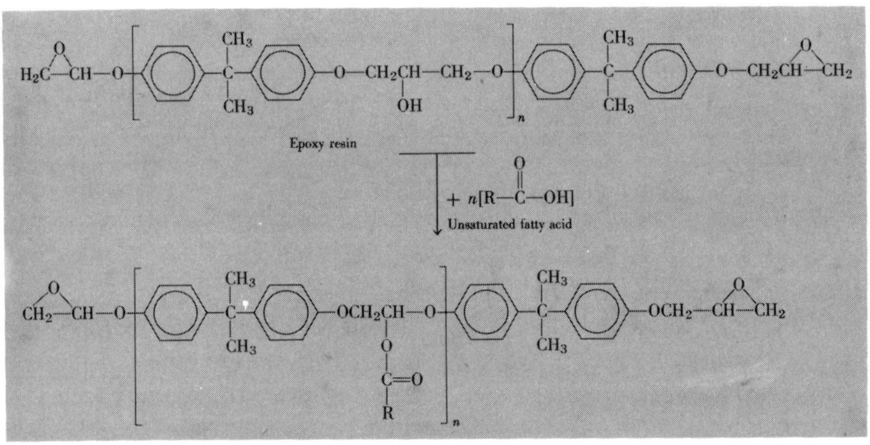

Figure 1-8: Preparation of esterified epoxy resins.

One of the Epon 1000F series epoxies is usually employed as the epoxy to be esterified with soybean oil, linseed oil, castor oil, or other unsaturated fatty acids. Driers such as the metal naphthenates, conventionally used for alkyd coatings, are also used in the esterified-epoxy fomulations. Hence both air-drying and baking-type coatings may be formulated. Baking-type enamels cure in 20 to 60 min at temperatures from 60 to 180°C. The incorporation of unsaturated chains has the advantage of imparting to the resin the more classical air-oxidation curing mechanism, lowering its cost, and achieving properities intermediate between those of nonesterified types and the alkyd esters. One of the early patents on esterified epoxies was issued to Greenlee in 1948.[28]

100-percent-solids Coatings. These are two-component nonester types that are fomulated without solvents or other volatile ingredients—hence their name, 100 percent solids. The resin is chosen of low enough viscosity to allow it to function both as a vehicle and as a reactive component.

Properties of Epoxies

Epoxies possess a better combination of properties than other coating types, which often renders them the sole candidates for electrical applications. For example, they are superior to phenolics, polyurethanes, alkyds, and silicones in their adhesion to a wide variety of substrates. They have excellent moisture and salt-spray resistance, with moisture-permeability rates three to four times lower than those of polyurethanes. In addition, substrate adhesion is often not affected by exposure to moisture. Epoxies are also superior to alkyds and silicones in solvent and chemical resistance. Electrical properties are good and remain fairly constant under 95 to 100 percent relative humidity and temperatures up to 150°C.

Thus the three properties which single out epoxies from other coating types are:

1. Excellent adhesion to a wide variety of substrates under many environmental conditions
2. Excellent resistance to moisture, salt spray, organic solvents, and chemicals
3. Good electrical characteristics and stability of these parameters under various environmental conditions, notably moisture and temperature

These desirable properties are primarily attributable to the thermosetting and highly cross-linked nature of the epoxy structure. However, this same structure renders epoxies less flexible than other coating types (such as polyurethanes), and problems of stress dissipation, shrinkage, and brittleness may result. Flexibility may be achieved by the use of flexible hardeners or plasticizers, but usually at the expense of other important properties, such as electrical insulation and moisture resistance. Often a compromise of properties must be accepted for lack of an ideal insulation coating.

Adhesion. Epoxies are noted for their excellent adhesion to many diverse substrates and surface conditions. This is generally attributed to the highly polar and surface-active nature of the epoxy structure, which provides both a chemical and mechanical interlocking with the substrate. The numerous ether linkages, with their available electron pairs (from oxygen atoms), provide sites for hydrogen bonding, with the hydrogen atom provided by the metal or plastic surfaces.* The hydroxyl groups present in an epoxy structure provide labile hydrogen atoms and also take part in hydrogen bonding. Each hydrogen bond represents a van der Waals force of attraction of about 5 kcal/mole, and the large number of such bonds which form account for the very high total force of attraction. In addition to these modes of attraction, the epoxy ring itself can chemically interact and interlock with

*The aromatic ether structure of the bisphenol portion of the resin augments this ability to hydrogen bond because the oxygen becomes more electronegative owing to resonance effects.[29]

active hydrogen atoms on the surface. For these reasons epoxies are widely employed as adhesives, yielding lap shear strengths of 1,000 to 5,000 psi on a wide variety of metal or plastic substrates.[30]

Many of the outstanding properties of epoxies as coatings, such as resistance to abrasion, weathering, humidity, salt spray, and impact, are attributable to the adhesive characteristics just described. Epoxies adhere well to aluminum, magnesium, steel, titanium, beryllium, copper, most alloys, and plastics. The sensitivity of adhesion to rough surfaces, contaminants and moisture, varies with the curing agent employed. Polyamide-cured epoxies are less sensitive to these factors than epoxies cured with other agents. Polyamide-cured epoxy coatings have been applied in environments of high relative humidity without subsequent loss of adhesion, and one ablative coating composition containing cork as a filler was found to adhere well even to aluminum surfaces containing oily residues.[31]

Water and Salt-spray Resistance. Another outstanding property of epoxy coatings is excellent resistance to water absorption and permeation. Water-absorption values for epoxies cured with a large variety of anhydrides, aliphatic amines, aromatic amines, or boron trifluoride complexes lie in the narrow range of 0.1 to 0.25 percent. Properly cured epoxies are stable even in boiling water and, in comparison to other polymer types, absorb only minimal amounts. Of 38 epoxy formulations in which both the resin and curing-agent proportions were varied, the weight gain after 2 hr of immersion in boiling water ranged from 0.13 to 2.3 percent, with typical values of 0.5 to 1 percent.[32] Water-absorption values may be greater than 1 percent, however, when nonstoichiometric compositions are used, that is, compositions with a large excess of either resin or curing agent, such that they remain unreacted in the cured coating. High values will also be obtained if the coating is undercured. The salt-spray and salt-water resistance of epoxies is also superior to that of many other coating types. Epoxies are being used effectively for the protection of metal enclosures and equipment used in sea water. Equipment located outside the hull of underwater vehicles, for example, may be adequately protected with epoxy coatings. Laboratory tests using aluminum and magnesium panels coated with both an air-curing and a baking-type epoxy showed no signs of corrosion or coating deterioration after being tested 800 hr in salt spray.

Electrical Properties. Epoxies are also noted for their electrical properties, rendering them suitable for many high-stress insulation applications. These properties are fairly stable under a variety of environmental stress conditions, such as temperatures up to 150°C and relative humidities of 95 to 100 percent.

Volume resistivity. Volume resistivities at room temperature are high, generally 10^{12} to 10^{15} ohm-cm. Values are temperature dependent and may drop sharply with increases in temperature; however, for many insulation uses a drop of even six decades is not considered serious. Changes in volume resistivity with temperature for a bisphenol A-epoxy formulation cured with diethanolamine are given in Table 1-20. These data are typical of what occurs with other epoxy formulations. Volume resistivity is also dependent on moisture content and humidity exposure, with values deteriorating upon longer exposures at high humidity. The electrical properties of epoxies are, however, much more stable than those for polyesters and alkyds under the same conditions, as has been demonstrated by Graves and Pizzino.[34] Flat specimens 2 by 2 by 0.125 in. were suspended over water at 70°C in a sealed container for varying periods of time, removed, and checked electrically. Initial values for some epoxies were greater than 3.2×10^{12} ohms and dropped to

Table 1-20: Variation of Volume Resistivity with Temperature for Epoxy-Amine Formulation[33]

Temperature, °C	*Volume resistivity, ohm-cm*
23	2.05×10^{14}
66	1.97×10^{13}
93	9.3×10^{10}
121	2.43×10^{9}
149	3.68×10^{8}

2.5×10^{12} ohms after 7 days of conditioning, 3.5×10^{11} ohms after 14 days, and 8×10^{10} ohms after 28 days. Other unfilled epoxies remained stable at values greater than 10^{12} ohms after 28 days of exposure. Conversely, both filled and unfilled polyesters showed very poor performance, some even at the end of 4 days. Similar trends were noted in specimens prepared by embedding 50-mil-diameter copper wires at 0.125 in. between centers.

A comparison of the electrical values for bisphenol A-epoxy formulation with those for a novolac-epoxy formulation showed no significant differences when measured at room temperature (see Table 1-21).

Table 1-21: Electrical Values for Epoxy Formulations[22]

	. .Novolac Epoxy* . .		. Bisphenol A Epoxy** .	
Property	**Initial Value**	**Value After 24 hr in Water**	**Initial Value**	**Value After 24 hr in Water**
Dielectric constant				
60 hz	3.78	3.82	4.12	4.19
10^3 hz	3.74	3.80	4.07	4.15
10^6 hz	3.39	3.44	3.55	3.61
Dissipation factor				
60 hz	0.0027	0.0021	0.0035	0.0043
10^3 hz	0.012	0.012	0.015	0.016
10^6 hz	0.024	0.025	0.032	0.032
Volume resistivity, ohm-cm	3.8×10^{15}	1.83×10^{14}	1.81×10^{15}	2.31×10^{14}

*Cured for 1 hr at 95°C plus 16 hr at 180°C.
**Cured with MNA for 16 hr at room temperature plus 4.5 hr at 150°C.

Dielectric constant and dissipation factor. Dielectric constants and dissipation factors for epoxies are low, ranging from about 3 to 6 and 0.003 to 0.03, respectively, at room temperature and 60 to 1,000 hz. At higher frequencies, dielectric constants decrease slightly; for example, a value of 3.0 at 10^3 hz may drop to 2.7 at 10^6 hz.

Dielectric strength. Dielectric-strength values are high and similar to those for other coating types. Typical values range from 300 to 450 volts/mil for 0.125-in.-thick samples and may increase to 2,000 volts/mil or more for 1-mil-thick samples. Dielectric strengths are only slightly affected by fillers, regardless of the types or amounts used.[35]

Arc resistance. Arc resistances of unfilled epoxies may vary from 80 to 100 sec. Fillers have a beneficial effect, increasing values to 125–225 sec, depending on the nature of the filler and hardener used.[35] The fillers studied included silica, mica, zirconium silicate, and hydrated alumina. Other investigators have attained improved antitracking properties for phenolic and epoxy coatings by incorporating 45 to 60 percent by weight of hydrous magnesium silicate powder.[36] Further improvement in arc resistance may be achieved by roughening the surface or allowing filler particles to project partially from the surface.[37]

Effect of fillers on electrical properties. The effect of fillers on the electrical properties of epoxies depends on whether the filler is electrically conducting or non-conducting in nature. Nonconducting fillers result in only minor changes, whereas conducting fillers cause a decrease in electrical insulating properties. This is true not only for epoxies, but also for other coating types. Iron oxide, in concentrations as low as 2 phr, reduces the volume resistivity of an epoxy from 3.9×10^{15} to 3.9×10^{14} ohm-cm.[38] Aluminum oxide or silicon dioxide has little effect; and surprisingly, aluminum powder may be used in rather high concentrations without significant reduction of volume resistivity (Table 1-22). The dielectric constants, however, increase appreciably with aluminum concentration.

Thermal Stability. Epoxies are generally stable up to 150°C, but some anhydride- and aromatic amine-cured formulations are stable to 200°C. Above these temperatures decomposition and large changes in electrical and mechanical properties occur. The thermal stability of epoxies is higher than that of polyurethanes (about 135°C) and alkyds (145 to 175°C) but lower than that of silicones (about 260 to 315°C). The percentage of outgassing products on aging at elevated temperatures is a good measure of thermal stability. Figure 1-9 shows the superiority of epoxies over alkyd polyesters and asphalt when vacuum-cycled at 150°C. The amount of outgassing from the epoxy stabilized within 4 to 5 hr, whereas volatiles continued to evolve from the others even after 55 hr.[40]

Figure 1-10 compares the outgassing properties at 150°F for an anhydride-cured epoxy versus an amine-cured epoxy. The samples were preconditioned at 100°C in vacuum for 7 hr prior to exposure at 150°C. In a 5-hr cycle the amine-cured epoxy evolved less than 0.2 ml of gas. In other tests polyesters evolved over 2.5 ml and phenolics and alkyds about 1.5 ml under essentially the same conditions.[40]

Pyrograms and thermograms, determined by thermogravimetric analysis and

Table 1-22: Electrical Properties of Epoxy with Aluminum Fillers[39]

Aluminum filler, phr	Volume resistivity		Dielectric constant at 10^3 hz, 23°C
	At 23°C, ohm-cm $\times 10^{14}$	At 93°C, ohm-cm $\times 10^{9}$	
No filler	1.0	6.3	4.4
10	1.1	6.4	4.7
30	0.89	6.1	6.5
50	0.81	5.6	8.7
100	0.45	5.3	10.0

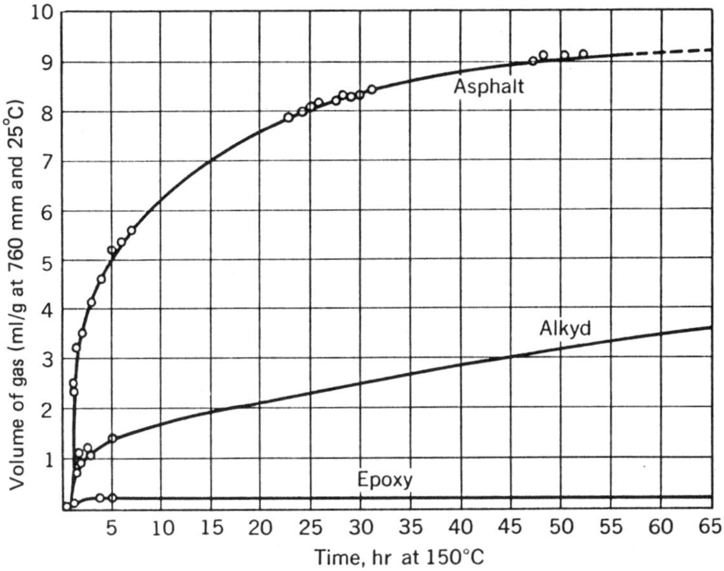

Figure 1-9: Gas evolution at extended vacuum cycles.[40]

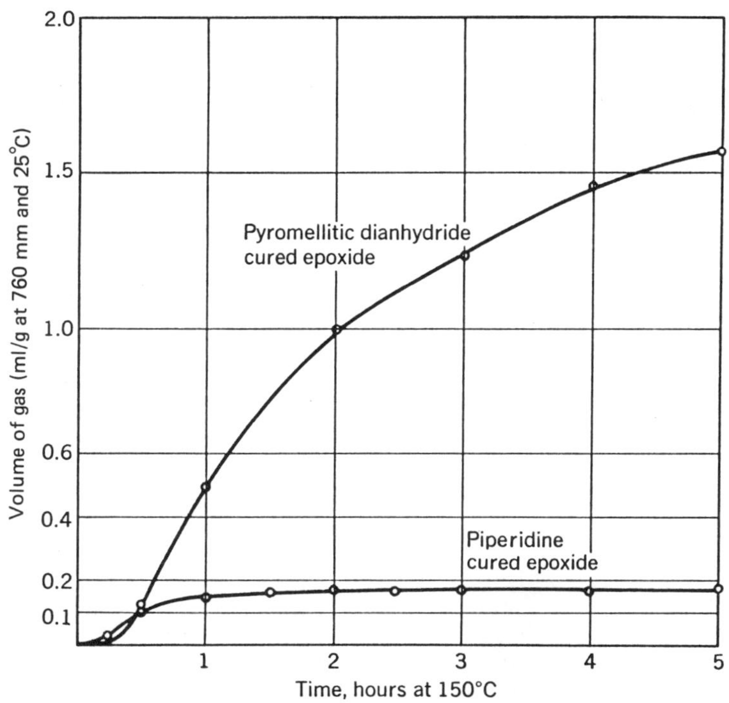

Figure 1-10: Gas evolution of anhydride-cured versus amine-cured epoxies.[40]

differential thermal analysis, respectively (see Chap. 8), are also useful in assessing the thermal stabilities of epoxies and other coatings. Pyrograms for several anhydride-cured epoxies are given in Figs. 1-11 and 1-12. The pyrograms for many other epoxies have been reported by Ehlers.[41]

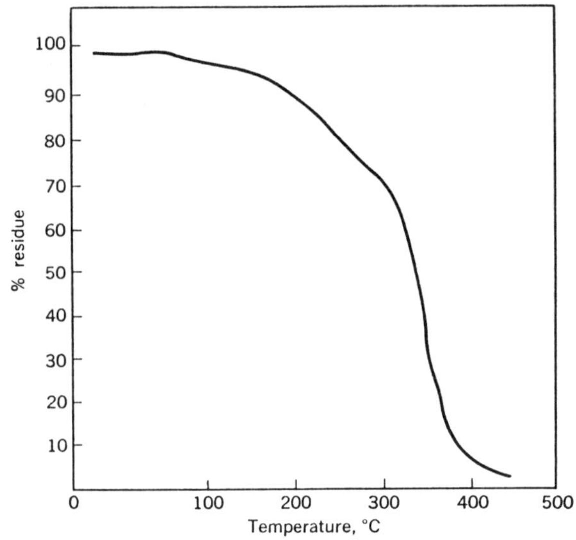

Figure 1-11: Pyrogram of Ciba Araldite 6010 (100 pbw), methyl nadic anhydride (80 pbw), and benzyldimethylamine (0.5 pbw).

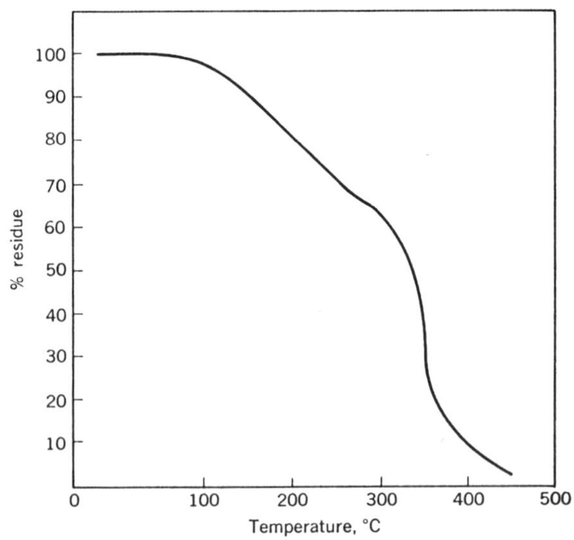

Figure 1-12: Pyrogram of Shell Epon 825 (100 pbw), methyl nadic anhydride (98 pbw), and benzyldimethylamine (0.5 pbw).

Properties of 100-percent-solids Coatings. The 100-percent-solids epoxy coatings have several advantages over their solution-coating counterparts. Because no solvents or volatile constituents are employed in their formulation, thicker layers (up to 60 mils) may be achieved in a single dip or spray operation , there is lower probability of pinholes arising owing to entrapped volatiles, and lower probability of outgassing products being released in a confined area. The low amounts of outgassing products released in comparison to solution coatings make them particularly desirable for applications in manned space vehicles or submarine crew compartments.

Among the disadvantages of 100-percent-solids coatings are their greater brittleness, their poorer shock and mechanical resistance, and the need for special spray equipment (because of normally higher viscosity and shorter pot lives). The problem of brittleness may be partially alleviated by the use of inherently flexible hardeners to cure the epoxy resin. Hence, instead of low molecular weight amine curing agents, polyamides or polysulfides may be used. The greater flexibility and shock-and vibration-damping properties of these epoxies are attributable to the long chain aliphatic portions of the polyamide or polysulfide structures. Varying degrees of flexibility and hardness may be achieved, depending on the proportions in which the epoxy is mixed with the curing agent. A typical formulation consists of 60 parts epoxy to 40 parts polyamide, by weight. The polyamide may be increased to 60 percent, yielding progressively more flexible and softer coatings. However, coatings with large amounts of polyamide, although advantageous from the flexibility standpoint, will normally have lower softening temperatures and display poorer electrical and moisture resistance properties. Hence a compromise among these properties needs to be established.

100-percent-solids coatings, both epoxies and other types, may be applied by dip, spray, brush, or casting. For spraying, special equipment adapted with dual-metered nozzles is often necessary, and thicknesses up to 125 mils may be achieved in one or two steps. As dip coatings, epoxies are widely used for insulating and impregnating electrical equipment, transformers, motors, etc. Applied by spraying or casting, they have been used as circuit-board coatings, insulation coatings, and ablative coatings. Epoxy-polyamide, epoxy-silicone-cork, and epoxy-phenol compositions have successfully been used as ablative coatings to protect missile and space electronic components from the intense heat generated during takeoff or reentry.

Properties of Esterified Epoxies. Esterified-epoxy coatings are not extensively used for electronic or electrical applications, except perhaps as protective coatings for chassis and housings. They are designed and formulated more as maintenance paints for storage-tank protection, appliance coatings, and other commercial uses. Their uses in electronic applications are mostly for environmental protection rather than electrical insulation. The main advantages of esterified epoxies are the corrosion and chemical and solvent protection they afford, combined with a much lower cost than the two-component or unmodified-epoxy coatings. Esterified epoxies are therefore intermediate in price and in properties between the alkyd esters and the two-component systems.

PHENOXIES

Phenoxy resins are structurally similar to epoxies and are prepared from dihydric phenols and epichlorohydrin in the presence of caustic.[42] Commercially available coatings utilize bisphenol A as the dihydric phenol, and their structure may be represented as[42]

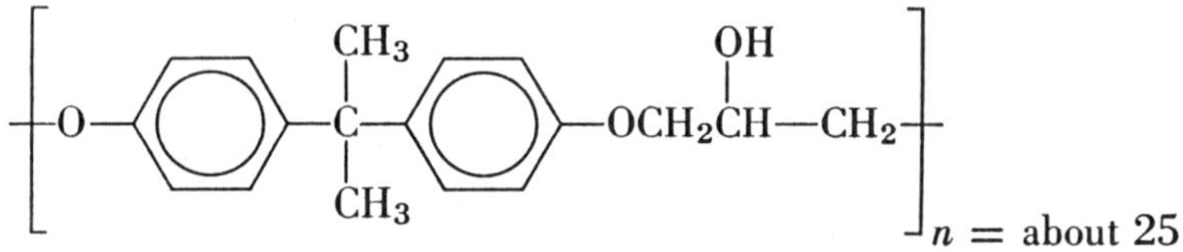

Typical phenoxy resin structure

Resins used for coatings are linear thermoplastic polymers with molecular weights from 80,000 to 200,000. The coatings are applied from solution, forming films 0.4 to over 2 mils thick upon evaporation of the solvent. Formulations with varying drying times are available, depending on the organic solvent employed. Coatings with methyl ethyl ketone solvent, for instance, are very fast drying, whereas those using Cellosolve acetate are slower. Intermediate drying times are obtained by using blends of these two solvents.[43]

Phenoxy coatings have not been used extensively for electrical or electronic applications. They have found widest use as coatings for can linings and as primers for epoxy, vinyl, or acrylic topcoats. Although the electrical and physical properties are good under ambient conditions, they deteriorate rapidly above 80°C. Phenoxies are therefore marginal coatings for most electronic applications, where operating or hot-spot temperatures are higher than 80°C. Still another problem which has limited the use of phenoxies is the difficulty of complete solvent release, causing rapid formation of a surface film and entrapment of solvent within the coating. Such entrapment of solvent, of course, reduces both physical and electrical properties. It is reported, however, that solvent release may be improved by introducing pigments or fillers in the coating or by force-drying.

Phenoxy resins may be cross-linked by the interaction of its many available hydroxyl groups with other resin polymers. New polymers resulting from interaction with polyurethanes, urea formaldehyde, melamine, and phenolic resins are reported to have improved physical and chemical resistance.[44]

SILICONES

Silicones, unlike other polymers discussed in this book are not entirely organic. Because of the presence of the silicon-to-oxygen structural backbone

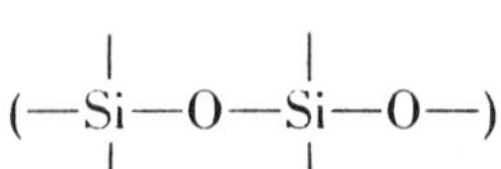

they are more accurately classified as *semiorganic*. The general structure for a linear silicone polymer may be represented as follows:

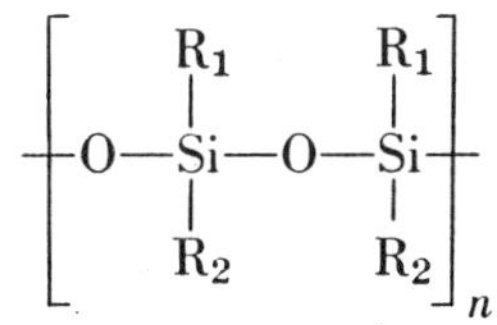

In this structure, R_1 may be the same as or different from R_2 and may represent any one or combination of organic groups such as methyl (CH_3), phenyl (C_6H_5), allyl ($—CH_2CH{=}CH_2$), or vinyl ($—CH{=}CH_2$). In the case of resins that cure through condensation, the polymer chain is normally terminated with hydroxyl groups.

The Chemistry of Silicones

Silicone resins are prepared in several steps, starting with the partial or complete hydrolysis of alkyl- or arylchlorosilanes by interaction with water. The starting materials may be mixtures of mono-, di-, and trichlorosilanes. The resulting hydrolysis products, referred to as *silanols* (Figure 1-13), are also highly reactive. They further condense with the elimination of water to give linear or cross-linked silicone resins. These resins are commercially available for further processing and application to electronic assemblies. The cross-linking of such linear polymers, or the further cross-linking of partially cross-linked resins using heat or catalysts or both, results in curing to rigid or elastomeric plastics or coatings.

There are essentially two cure mechanisms for silicone resins used as coatings. According to one mechanism, hydroxyl-terminated silicones polymerize by con-

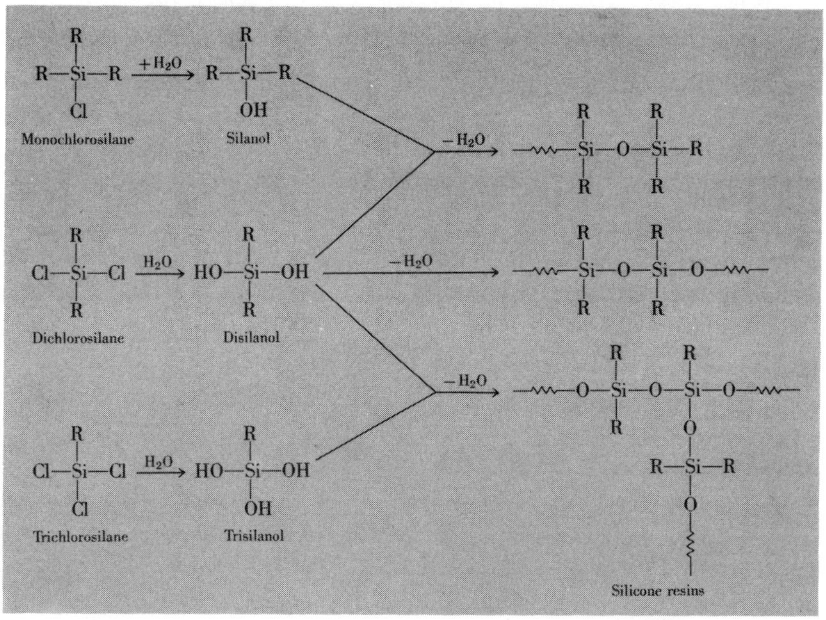

Figure 1-13: Typical reactions for preparation of silicone resins.

densation with other hydroxyl-containing silicone species or with alkoxy silanes. In this process water or an alcohol is eliminated as a by-product. The mechanism for the curing of a hydroxyl-terminated resin with propyl orthosilicate in the presence of a catalyst is given in Figure 1-14. Curing is effected at room temperature or at elevated temperatures in the presence of specific catalysts, such as dibutyltin dilaurate, zinc naphthenate, iron octoate, stannous octoate, or the metal salts of other organic acids. Amine catalysts, such as triethanolamine, may also be employed. Silicones prepared according to this procedure are often referred to as *room-temperature vulcanizing* (RTV) types. Their cures may be accelerated, however, by increasing the temperature.

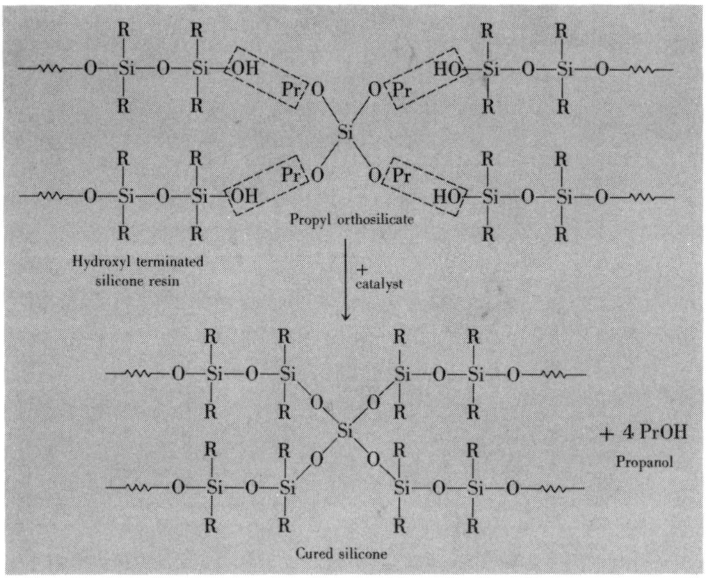

Figure 1-14: Condensation polymerization of silicones.

Dimethyl siloxane polymers may also cure by a free-radical mechanism involving the methyl groups. If ethylenic groups such as vinyl or allyl are also present in the silicone resin structure, polymerization is accelerated and the less active peroxide catalysts can be used. The catalysts used for polymerization consist of the usual free-radical-initiating types, such as benzoyl peroxide, dichlorobenzoyl peroxide, or di-tertiary butyl peroxide. The mechanism for the curing of a methyl vinyl silicone resin with dichlorobenzoyl peroxide catalyst is given in Figure 1-15. On heating, each molecule of the catalyst dissociates into two free-radical molecules (step 1). The dichlorophenyl free radical then adds to the vinyl group of the silicone resin (step 2), creating another free radical. The latter, on contacting a methyl group from a neighboring polymer chain, abstracts a hydrogen atom and creates a methyl free radical (step 3). Simultaneously the dichlorophenyl group is removed from the other polymer chain (step 4). There is now an ethyl free-radical side group on one polymer chain and a methyl free-radical group on another, which immediately

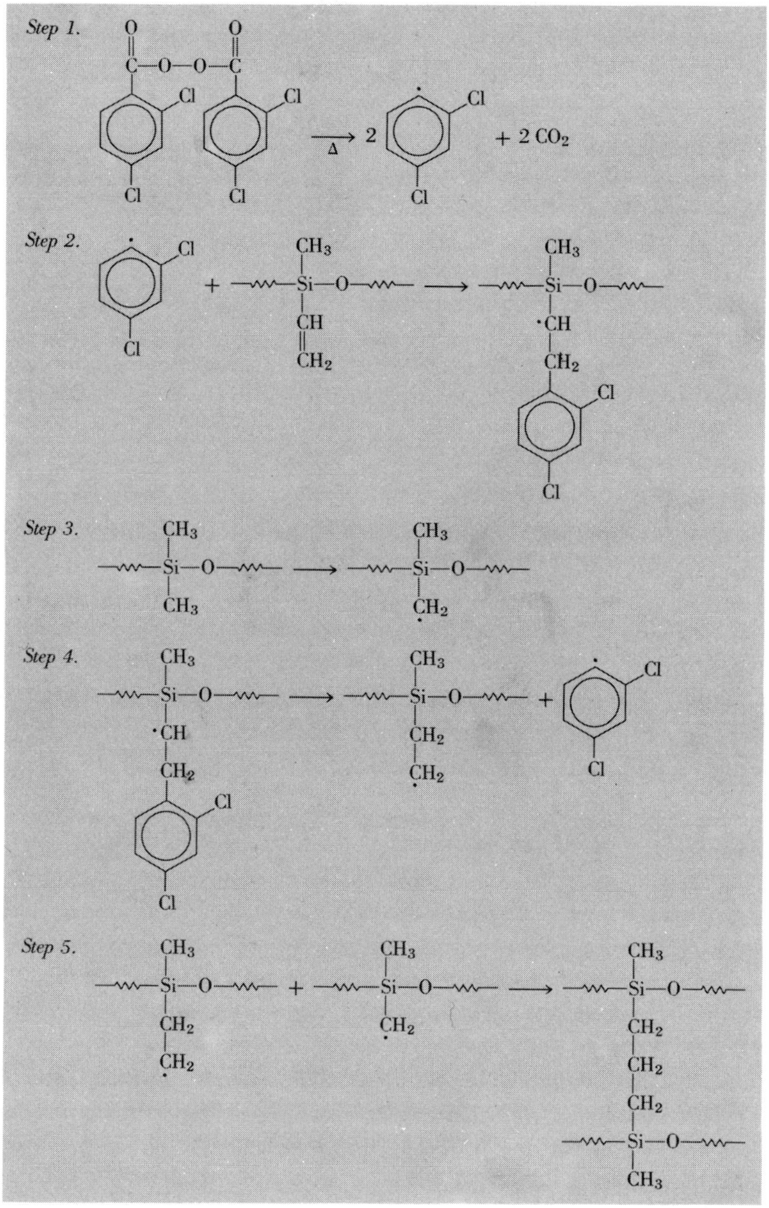

Figure 1-15: Free-radical addition polymerization of silicones.

couple to form a propyl cross-link between the polymer chains (step 5). The dichlorophenyl free radical has been regenerated and is, of course, available to repeat the process. Silicones cured in this way are referred to as *heat vulcanizing* types. They are of special interest for many electronic applications because no water or other contaminant by-products are released during cure.[45] The Dow Corning Sylgards are examples of this type. Details of the chemistry and cure mechanisms of these and other silicones are given in the literature.[46–49]

Silicone Alkyds. Silicone alkyds are resins in which the silicone structure has been chemically combined with an alkyd polymer chain. They are prepared by reacting glycerol with a silicone ester or a silanol instead of the fatty acids normally used in preparing alkyds:[46]

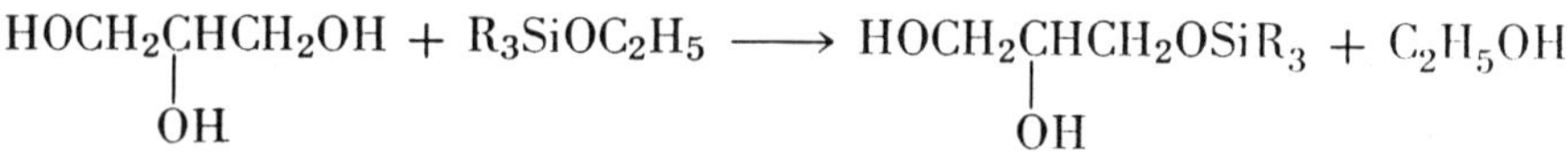

Other silicone diesters and hydroxyl-terminated silicones may be used to give structures such as

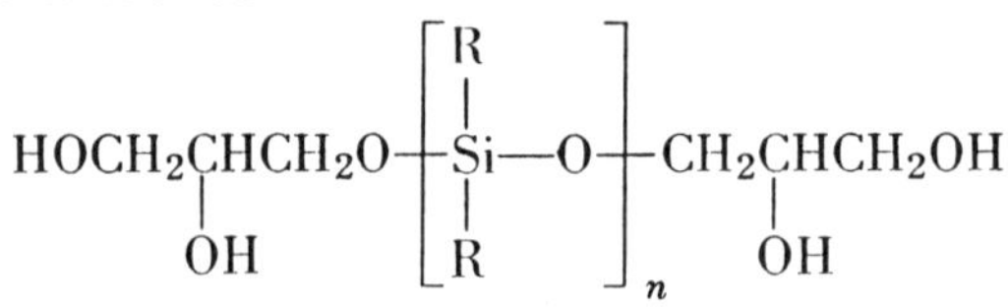

Phthalic anhydride is then condensed with this product and yields a high molecular weight, highly cross-linked silicone alkyd, of which a typical structure is:

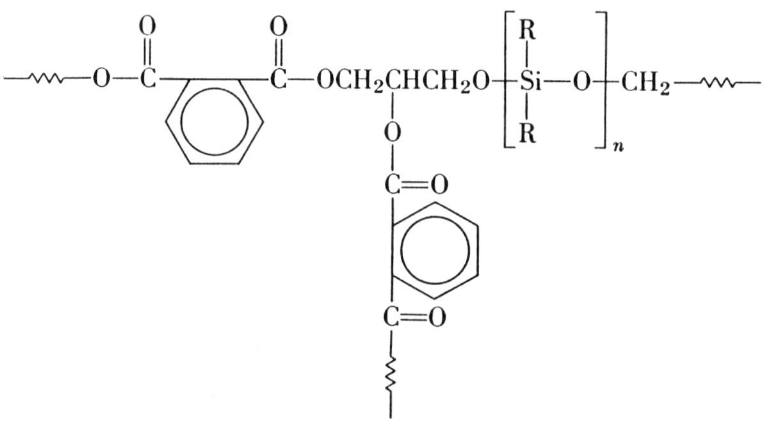

Typical structure for a silicone alkyd

It is apparent that numerous modifications of this basic chemistry are possible, giving rise to a multitude of proprietary silicone-alkyd coatings. Silicone alkyds provide coatings with greater flexibility, hardness, and thermal stability than alkyds, but naturally then would not possess the high thermal and oxidative resistance of the unmodified silicones and must therefore be considered intermediate in nature. Nevertheless, these intermediate properties are considered very important in

achieving a compromise between thermal stability and ease of handling for many applications. Silicone-alkyd coatings are widely employed as moisture-barrier and corrosion-protective coatings for electronic components such as resistors, transistors, and integrated circuits.

Halogenated Silicones. As in the case of epoxies, chlorinated or brominated silicones may be prepared, and the presence of such halogen groups renders these resins nonflammable, or less flammable than their nonhalogenated counterparts. Chlorinated diphenyl silicone, methyl chlorophenyl silicone, and many others have been prepared and characterized.[50]

Silicone Properties

Silicone formulations are available as solvent solutions, room-temperature-vulcanizing (RTV) rubbers, and solventless resins (Table 1-23). They have numerous and wide applications in the electrical and electronics industries. By far the most important use is as insulation in the form of extrusion or resin coatings for high-temperature high-voltage operating parts. For many applications silicones are in fact the only insulation that is functional. Some typical applications are:

1. Insulation for heating cable
2. Insulation for wire (hookup, aircraft, ignition)
3. Circuit-board coatings
4. Welded module conformal coatings (also referred to as *dip* or *freeze coatings*)
5. Semiconductor-junction coatings
6. Electronic-component protective coatings
7. Water repellents
8. Impregnating varnishes for coils (stator, rotor, generator, transformer)
9. Varnishes for flexible mica sheets, mica tape, glass cloth, and sleeving
10. Clear or pigmented corrosion-protective coatings
11. Ablative coatings
12. Insulating coatings for power supplies, connectors, relays and magnetic amplifiers
13. Gel coatings for integrated circuits

Some specific silicone coatings and their applications are given in Table 1-23.

The chief properties that render silicones particularly useful for all these applications and distinguish them from other coating types are the wide temperature range over which they are stable and their excellent electrical properties over an extreme range of environmental conditions. Silicones retain their high resiliency at temperatures as low as −115°C and are remarkably stable both electrically and mechanically at continuous temperatures as high as 260°C and intermittent temperatures of 315°C. The greater bond energy and stability associated with the Si-O

Table 1-23: Typical Silicone Coatings* and Their Uses

Coating	Properties and uses
Dow Corning 997	Coating and impregnating varnish for motors and transformers; reliable to 220 C.
Dow Corning 1-2577	Coating rigid and flexible circuit boards; resistant to ultraviolet light and dirt; meets MIL-I- 46058C.
Dow Corning 94-003	A fluorosilicone dispersion coating used to control corrosion.
Dow Corning 96-082	Impregnating high voltage coils and transformers; general purpose applications where long pot life is critical.
Dow Corning HIPEC 648	A 50-percent-solids resin in xylene for use in high temperature, high voltage applications; semiconductor passivation; alpha particle barrier.
✓Dow Corning HIPEC R-6101	Solventless coating for semiconductor junction passivation; protection of integrated and hybrid circuits, Led displays, optoelectronics; alpha particle barrier.
Dow Corning Sylgard 170	Impregnating high voltage transformers and coils; general purpose encapsulating and potting of electronic and electrical circuits.
✓Dow Corning Sylgard 182	Solventless resin with long pot life; general purpose coating.
Dow Corning Sylgard 527	Primerless gel used for sealing and protecting delicate electronic circuits and as a particle getter in hybrid circuits.
Dow Corning Q3-6636	Dielectric gel used for sealing and protecting microelectronic devices; working range of -50 to 200 C.
Dexter R62121	Silicone alkyd used as corrosion protective coating for integrated circuits and other electronic devices.
G.E. SR-17M	General purpose impregnating and dipping varnish for use on transformers, coils, flexible tapes, and rotating equipment.
G.E. SR-32	A 60-percent solids resin in toluene for use in flexible electrical tapes and an electrical insulating varnish; protection from -40 C to 220 C with 5 to 10 minute surges up to 370 C.
G.E. SR-80M	An air-drying varnish for use over 105 and 130 C insulating systems for added protection from humidity, corrosive chemicals, and dust.

(continued)

Table 1.23: (continued)

Coating	Properties and uses
G.E. SR-98	Varnish with good high-temperature strength, hardness, and abrasion resistance; used for bonding coils and insulating materials.
G.E. SR-112	A 50-percent-solids resin in xylene used for high heat- and weather-resistant coatings and in formulations for resistor coatings.
G.E. SR-125	A 50-percent-solids resin in xylene for use in formulations for resistor coatings and as a paint resin; has flexibility and excellent thermal-shock resistance.
G.E. SR-141	A 50-percent-solids resin in toluene for use in high temperature paint applications up to 590 C.
G.E. SR-224	A low-temperature-curing (50 to 100 C) flexible varnish for coating glass, mica, and polyester films.
G.E. SR-240	A 50-percent-solids resin in toluene for use in electrical applications or as an insulating varnish and in high temperature ceramic paint applications up to 760 C.
G.E. SR-882M	An 80-percent-solids resin in toluene for uses as a blending resin and as a general purpose paint resin.
Wacker Silicones SWS RSV S-7	A 100-percent-solids, radiation (Ultraviolet) stimulated vulcanizing (RSV) conformal coating.
SWS RSV S-10	A 100-percent-solids, RSV conformal coating for rigid and flexible printed circuit boards.
SWS SILGAN V-501 dispersion	A 80-percent-solids coating for use on printed circuit boards, transformers, capacitors, motors and pumps.
SWS MTV V-67	A high viscosity one component encapsulant for use in passivation and environmental protection of diodes, power transistors and hybrids.

*RTV types are listed in Tables 1-25 to 1-27

linkage, in comparison to the C-C bond found in completely organic polymers,[29] provides this higher continuous-service temperature and greater thermal stability. Silicones have been classified as Class H insulation; that is, they may be used safely up to 180°C for long periods of time (5 to 10 yr) (see Figure 1-16). Before the introduction of silicones, electrical applications were essentially limited to 130°C maximum operating temperature (Class B). Sufficient data are now available to show that the life expectancy of equipment insulated with silicones may be increased 100-fold over Class B insulated equipment. Extensive data for insulated motors are available.

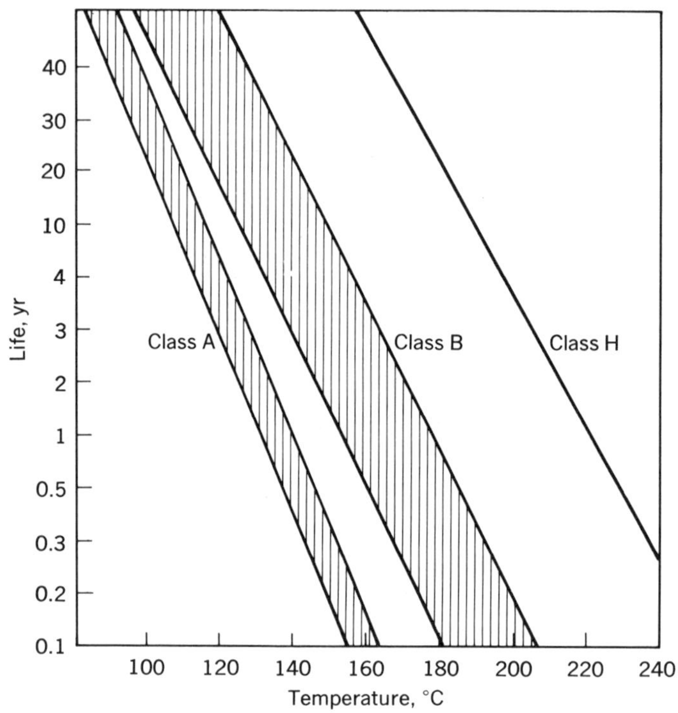

Figure 1-16: Insulation classes and their relationship to life expectancy.[51]

The electrical properties of silicones are superior to those of most polymer types, including many epoxies. For example, dielectric constants at room temperature and 100 hz for a large number of commercial formulations lie in the rather narrow range of 2.8 to 3.8. At higher frequencies, temperatures, and humidities, very little change occurs in either the dielectric constant or the dissipation factor. In fact, unlike many plastics, values may even decrease with increasing temperature (Figures 1-17 and 1-18). The *k* value for Sylgard 183, for instance, drops from 3.3 at room temperature to 2.6 at 200°C.[53] Silicones have been aged at 300°C for 1,000 hr with very little change in dielectric constant. Dielectric-breakdown voltages for silicones are high, ranging from 500 to 2,000 volts/mil, depending on the formulation and thickness of film. Volume resistivities are all 10^{12} ohm-cm or greater.

Like other polymer types, silicones have some limitations, which may be offset if they are taken into consideration. Some silicones have a tendency to migrate to other parts of an assembly, and because of their inherently good mold-release properties they may degrade the adhesion of coatings, encapsulants, adhesives, or other plastic materials. Many users resolve this difficulty by performing their silicone operations last and keeping these operations separate from other plastic operations. Separate ovens are advisable for drying or curing silicones. Besides jeopardizing the adhesion of other materials, the adhesion of silicones themselves to surfaces is often poor, or at most marginal, but this may be remedied by the use of primers.

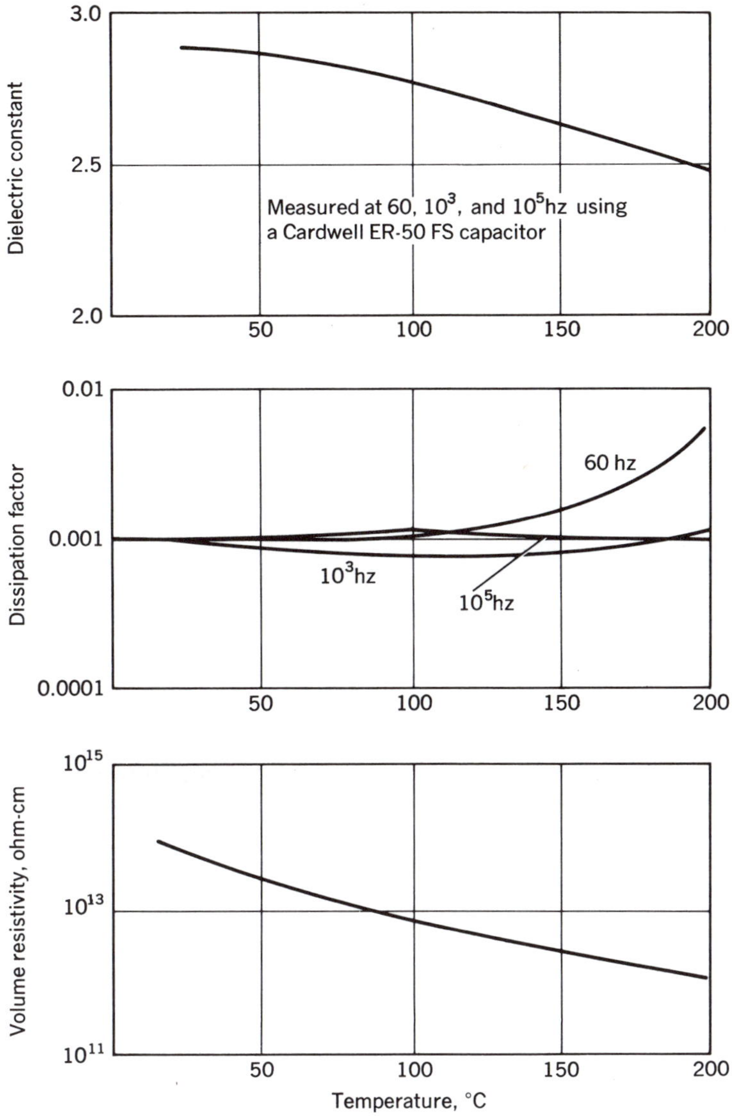

Figure 1-17: Electrical properties of cured Dow Corning Sylgard 182 resins at elevated temperatures.[52]

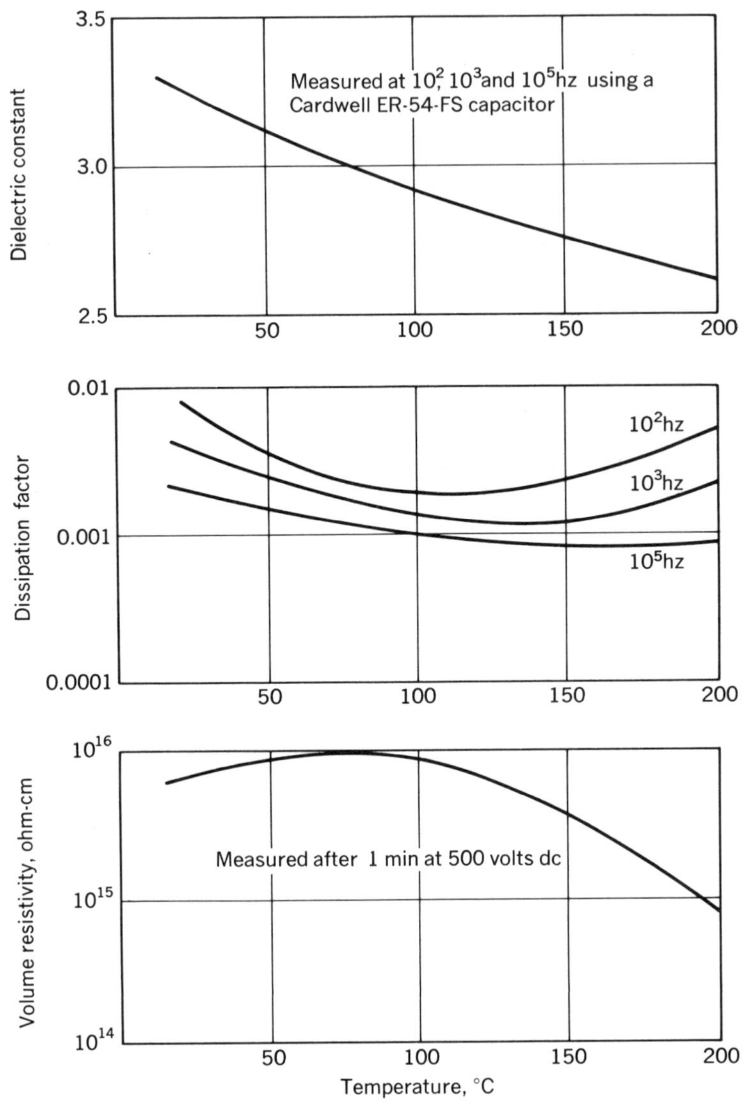

Figure 1-18: Electrical properties of cured Dow Corning Sylgard 183 resins at elevated temperatures.[53]

RTV (Room Temperature Vulcanizing) Silicones

Early silicone formulations required cures of more than 24 hr at temperatures of 120°C or higher. The introduction of the RTV silicones was a great step forward in reducing these severe curing schedules and expanding the silicone technology. The RTV silicones may be cured at room temperature or slightly above, obviating the need to expose sensitive electronic assemblies to high curing temperatures. The use of silicones was therefore extended to many new applications in which high curing temperatures could not be tolerated.

RTV silicones are available in a wide range of viscosities from several manufacturers. They may be one- or two-component systems. In the two-component systems the curing agent consists of a liquid-metal soap (metallic salt of a long-chain fatty acid). The commonly used catalysts (Table 1-24) are used in concentrations of 0.1 to 1.0 percent. Higher catalyst concentrations provide faster cures but also reduce the pot life of the mixture. Besides their ease of application and curing, RTV silicones have the advantage of easy removability and repair. Since they are rubbery and soft, they may be cut away or softened with solvents, defective parts may be replaced, and the area recoated. This is an important economic advantage for expensive electronic assemblies.

As with other silicones, the electrical properties of the RTV silicones are extremely good at room temperature, elevated temperatures, and after thousands of hours of aging at 200°C (Figure 1-19). The physical and electrical properties for RTV silicones from three manufacturers are reported in Tables 1-25 to 1-27. Some higher-viscosity RTV silicones not included in these tables may, if thinned properly with solvents such as xylene or toluene, also be used as coatings.

Table 1-24: Commonly Used RTV Silicone Catalysts

Chemical type	Suppliers
Dibutyltin dilaurate	Anderson Development, Ferro
Lead 2-ethyl hexoate	Shepherd Chemical
Stannous 2-ethyl hexoate	M & T Chemical

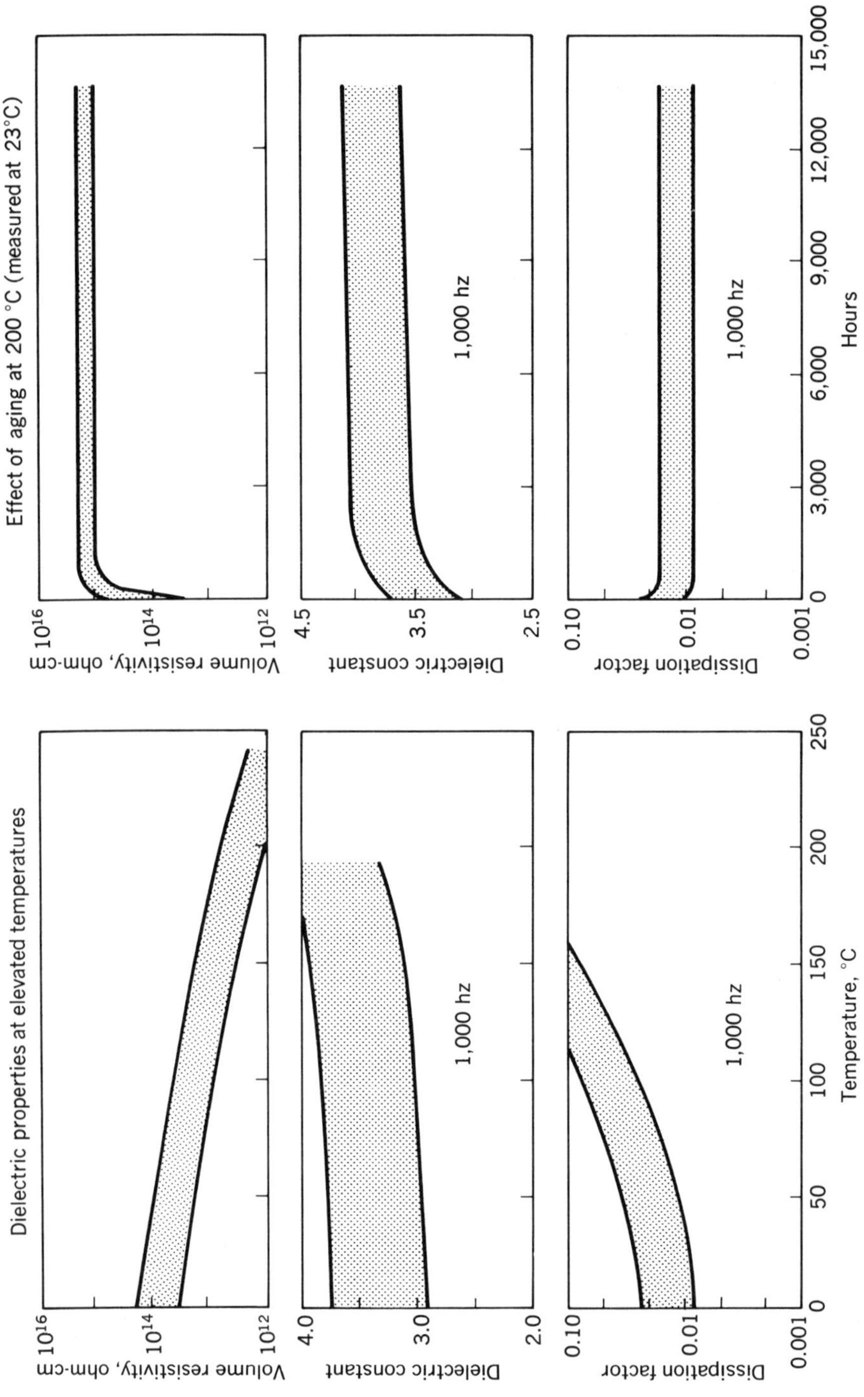

Figure 1-19: Typical electrical properties of Dow Corning RTV silicones.

Table 1-25: Properties of RTV Silicones (Dow Corning)

Property	Sylgard 170[a]	Sylgard 184[b]	Sylgard 527[c]	3140[d]	1-2577[d]	SILASTIC E[a]
Viscosity at 25 C, poises	25[e]	39[e]	3.3	35	800	600[e]
Specific gravity at 25 C	1.33	1.05	0.97	1.05	1.11	1.12
Pot life at 25 C, hr	1/4	2	2	na	na	2
Cure schedule, hr/C	8/25	24/25	24/25	72/25	24/25	24/25
Useful temperature range, C	-55 to 225	-55 to 200	-55 to 180	-55 to 200	-65 to 200[f]	-55 to 200
Radiation resistance, megarads	200	200	20	100	----	200
Thermal conductivity (25 to 100 C), gm cal/cm sec°C	7.5	3.5	na	2.91	2.9	4.53
Arc resistance,sec	120	115	182	50	na	121
Dielectric constant						
100 Hz	3.15	2.66	2.95	2.64	2.68	2.81
100 KHz	3.10	2.65	2.95	2.63	2.71	2.82
Dissipation factor						
100 Hz	0.008	0.0009	0.0025	0.0016	0.00067	0.006
100 KHz	0.002	0.001	0.00015	0.0004	0.0014	0.002
Dielectric strength, volts/mil	450	450	450	500	1100	500
Volume resistivity, ohm/cm, x 10^{14}	10	20	23.3	5.0	15	5.0

Note: All values are typical of production materials and are not intended for use in preparing specifications.

a Dow Corning 1200 RTV prime coat
b Sylgard primer
c no primer required
d Dow Corning 1204 primer
e Measured after catalyzation
f Tested according to MIL-I-46058

Table 1-26: Properties of RTV Silicones (General Electric)

Property	RTV-11	RTV-12	RTV-21	RTV-60	RTV-615	RTV-670	RTV-8111
Viscosity at 25 C, centipoises	12,000	1,900	25,000	50,000	3,500	5,000	12,000
Specific gravity at 25 C	1.18	0.99	1.31	1.47	1.02	1.06	1.18
Pot life at 25 C, hr	1-4	2	1-2	1-2	4	5	--
Cure schedule, hr/C	72/25	72/25	72/25	72/25	1/100	1.5/100	--
Useful temperature range, C	-54 to 204	-60 to 204	-54 to 204	-54 to 260	-60 to 204	-60 to 204	-54 to 204
Thermal conductivity (25 to 100 C), gm cal/cm^2/C/sec/cm x 10^{-4}	6.9	--	7.4	7.4	4.5	4.5	7.0
Dielectric constant 100 Hz	3.6	3.0	3.4	3.7	3.0	2.8	3.6
Dissipation factor 100 Hz	0.019	0.002	0.01	0.020	0.001	0.001	0.019
Dielectric strength, volts/mil	500	400	500	500	500	500	500
Volume resistivity, ohm/cm, x 10^{14}	6	0.1	8.8	1.3	10	40	6

Table 1-27: Properties of RTV Silicones (Wacker Silicones Corp.)

PROPERTY	T-67	T-81	SILGAN J-500	SWS-931	SWS-934	SWS-960
VISCOSITY AT 25°C, CENTIPOISES	80	1,000	300	--	--	4,000
SPECIFIC GRAVITY AT 25°C	0.82	0.91	--	1.04	2.4	0.93
ELONGATION, PERCENT	425	300	250	300	150	300
USEFUL TEMP. RANGE, °C	--	--	-70 TO 120	-70 TO 200	--	--
THERMAL CONDUCTIVITY (25 TO 100°C), cal/cm sec °C	--	--	--	--	15.5	5.25
DIELECTRIC CONSTANT						
100 Hz	2.61	--	2.91	2.65	--	--
1000 Hz	2.61	2.76	--	--	3.97	2.73
DISSIPATION FACTOR						
100 Hz	0.001	--	0.007	0.0003	--	--
1000 Hz	0.001	0.0003	--	--	0.006	0.0003
DIELECTRIC STRENGTH, VOLTS/MIL	750	425	500	500	360	425
VOLUME RESISTIVITY, ohm/cm, x 10	7×10^{15}	1×10^{15}	1.6×10^{15}	1.5×10^{14}	4.6×10^{14}	1×10^{15}

REFERENCES

1. Kline, G. M.: Plastics in Germany, 1939–1945, *Mod. Plastics,* **23:** 152A–P (1945).
2. Patton, T. C.: *Official Digest,* **34** (446): 342 (1962).
3. Urethane Coatings, A New Approach to the Old Problem of Preventive Maintenance, *Mobay Chem. Co. Prod. Information Bull.*
4. Stanton, J. M., and D. M. Rohalt: Oil Modified Urethanes in Industrial and Maintenance Finishes, *Paint Ind. Mag.,* July, 1961.
5. *Spencer Kellogg Tech. Bull.* TS-6821, 1968.
6. Roberts, J. D., and M. C. Casserio: "Basic Principles of Organic Chemistry," Benjamin, New York, 1964.
7. Coatings for Metals, *Mobay Chemical Corporation.*
8. *Conap, Inc., Product Bulletin* P-172.
9. *Conap, Inc., Product Bulletin* C-120a.
10. Hostettler, F., and E. F. Cox: *Ind. Eng. Chem.,* **52:** 609 (1960).
11. Armitage, F., and W. T. C. Hammond: *Chem. Ind. (London),* **1951:** 1082.
12. Martin, S. R. W.: Convertible Coatings, in C. J. A. Taylor and S. Marks (eds.), "Paint Technology Manuals," part 3, Reinhold, New York, 1962.
13. Farmer, E. H., et al.: *Trans. Faraday Soc.,* **38:** 348 (1942).
14. Farmer, E. H., and D. A. Sutton: *J. Chem. Soc.,* **1942:** 139.
15. Saunders, J. H., and K. C. Frisch: "Polyurethanes: Chemistry and Technology." part II, Interscience, New York, 1964.
16. Patton, T. C.: Fundamentals and Advances in Urethane Coating Technology, *J. Paint Technol.,* vol. 39, no. 512, September, 1967.
17. Wells, E. R., G. A. Hudson, J. H. Saunders, and E. E. Hardy: The Properties and Applications of Urethane Coatings, *Official Digest, Fed. Soc. Paint Technol.,* vol. 31, no. 416, September 1959.
18. Lee H., and K. Neville: "Epoxy Resins," McGraw-Hill, New York, 1957.
19. Lee H., and K. Neville: "Handbook of Epoxy Resins," McGraw-Hill, New York, 1967.
20. Skeist, I.: "Epoxy Resins," Plastics Application Series, Reinhold, New York, 1962.
21. Erich, W.: Literature Survey on the Synthesis of Epoxy Resins from Bis-phenols, AD158199, Picatinny Arsenal TN-12, March 1958.
22. Applegath, D. D., R. F. Helmreich, and G. A. Sweeney: Epoxy Novolacs, *SPE J.,* January, 1959.
23. Koroly, J. E.: Expanded Cellular Resinous Products Containing Polyepoxides and Aconitic Acid, U.S. Patent 2623023, Dec. 23, 1952.
24. Wear, R. L., W. St. Paul, and R. L. Bowman: Heat Curing Epoxy Resin Composition. U.S. Patent 3052650, September, 1962.
25. Floyd, D.E.: "Polyamide Resins," Plastics Application Series, Reinhold, New York, 1962.
26. Saunders, T. F., M. F. Levy, and J. F. Serino: A Mechanistic Study of the Dicyandiamide Epoxy Curing Process, *Proc. Am. Chem. Soc., Div Organic Coatings Plastics,* vol. 26; no. 1 (March, 1966).
27. Belanger, W. J.: Epoxide Resin-Amide Compositions, U.S. Patent 2928811, March 15, 1960.
28. Greenlee, S. O.: Synthetic Drying Compositions, U.S. Patent 2456408, Dec. 14, 1948.
29. Pauling, L.: "Nature of the Chemical Bond," Cornell University Press, Ithaca, N.Y., 1960
30. Licari, J. J.: High Temperature Adhesives, *Prod. Eng.,* Dec. 7, 1964.
31. Dittman, D. O., J. J. Licari, and M. Glovak: Ablative Thermal Insulation Composition of a Silicone, Epoxy Resin, Cork, and a Curing Agent, U.S. Patent 3380941, April 30, 1968.
32. Epocasts, *Furane Plastics Corp. Tech. Bull.*
33. Quant: A Low Density Potting Compound, *Sandia Corp. Preprint* SCR-417, 1961.
34. Graves, H. K., and M. A. Pizzino: Effects of High Humidity on Dielectric Properties of Casting Resins, *Elec. Mfg.,* April, 1956.

35. Comparative Properties of Filled Epoxy Electrical Potting Compounds, *Union Carbide Tech. Bull.,* August, 1959.
36. Stringfellow A. E., and D. H. Reighter: Circuit Interrupter Having Track-resistant Coating Thereon, U.S. Patent 3192347, June 29, 1965.
37. Radley, G. R.: Electrical Apparatus with a Thermal and Electrical Insulating Coating, U.S. Patent 2862838, Dec. 2, 1958.
38. Ringwood: Epoxy Resins for Encapsulation of Electric Components, *SPE J.,* January, 1960.
39. Review of Technical Papers, *Electro-Technol.,* December, 1960.
40. Torossian, K. A., and S. L. Jones: A Study of Polymers Applicable to Insulation Systems, *Seventeenth Ann. Tech. Conf. SPE,* Washington, D.C., January, 1961.
41. Ehlers, G. F. L.: Thermal Gravimetric Analysis of Polymers, *TD Rept.* ASD-TR-61-622, 1961.
42. Lee, H., D. Stoffey, and K. Neville: "New Linear Polymers," McGraw-Hill, New York, 1967.
43. Eponol 55 Resin, *Shell Chem. Co., Plastics Resins Div. Bull.* SC:61-160, October 1961.
44. Eponol 53-B-40 Resin Solution, *Shell Chem. Co., Plastics Resins Div. Bull.* SC:64-51, September, 1964.
45. Lockhart, F. J.: The Role of Silicone Packaging Materials in the Semiconductor Industry. *Proc. Natl. Electronic Packaging Prod. Conf., Long Beach, Calif.,* January, 1967.
46. McGregor, R. R.: "Silicones and Their Uses," McGraw-Hill, New York, 1954.
47. Rochow, E. G.: "An Introduction to the Chemistry of Silicones," Wiley, New York, 1951.
48. Eaborn, C.: "Organo-Silicon Compounds," Butterworth, London, 1960.
49. Noll, W.: "Chemistry and Technology of Silicones," Academic Press, New York,1968.
50. Rochow, E. G.: Halogenated Aryl Silicones, U.S. Patent 2258219, October, 1941.
51. Meals, R. N., and F. M. Lewis: "Silicones," Plastics Application Series, Reinhold, New York, 1961.
52. *Dow Corning Tech. Bull.* 07-157, March, 1965.
53. *Dow Corning Tech. Bull.* 07-103, December, 1963.

2

Chemistry and Properties of Coatings

Polyimides—Fluorocarbons—Polyxylylenes

POLYIMIDES

Polyimides were developed by Du Pont and first introduced on the market in 1961. They are often confused with polyamides because of the similarity in names, but in fact they are quite different, both structurally and in other respects. Polyimides possess a highly aromatic ring structure and as such have exceptional thermal stability even at temperatures of 500 to 600°C for short periods of time. They belong to a new group of plastics with the highest thermal stability developed to date. This group also includes the polybenzimidazoles, polyquinoxalines, and polybenzothiazoles.[1]

The Chemistry and Structure of Polyimides

Polyimides can be divided into two types, those that are synthesized by condensation polymerization and those that are synthesized by addition polymerization. Both types are initially formed by the condensation polymerization of dianhydrides with aromatic diamines and are characterized by the repeating phthalimide unit.

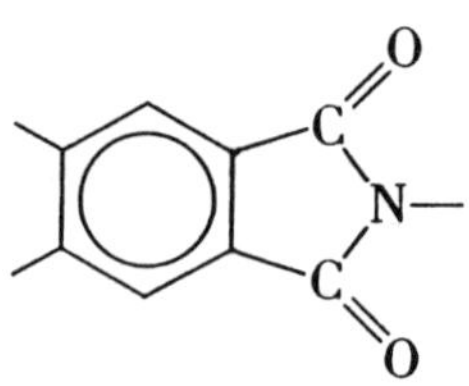

Phthalimide unit

Condensation-type Polyimides. A common procedure is to condense pyromellitic dianhydride with bis(4-aminophenyl)ether, as shown in Figure 2-1. The reaction proceeds according to two steps:

1. The formation of an intermediate polyamic acid
2. Ring closure through the elimination of water to yield the high molecular weight polyimide

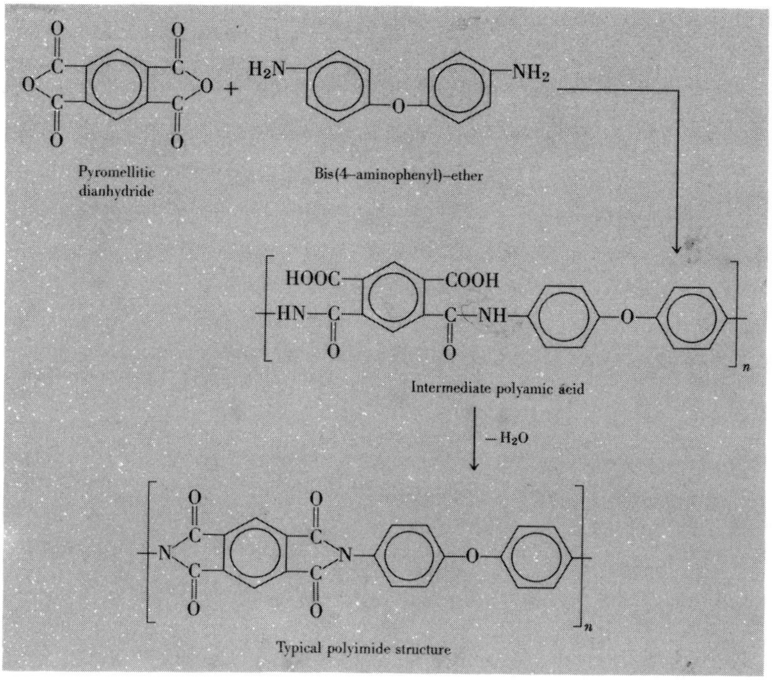

Figure 2-1: Preparation of a typical polyimide polymer.

The commercially available enamels and varnishes are solutions of polyamic acids which convert on baking to the more thermally stable polyimides. Minimum temperatures of 120 to 150°C are necessary for cyclization (ring closure), but quantitative conversion requires temperatures of 300 to 400°C for several minutes to 1 hr. The mechanism and kinetics of polyimide formation have been studied in detail.[2] The high temperatures needed for the complete conversion of polyamic acid to polyimide limits the use of these polymers to applications where such temperatures can be tolerated. However, some formulations employing benzoguanamines and *s*-triazines have been developed which have more moderate baking temperatures of 125 to 205°C.[3]

An interesting and useful method for producing precisely-defined patterns of polyimide insulation makes use of the two-step cure process and photolithography. The polyimide is first applied to the substrate and soft-baked (B-staged) to remove

solvent, leaving the coating as the polyamic acid. Then photoresist is applied, exposed, and developed and the exposed polyimide is dissolved with an alkaline solution such as ammonium hydroxide. Finally, the photoresist is removed and the patterned polyimide is fully cured to the polymer (C-staged). By this process, insulation layers with precisely-defined vias can be formed and may be used in producing thin-film multilayer interconnect substrates.[4]

Many variations of the polyimide structure are possible, and varying properties have been obtained, depending on the diamine and anhydride used. For instance, polyimides derived from bis(4-aminophenyl)thioether are stable for 3 months in boiling water, whereas those derived from meta- or paraphenylenediamine become brittle after only a week under the same conditions. A wide range of dianhydrides and diamines have been explored as intermediates for polyimides. Some diamines that have been used are shown in Figure 2-2.

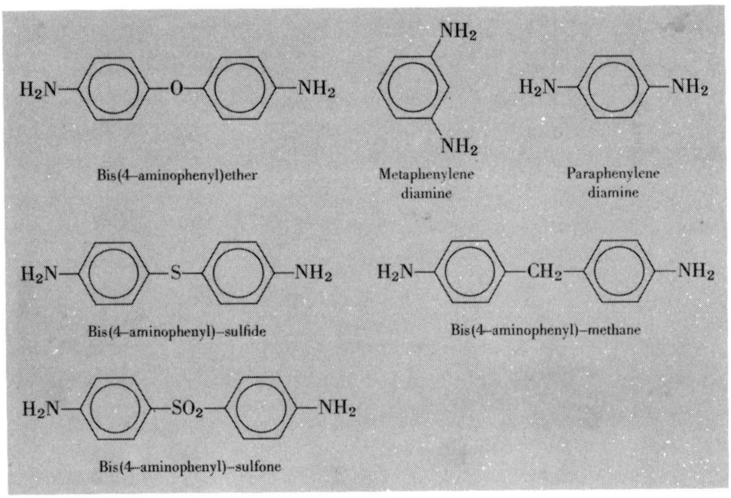

Figure 2-2: Diamines commonly used as polyimide intermediates.

During condensation polymerization large amounts of water are released as a byproduct and can remain in the polymer yielding a porous structure. Also, the water itself can cause failure of devices and circuits through electrical leakage or corrosion if other contaminants are present. To avoid this, polyimides were developed which polymerize through an addition process.

Addition-type Polyimides. The first addition-type polyimides were developed by Hughes Aircraft under U.S. Air Force contracts.[5] These polyimides are polymerized through unsaturated end groups such as acetylene, nadic anhydride, cyclohexene, and others (Figure 2-3).

Imidized prepolymers, called oligomers, are formed by reacting a dianhydride with a diamine, followed by the addition of the modified amine which contains the reactive terminal group (Figure 2-4). The reaction is performed in a high boiling solvent such as N-methylpyrrolidone (NMP) or dimethyl formamide (DMF) so that the water evolved during condensation is readily boiled off. The oligomer is col-

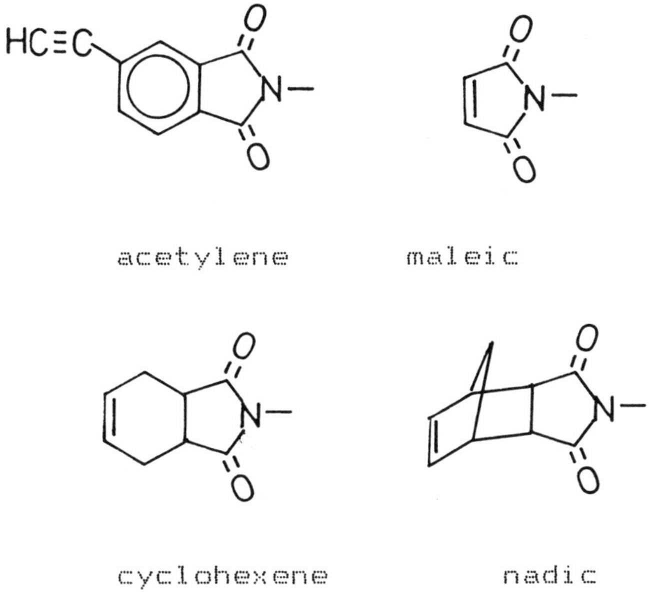

Figure 2-3: Crosslinking end groups for addition type polyimides.[6]

lected and dissolved in a polar solvent such as tetrahydrofuran (THF) or acetone. The resulting polymers are thermosets which cure into a highly cross-linked network. They differ from condensation polymers which result in linear structures. Cross-linking improves overall strength; however, the coatings produced are often brittle.[6]

Addition-type polyimides are cured at temperatures of 200°C or above giving materials with high glass-transition temperatures (Tg). These polymers are currently used as high-temperature-stable structural adhesives,[6] wire coatings,[7] and, with the addition of a filler, as molding compounds.[8]

Mechanical strength and flexibility depend on the size of the oligomer which is indicative of the amount of cross-linking that will occur during cure. For instance, the larger the oligomer, the lower the degree of cross-linking and therefore the greater the flexibility. The structures and properties of fourteen acetylene-substituted polyimides are described by Bilow.[7]

The thermal and electrical properties of addition-type polyimides are comparable to those of condensation types. Dielectric constants range from 2 to 3 and long term thermal stabilities from 300 to 400°C.[6,7]

Solvent-Soluble Polyimides. Polyimides are inherently insoluble in organic solvents. This limits their usefulness for microelectronic applications such as hybrid circuits or printed circuit boards where easy removal of the coating is required to rework and repair components. Variation of the diamine, the anhydride, or both has resulted in solvent-soluble polyimides, some examples of which are shown in Figure 2-5. The approach has been to incorporate bulky groups such as trifluoromethyl or sulfone, bridging groups, and meta-linked diamines to reduce charge-transfer complexing in aromatic polyimides and thereby facilitate solubility.[9]

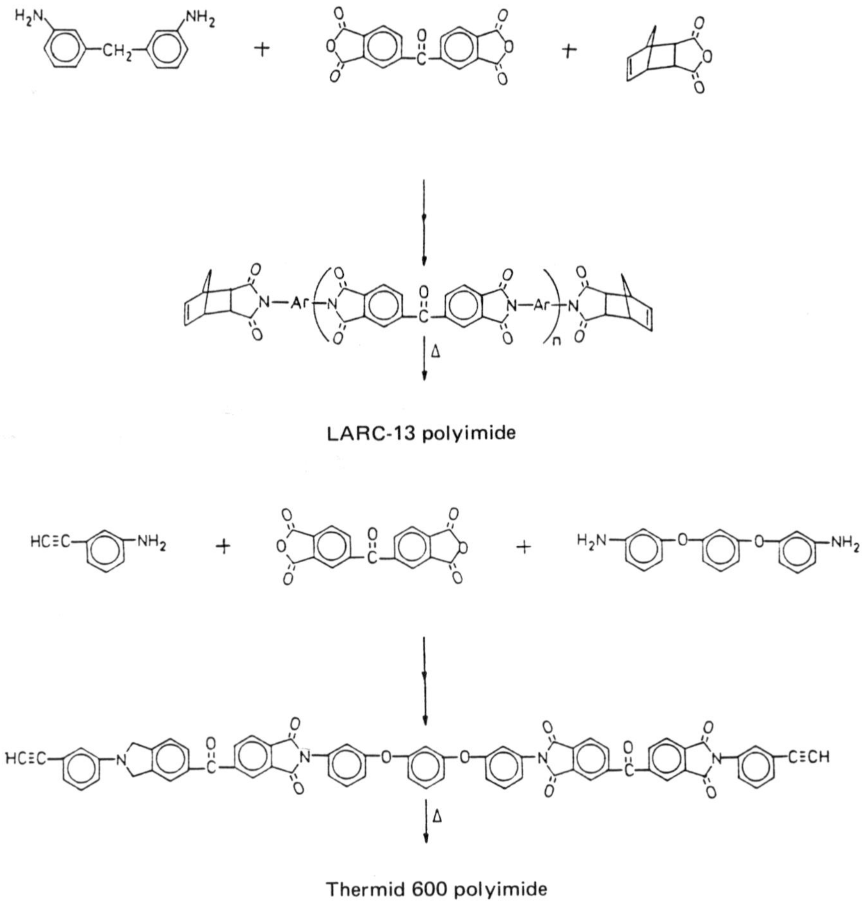

Figure 2-4: Starting materials and imide prepolymers for two addition-type polyimides.[6]

Although these polyimides are condensation types they are supplied fully imidized (prepolymerized) and are therefore stable at room temperature and have long shelf lives. Because they are preimidized no water is released during the cure process. Some solvent-soluble polyimides can be thermally cross-linked (at temperatures ranging from 200 to 350°C) to form a dense coating having little or no shrinkage.[10] Once cross-linked, the polyimide coatings are chemically resistant to most commercially available organic solvents.[10]

The ability to precisely control the viscosity of the polyimide allows the application of very thin coatings (0.5 to 5 microns). This has resulted in their use as passivation coatings,[11,12] interlayer dielectrics,[13–15] x-ray masks,[16] planarization coatings,[17,18] and alpha-particle barriers[19,20] for integrated circuits. The physical and electrical properties of one solvent-soluble polyimide coating are shown in Table 2-1.

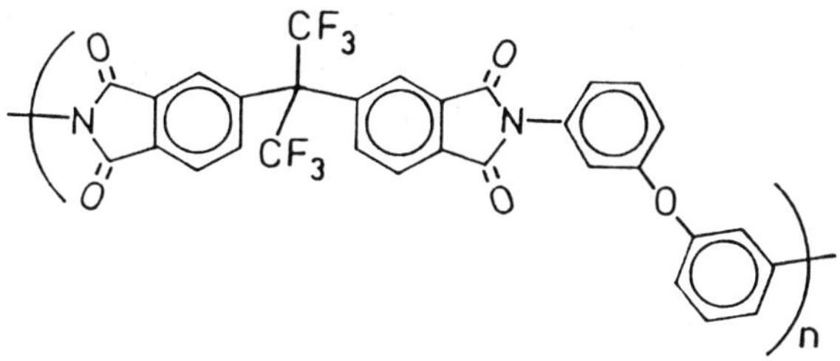

Polyimide derived from 2,2-bis(3,4-dicarboxyphenyl)hexafluoropropane dianhydride + 3,3'-oxydianiline diamine[9].

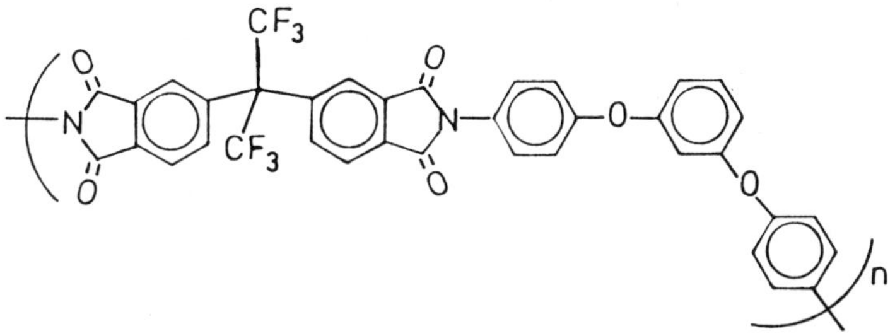

Polyimide derived from 2,2-bis(3,4-dicarboxyphenyl)hexafluoropropane dianhydride + 1,3(4-aminophenoxy) benzene diamine[9].

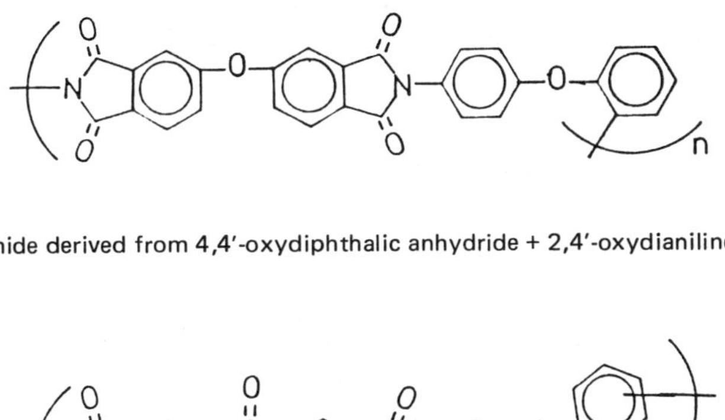

Polyimide derived from 4,4'-oxydiphthalic anhydride + 2,4'-oxydianiline diamine[9].

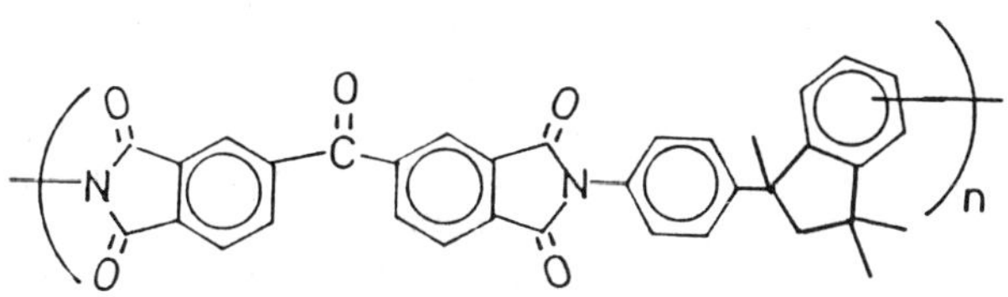

Polyimide derived from 2,2-bis(3,4-dicarboxyphenyl)carboxy dianhydride + diaminophenyl trimethyl indane[21].

Figure 2-5: Examples of solvent-soluble polyimides.

Table 2-1: Physical and Electrical Properties of Solvent-Soluble Polyimide, Probimide Series 200 (Ciba-Geigy)

Physical Properties	
Glass transition temperature (Tg)	320°C
Coefficient of expansion, 20°C	28ppm/°C
Thermal conductivity	39 × 10^{-5} cal/cm-sec-°C
Electrical Properties	
Dielectric constant, 25°C, 1kHz	3.3
Dissipation factor, 25°C, 1kHz	0.003
Dielectric strength	220v/micron
Volume resistivity	4.4 × 10^{16} Ohm-cm

Photosensitive Polyimides. A variation on solvent-soluble polyimides is to use a photosensitive side group in the polyimide structure, thereby making the polyimide photoimageable.[21] These polyimides are discussed in detail in Chapter 9.

In addition to the polyimides, a number of amide-imide and ester-imide copolymers have been synthesized. The amide-imide polymers are formed by the reaction of dianhydrides with amide-modified aromatic diamines.[22]

Properties of Polyimides

The outstanding properties of polyimides include:

1. High-temperature oxidation stability and low weight loss
2. Temperature stability from -190 to 600°C
3. Flame resistance (since polyimide will not support combustion in air and will ignite only at temperatures above 400°C)
4. High wear resistance
5. High radiation resistance
6. High chemical and solvent resistance

Thermal Stability. By far the most valuable property of the polyimide polymers, and the reason they are replacing many other materials, is their high-temperature stability and their retention of physical and electrical properties at high temperatures. Practical continuous-service temperatures range from 150 to 370°C, with stability up to 600°C for short periods of time. Softening points are generally above 700°C, although some investigators report that the material only decomposes gradually with charring. The polyamide-imide copolymers, however, are reported to have distinct melting points at about 400°C.

Another indication of the high thermal stability of polyimides is their low weight loss. Extensive weight-loss data at various temperatures and times are reported in the literature.[23–25]. Typical weight-loss values for the polyimide enamel Pyre-M.L. (trade name of Du Pont) aged in air are given in Tables 2-2 and 2-3. Weight loss for thin polyimide films in air at 320°C is low in comparison to that for phenolic, modified phenolic, and amide-imide films but slightly higher than for silicones (Figure 2-6). Weight losses in an inert environment are much lower: for instance, 1.5 percent at 400°C and 7 percent at 500°C after 15 hr.

Table 2-2: Weight Loss of Pyre-M.L. Enamel at Elevated Temperatures[27]

Time (hr)	Percent Loss
At 220°C	
2,300	3.3
6,900	4.8
13,000	5.6
At 240°C	
2,300	5.0
6,200	9.0
12,500	13.4
At 260°C	
2,200	10.5
6,500	23.2
11,500	27.1
At 300°C	
5	0.75
144	4.1
195	5.1
404	7.0
500	8.3
645	8.3
808	8.3

Table 2-3: Weight Losses for Pyre-M.L. Resin Coatings[26]

Solids Content (%)	Solvent	Viscosity (poises)	Temperature (°C)	Time (hr)	Weight Loss (%)
16.5	*N*-Methyl pyrrolidone/ dimethylacetamide	70	300	50	1
			400	3.3	11
			450	3.3	20
25	Dimethylformamide	35	300	50	1
			400	3.3	12
			450	3.3	35
25	Dimethylformamide	8	300	50	4
			400	3.3	11
			450	3.3	46
60	*N*-Methyl pyrrolidone/ xylene	80	300	50	5
			400	3.3	12
			450	3.3	31

The thermal stability of polyimides has also been demonstrated by the use of both *thermal gravimetric analysis* (TGA) and *differential thermal analysis* (DTA). Here again, the higher thermal stability of samples heated in an inert (argon) atmosphere versus air is apparent (Figure 2-7). A comparison of the TGA curves for a number of aromatic polymers with those for polyimides is shown in Figure 2-8.

The coefficients of thermal expansion (CTE) for polyimide coatings and varnishes lie between 2×10^{-5} and 4×10^{-5} in./(in.)(°C) and in this respect are in the same range as many other polymeric materials, including epoxies and fluorocarbons.[26] However, both DuPont and Hitachi have introduced highly densified polyimides that have expansion coefficients of 3 to 5×10^{-6} in./(in.)(°C) closely matching the CTE of silicon. Thermal-conductivity values at 32° to 300°C lie between 3.72×10^{-4} and 4.51×10^{-4} cal/(sec)(cm)(°C).[21] The thermal conductivity of Pyre-M.L. was also reported by another source[26] as 3.5×10^{-4} cal/(sec)(cm)(°C).[21]

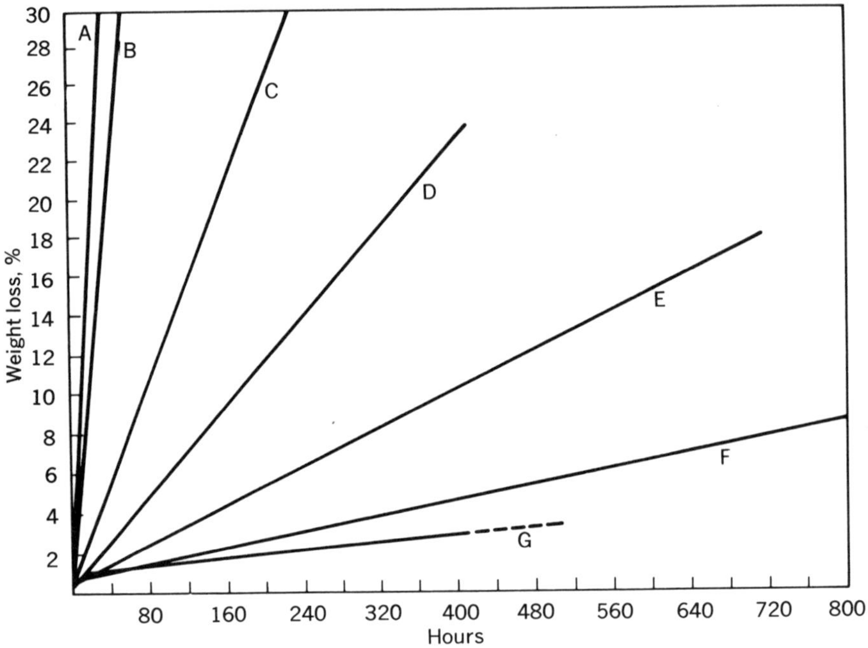

Figure 2-6: Comparative weight loss of thin films as a function of time, in air at 320°C: (A) phenolic, (B) modified phenolic, (C) aromatic amide-imides, (D) aromatic amide-imides, (E) aromatic polyimides, (F) aromatic polyimides, (G) silicone.[28]

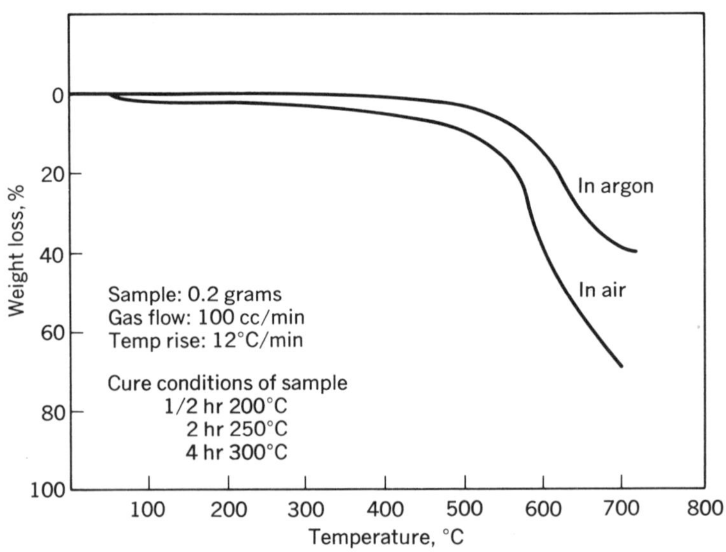

Figure 2-7: TGA curves for polyimide resin solutions.[29]

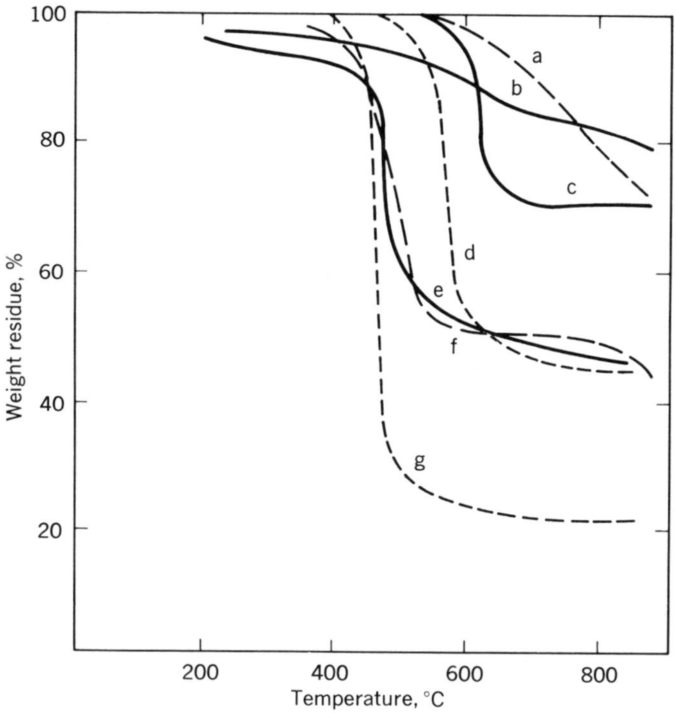

Figure 2-8: TGA curves for polyaromatic polymers: (a) polybenzimidazole, (b) polyphenylene, (c) polyimide, (d) polyphenylether, (e) polyoxadiazole, (f) polyphenyl sulfide, (g) polycarbonate.[30]

Solvent and Chemical Resistance. The resistance of polyimide varnishes and coatings to organic solvents, moisture, and synthetic lubricants and greases is very high. Only slight deterioration occurs when polyimides are immersed up to 4 weeks in organic solvents, including ketones, alcohols, aldehydes, chlorinated solvents, benzene, and naphtha. The coatings are also not affected by exposure to standard synthetic greases, either for 3 months at room temperature or for 6 hr at 285°C. The amide-imide enamels also have excellent solvent resistance. A comparison of this type of enamel with polyimide and with a standard magnet-wire coating, is given in Table 2-4.

The resistance of polyimides to strong acids and bases is, however, not as good as its resistance to organic solvents; deterioration of some form or other occurs. In concentrated sulfuric acid the coating dissolves after 1 day immersion, and in a 5 percent sulfuric acid solution it becomes stiff in 3 weeks. In concentrated hydrochloric acid the coating becomes stiff and cloudy in 3 weeks, and in a 5 percent hydrochloric acid solution it stiffens slightly in 4 weeks. In concentrated nitric acid embrittlement and deterioration occur in only 1 day. The coating also stiffens on immersion in concentrated and dilute alkalis such as sodium hydroxide and ammonium hydroxide.[31]

Table 2-4: Comparative Properties of No. 18 AWG Magnet Wires[32]

Chemical Properties	Wire Coatings Amide-Imide*	Standard**	Polyimide***
Solvent shock, 15% prestretch 3X wrap in boiling xylol	OK	OK	OK
Solvent resistance, 24 hr immersion followed by cheesecloth wipe			
Toluol	OK	OK	OK
Xylol	OK	OK	OK
5% H_2SO_4	OK	OK	OK
Perchloroethylene	OK	OK	OK
Ethanol	OK	OK	OK
1% KOH	OK	OK	OK
Methylene chloride	OK	Fails	OK
Scrape abrasion under liquid R-22, strokes to failure	1,000+	225	90
Soxhlet extraction in toluol methanol, %	0.5	1.4	0.5
Soxhlet extraction in R-22, %	0.2	0.3	0.3
Blister test in R-22	OK	Blisters	OK

*Standard A.I.-220.
**Heavy Isonel-200 17.
***Heavy Pyre-M.L.

Moisture Resistance. The percentage of water absorption of polyimides after 24 hr of immersion at room temperature ranges from 0.5% to several percent depending on the formulation and degree of cure (see Table A-7). Nevertheless, hydrolytic stability is reported to be excellent. Polyimide-coated glass fabrics sealed in a tube with water showed no deterioration after 7 weeks at 200°C. Du Pont also reports excellent protection of polyimide-coated steel and aluminum after 4 months at 100 percent relative humidity and 38°C. In addition, fully cured free films have withstood boiling water for a 1-year period with no deterioration.

Electrical Properties. Considerable data are available on the effect of high-temperature and long-term thermal aging on the electrical properties of polyimides. In general, electrical properties are very stable over a wide temperature range, as illustrated by the dielectric-constant and dissipation-factor values in Table 2-5. These values will vary appreciably with the degree of cure; in fact, changes in dissipation factor have been utilized for the quality control of wire insulation during production.

The dielectric strength of polyimide coatings can range from 3,000 volts/mil to 6000 volts/mil, depending on the formulation, thickness of the sample, and many other factors. Dielectric strengths for polyimide-amide coatings are also high, ranging from 4,400 to 4,700 volts/mil. Exposure to water for 3 months at 70°C affected the breakdown voltages only slightly, reducing them to 4,000 to 4,300 volts/mil, respectively.[25]

FLUOROCARBONS

The Chemistry of Fluorocarbons

The two main types of fluorocarbon coatings available commercially are Teflon TFE (tetrafluoroethylene), produced by Du Pont and also known as PTFE (polytetrafluoroethylene), and Teflon FEP (fluorinated ethylene propylene). Teflon TFE is

Table 2-5: Electrical Data for Polyimide Pyre-M.L. Enamel[27]

Frequency, hz	Room temp.	150°C	200°C	225°C	250°C	285°C
	Dissipation factor					
10^2	0.0080	0.0087	0.0095	0.0088	0.0114	0.0192
10^3	0.0075	0.0093	0.0101	0.0099	0.0105	0.0140
10^4	0.0096	0.0125	0.0137	0.0131	0.0119	0.0162
10^5	0.0115	0.0114	0.0133	0.0141	0.0139	0.0168
	Dielectric constant					
10^2	3.99	2.29	2.30	2.31	2.36	2.29
10^3	3.95	2.25	2.27	2.28	2.30	2.24
10^4	3.91	2.22	2.24	2.25	2.28	2.20
10^5	3.84	2.17	2.18	2.20	2.23	2.15
10^6	3.67	2.28	2.25	2.29	2.18	2.25

prepared by the polymerization of tetrafluoroethylene, a gaseous monomer, whereas Teflon FEP is formed by the copolymerization of tetrafluoroethylene and hexafluoropropylene. Structures and reaction steps are given in Figure 2-9.

Other fluorocarbon resins include polychlorotrifluoroethylene (CTFE), fluorinated acrylate polymers such as the Fluorad coatings (trade name of 3M Corporation), and vinylidene fluoride, discussed under vinyl polymers in Chapter 3.

Fluorocarbon coatings are used as protective coatings,[33,34] release coatings for photolithographic processes,[35] and barrier coatings to deter lubricant migration.[36]

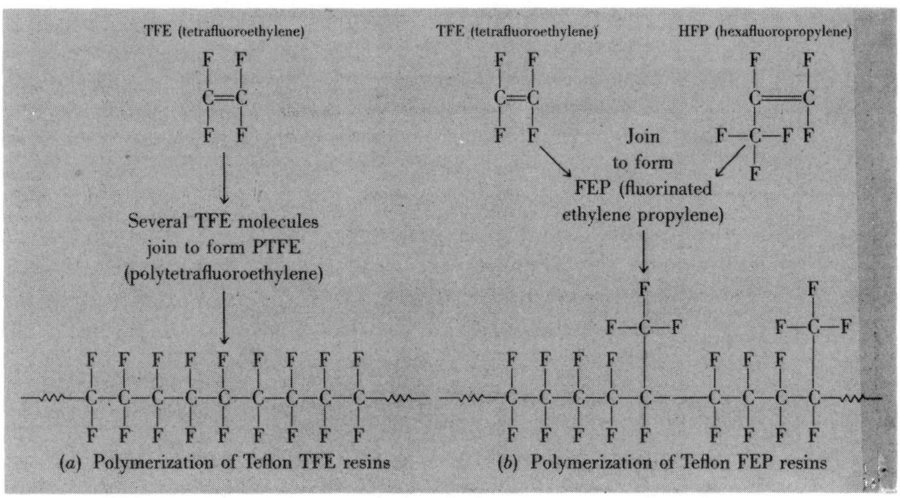

Figure 2-9: Polymerization diagrams for fluorocarbon polymers.

Properties of Fluorocarbons

Regardless of the formulation or method of application, both TFE and FEP coatings have many outstanding properties in common:

1. High thermal stability; continuous-service temperature 200 to 260°C
2. Excellent electrical insulating and dielectric properties up to 260°C
3. High inertness to chemicals and solvents
4. Extremely high degree of purity; almost no side reactions, byproducts, or contaminants
5. Lowest coefficient of friction of any solid material

All these unique properties may be attributed to the fluorocarbon polymer structure. The molecules consist of closely packed long linear chains of carbon atoms protected by a tightly held sheath of fluorine atoms. Since the carbon-to-fluorine bond strength is one of the highest among atom combinations, fluorocarbon resins are inert to both chemical attack and high-temperature exposure.

Because of their properties, fluorocarbon coatings have found many uses in the electronics and electrical industries as capacitor-layer insulation; coatings for glass fabrics and yarns for electrical insulation of motors, generators, transformers, and wire; enamel for magnet wire insulation; and as low-friction, corrosion-protective surface coatings.[37]

Electrical Properties. Of all the known polymers, Teflons TFE and FEP have the best electrical properties and the best retention of electrical properties up to 260°C. The dielectric constant of Teflon is 2.0 to 2.2, among the lowest for polymeric materials (see Appendix). The dielectric constant of Teflon FEP remains essentially unchanged from room temperature to 205°C, and that of Teflon TFE remains unchanged to 260°C, in the frequency range of 60 to 3 X 10^9 hz. These fluorocarbons also exhibit one of the lowest dielectric power losses over the same temperature and frequency ranges. Dissipation factors range from 0.001 to 0.0001 (Table 2-6), with Teflon TFE having somewhat lower values than Teflon FEP (Figure 2-10).

Table 2-6: Dielectric Constant and Dissipation Factor, for Teflon FEP, per ASTM D 150-59T, 73° to 100°F[38]

Frequency	Dissipation factor	Dielectric constant
100 hz	0.0003	2.1
1 khz	0.0002	2.1
10 khz	0.0002	2.1
100 khz	0.0005	2.1
1 Mhz	0.0007	2.1
50 Mhz	0.0006	2.1
3,000 Mhz	0.0004	2.1

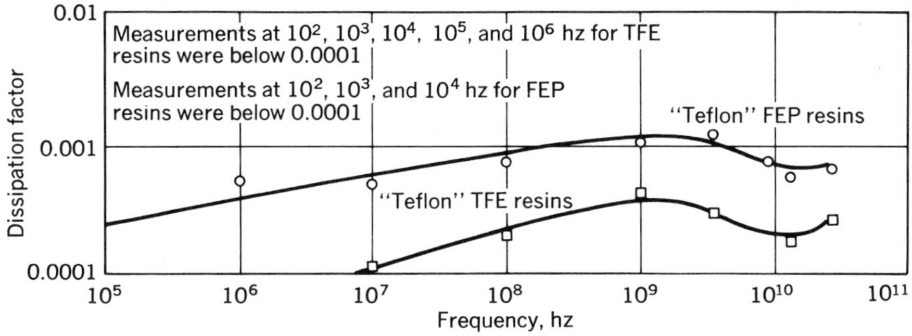

Figure 2-10: Dissipation factor versus frequency for Teflon resins (at 23°C, according to ASTM D 150-54T).[38]

Surface and volume resistivities are very high for both TFE and FEP. Surface resistivity, measured according to ASTM D 257–52T, is greater than 10^{15} ohms, and volume resistivity, according to the same procedure, is greater than 10^{18} ohm-cm. In both cases there is little variation over the temperature range of -40 to 230°C. Dielectric-strength values are also very high and are given in Table 2-7 as a function of film thickness from 1 to 100 mils.

Table 2-7: Dielectric Strength for Teflon FEP, per ASTM D 149-44, Short-Time Test[38]

Thickness, mils	*Volts/mil*
1	4,000
2	3,500
5	2,700
10	2,100
15	1,750
50	840
100	500

Thermal Properties. Teflon TFE has a continuous-service temperature of 260°C and may be used for short periods of time up to its transition temperature of 325°C. Specimens have been heat-aged at 300°C for 6 months with no change in dielectric constant. Teflon FEP has a lower continuous-service temperature of 205°C but may be used for short periods of time up to its melting temperature of 290°C. Specimens have been heat-aged at 260°C without any change in dielectric constant. Although the maximum service temperature of FEP is lower than for TFE, it is considerably easier to process and apply.

Other thermal properties for these two fluorocarbons may be found in Table 2-8. Note that Teflon TFE shows a very low weight-loss rate even at temperatures above 205°C (Table 2-9).

Mechanical Properties. Two outstanding properties of Teflons are their low coefficients of friction—the lowest for any known plastic materials—and their high

Table 2.8: Comparative Thermal Properties of Fluorocarbons

Property	Teflon FEP	Teflon TFE
Melting point	250°–285°C	325°C*
Thermal conductivity (ASTM C 177), cal/cm sec °C	6×10^{-4}	7.0×10^{-4}
Specific heat, Btu/(lb) (°F)	0.28	0.25
Coefficient of linear expansion (−100° to 160°F, per ASTM D 696), cm/(cm) (°F)	$4.6 - 5.8 \times 10^{-5}$	5.5×10^{-5}

*Changes to a rubbery, noncrystalline state.

Table 2.9: Weight-Loss Rate Data for Teflon TFE[39]

Temperature (°C)	Weight Loss (%/hr)
205	0.0002*
260	0.0002*
315	0.0002*
360	0.001**
390	0.006**
420	0.09**

*Test on granular molding powder.
**Test on molded sheet.

abrasion resistance. Some experimental values for these properties are given in Tables 2-10 and 2-11. The friction properties of the Teflons are so good, in fact, that they are often used as fillers in other plastics (such as epoxies and acetals) to lower the coefficients of friction and improve wear properties.

Chemical and Solvent Resistance. Fluorocarbon polymers are resistant to the widest variety of organic and inorganic acids and bases, solvents, and gases. They therefore provide excellent corrosion protection for metals. Their excellent compatibility with numerous chemicals, acids, bases, and solvents is evident from the data in Tables 2-12 to 2-14. In fact, the only chemicals to which fluorocarbons are not resistant are molten alkali metals (such as sodium and potassium), free-fluorine gas, and fluorine precursors (such as oxygen difluoride and chlorine trifluoride) at elevated temperatures. Even when there are weight increases and slight absorption of chemicals no significant loss of physical or electrical properties is reported.

Table 2-10: Coefficient of Friction* for Teflon FEP (Against Steel, 23°C)[38]

Load, psi	Static coefficient	Dynamic coefficient, 25 ft/min
1	0.14	0.425
10	0.11	0.415
100	0.08	0.350

Table 2-11: Abrasion Resistance of Teflon TFE and FEP Coatings[38]

Test	Weight loss	
	FEP	TFE
Armstrong (200 cycles, per ASTM D 1242-56), g/in.^{2}.	0.174	0.337
Taber (1,000-g load, CS-17F Calibrase Wheel), mg		
1,000 cycles.	7.5	8.9
2,000 cycles.	13.2	13.4

Table 2-12: Chemical Compatibility of TFE and FEP Resins

Chemical	Resistant at °C*	Chemical	Resistant at °C*
Abietic acid	180	Hydrochloric acid, 0 to 100%	bp
Acetic acid, 0 to 100%	bp	Hydrofluoric acid, 0 to 100%	bp
Acetone	bp	Hydrogen peroxide, 90%	65
Acetophenone	bp	Methyl ethyl ketone	bp
Aniline	bp	Naphthols	bp
		Nitric acid, 0 to 100%	bp
Benzyl alcohol	bp		
n-Butyl amine	150	Nitrogen tetroxide	bp
Butyl acetate	125	Oils, animal and vegetable (TFE)	260
		Oils, animal and vegetable (FEP)	205
Carbon disulfide	45	Ozone	25
Cetane	50	Perchloroethylene	bp
Chloroform	60		
Cresols	bp	Phenol	bp
Cyclohexane	150	Piperidine	105
Cyclohexanone	bp	Potassium hydroxide, 60%	bp
Dibutyl phthalate (TFE)	260	Pyridine	bp
Dibutyl phthalate (FEP)	205	Soap and detergents	bp
Dibutyl sebacate	bp	Sodium hydroxide, 0 to 100%	bp
Diethyl ether	bp	Sodium hypochlorite, 20%	bp
Diisobutyl adipate	bp	Sodium peroxide	bp
Dimethyl formamide	150		
Unsym. dimethyl hydrazine	bp	Solvents, aliphatic and aromatic (TFE)**	260
Dioxane	bp	Solvents, aliphatic and aromatic (FEP)**	205
Ethyl acetate	bp	Stannous chloride (TFE)	245(mp)
Ethyl alcohol	205	Sulfuric acid, 0 to 100% (TFE)	260
Ethylene bromide	100	Sulfuric acid, 0 to 100% (FEP)	205
Ethylene glycol	150	Tetrabromoethane	bp
Furane	bp		
Gasoline	95	Trichloroethylene	bp
Hexachloroethane	bp	Tricresyl phosphate	150
		Xylene	135
Hydrazine	bp		

*Both TFE and FEP, except as indicated.

**Some halogenated solvents will cause moderate swelling.

Note: By permission of E.I. Du Pont de Nemours & Co.

Table 2-13: Exposure of Teflon Resins to Common Acids and Bases

Reagent	Exposure Temp., °C	Exposure time	Weight increase, %
Hydrochloric acid, 10%........	25	12 months	0
	50	12 months	0
	70	12 months	0
Hydrochloric acid, 20%........	100	8 hr	0
	200	8 hr	0
Nitric acid, 10%..............	25	12 months	0
	70	12 months	0.1
Sulfuric acid, 30%............	25	12 months	0
	70	12 months	0
	100	8 hr	0
	200	8 hr	0.1
Sodium hydroxide, 10%........	25	12 months	0
	70	12 months	0.1
Sodium hydroxide, 50%........	100	8 hr	0
	200	8 hr	0
Ammonium hydroxide, 10%....	25	12 months	0
	70	12 months	0.1

* Weight changes less than about 0.2 percent are not experimentally significant.

By permission of E. I. du Pont de Nemours & Company.

Table 2-14: Exposure of Teflon Resins to Common Solvents

Solvent	Exposure temp., °C	Exposure time	Weight increase, %
Acetone	25	12 months	0.3
	50	12 months	0.4
	70	2 weeks	0
Benzene	78	96 hr	0.5
	100	8 hr	0.6
	200	8 hr	1.0
Carbon tetrachloride..........	25	12 months	0.6
	50	12 months	1.6
	70	2 weeks	1.9
	100	8 hr	2.5
	200	8 hr	3.7
Ethyl alcohol, 95%...........	25	12 months	0
	50	12 months	0
	70	2 weeks	0
	100	8 hr	0.1
	200	8 hr	0.3
Ethyl acetate.................	25	12 months	0.5
	50	12 months	0.7
	70	2 weeks	0.7
Toluene	25	12 months	0.3
	50	12 months	0.6
	70	2 weeks	0.6

By permission of E. I. du Pont de Nemours & Company.

Fluorocarbon Dispersion Coatings

Both Teflon TFE and Teflon FEP are available for coating applications as colloidal aqueous dispersions. These consist of suspensions of hydrophobic negatively charged particles (average size about 0.2 microns) in water. Dispersions contain from 32 to 62 percent resin by weight and may be stabilized with an immiscible oil, nonionic wetting agents, or mixtures of ionic and nonionic wetting agents. A commonly used nonionic wetting agent is Rohm and Haas Triton X-100 used in concentrations from 6 to 10 percent by weight of solids.

Dispersion coatings may be applied by spraying, dipping, flowing, or casting. To obtain a continuous integral coating, the coating must first be baked at about 120°C to remove water and then sintered at 360 to 400°C to fuse the Teflon particles together. The sintering temperature and time are very critical in obtaining optimum properties. It is reported that exposures of 5 min at 370°C provide the best mechanical and electrical coating properties (Table 2-15), and that these properties are equivalent to those obtained from Teflon parts fabricated by the molding technique.[40] In addition to the fusion of Teflon particles, high sintering temperatures are necessary to decompose and volatilize the wetting agent. Some discoloration may be evident after sintering, owing to incomplete removal of carbonized residues from the wetting agent. Sintering cycles of 8 to 10 min. are recommended for obtaining high-purity coherent films of 1 mil thickness. The maximum thickness deposited in any single coating application should be limited to 0.5 mil; thicker coatings tend to crack or craze, and these cracks cannot be eliminated by fusing, even with prolonged baking. Thick films may be built up, however, by multiple coating and baking operations without crazing.

Table 2-15: Properties of 2-mil Cast Film of Teflon as a Function of Baking Temperatures[41]

Property	Method	Baking temperature, °C				
		350	370	390	410	430
	Mechanical properties					
Tensile strength, psi.	ASTM D 412	3,810	4,010	3,490	2,420	1,710
Elongation, %.	ASTM D 412	330	370	390	420	100
Elmendorf test g/mil	. . .	6	8	9	14	
	Electrical properties					
Dielectric constant (1,000 hz)	ASTM D 150-47T	2.14	2.14	2.14	2.14	2.14
Dissipation factor (1,000 hz)	ASTM D 150-47T	0.0002	0.0003	0.0002	0.0002	0.0003
Volume resistivity, ohm-cm.	ASTM D 257-46	10^{14}	10^{14}	10^{14}	10^{14}	10^{14}
Dielectric strength (short-time), volts/mil.	ASTM D 149-44	3,740	4,090	2,600	2,250	3,160

On completion of the sintering step, the coatings may be cooled rapidly by quenching in water to impart maximum toughness, or cooled slowly to provide easier wettability for succeeding steps. For multiple coatings it is important to cool the film slowly after each application except the last one, which may be quenched rapidly.[41]

The properties of carefully prepared thin films cast from a TFE dispersion formulation are given in Table 2-16, and some commercially available coatings and their properties are given in Table 2-17.

In addition to application by colloidal dispersion, Teflons may be applied by extrusion, fluidized-bed techniques, or electrostatic spraying. In all cases Teflon FEP is somewhat easier to process because it has a softening temperature about 100°F lower than Teflon TFE.

Table 2-16: Typical Properties of Thin Films Cast from Teflon TFE Fluorocarbon Resin Dispersion

Property	Methods	Value*
Specific gravity	ASTM D 792-64T	2.15–2.20
Tensile strength at 25°C, psi	ASTM D 882-56T	3,000–4,000
Elongation at 25°C, %	ASTM D 882-56T	300–500
Tear strength, lb/in	ASTM D 624-48	400–800
Stiffness at 25°C, psi	ASTM D 747-48T	40,000–90,000
Moisture permeability (0.001 in thickness), g/100 in^2/day	–	0.2
Dielectric constant at 1,000 hz	ASTM D 150-54T	2.0–2.2
Power factor at 1,000 hz	ASTM D 150-54T	<0.0003
Volume resistivity, ohm-cm	ASTM D 149-44	$>10^{15}$
Dielectric strength 1- to 4-mil films, volts/mil	ASTM D 149-44	3,000–4,500
Electrical flaws at 800 volts/mil, flaws/100 ft/½ in width	ASTM D 1389-56T (modified)	0–8**
Water absorption, %	ASTM D 570-54T	<0.1
Brittleness temperature, °C	–	<-70

*Data shown are average ranges and should not be used for specifications.
**Varies with film thickness.

Note: By permission of E.I. Du Pont de Nemours & Co.

POLYXYLYLENES

Chemistry

Poly-*p*-xylylene films and coatings may be prepared by the pyrolytic dehydrogenation of gaseous *p*-xylene. At temperatures of 700 to 1100°C and reduced pressure of 1 to 5 torr, *p*-xylene loses a molecule of hydrogen and converts to an unstable quinoid structure. On quenching, this structure polymerizes rapidly to the desired poly-*p*-xylylene:[42]

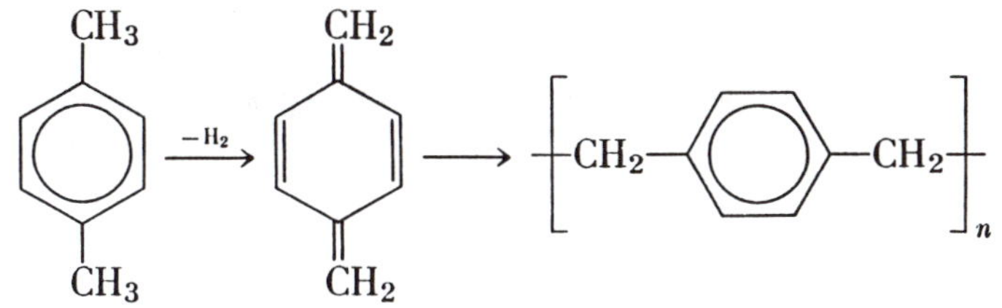

p–Xylene Quinoid structure Poly–*p*–xylylene

Table 2-17: Fluorocarbon Dispersion Coatings

Coating	*Characteristic features*
Du Pont Teflon 42	A water dispersion of 32 to 35 percent; TFE particles have a viscosity of 4 to 5 centipoises and a pH of 10; available for coating, casting, or impregnating.
Du Pont Teflon 30, and 30B	A water dispersion of TFE, 60 percent resin with 6 percent Triton X-100 (nonionic) wetting agent; viscosity about 15 centipoises, pH 10.
Du Pont Teflon 120	A water dispersion of about 50 percent, FEP useful as surface coating, wire and cable coating, and impregnant; contains 5 to 7 percent of mixed anionic and nonionic wetting agent; viscosity about 25 centipoises, pH 10.

Early work on this method was reported by Szwarc.[43,44] Poly-*p*-xylylene prepared by this method, however, is an infusible, intractable cross-linked polymer which cannot be fabricated. Subsequently, a simplified process for preparing a series of poly-*p*-xylylenes which are completely linear noncross-linked polymers was developed by Union Carbide Corporation. These polymers, called *parylenes,* were first introduced in 1965. The basic member of the series, poly-*p*-xylylene, is prepared by converting *p*-xylene to di-*p*-xylylene (a white solid dimer) and then pyrolyzing this dimer at about 650°C in a vacuum of 0.1 to 0.5 torr.[45] The dimer is believed to dissociate into a monomer, *p*-xylylene, which immediately polymerizes on contacting a surface. Items to be coated may be kept at room temperature or temperatures as high as 150°C. The reaction steps are summarized in Figure 2-11.[46]

The two key members of the parylene series are parylene N and parylene C. The latter is structurally similar to parylene N, except that one chlorine atom is contained in each benzene ring:

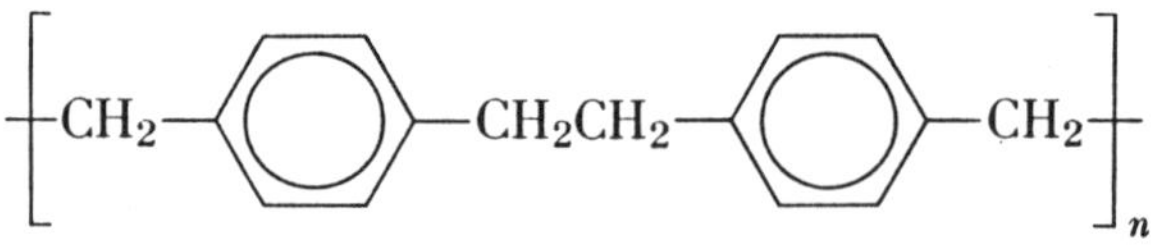

Parylene N

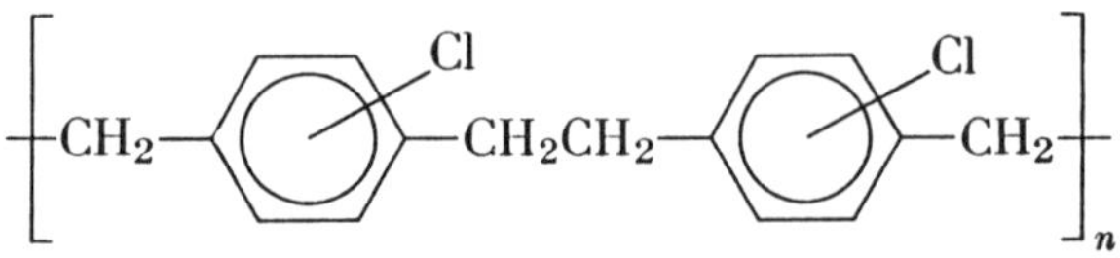

Parylene C

In addition to parylenes N and C, more than 40 other derivatives have been synthesized. The solid dimers, polymers, and apparatus for vapor deposition were covered by a series of Union Carbide patents.[47–50]

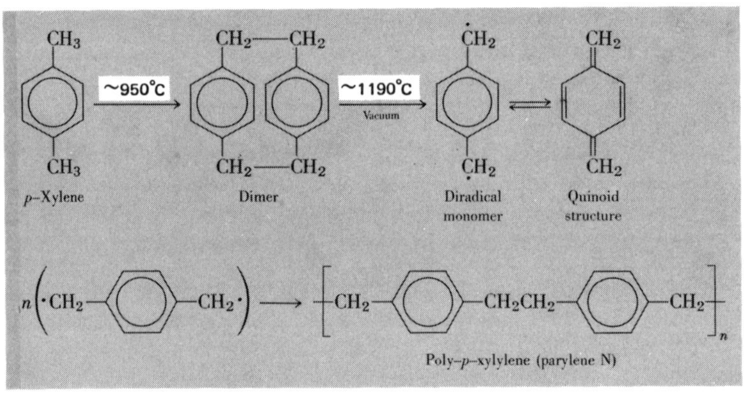

Figure 2-11: Parylene synthesis reactions.

Properties of Polyxylylenes

The unique and superior properties of parylene coatings are attributed largely to the process by which they are deposited and polymerized. The parylenes were among the first coatings to be successfully applied in the vapor phase without requiring either high surface temperatures, high vacuum, or external activation such as ultraviolet irradiation or gaseous discharge. Although the decomposition temperature of the dimer is relatively high (650 to 680°C), the actual temperature of the part may be maintained at room temperature to about 150°C, an essential requirement for many electronic assemblies. Because of the method of application, high-purity, ultrathin, pinhole-free coatings are obtained. Coatings may be applied in thicknesses less than 1,000 Å up to several mils. The starting material is a purified dimer, which dissociates, then repolymerizes in such a way that no side

Table 2-18: Electrical Properties of Parylenes Compared to Other Coating Types

Properties*	Method or conditions	Parylene C	Parylene N	Epoxies†	Silicones†	Urethanes†
Dielectric strength						
Short-time, volts/mil						
1-mil films........	ASTM D 149-64	5,600	7,000			
Corrected to ⅛ in...	ASTM D 149-64	590	700	400–500	550	450–500
Step-by-step						
1-mil films........	ASTM D 149-64	4,700	6,000			
Corrected to ⅛ in...	ASTM D 149-64	550	550	380	550	450–500
Volume resistivity (23°C, 50% humidity, ohm-cm	ASTM D 257-61 (1-in.^{2} mercury electrodes)	8.8×10^{16}	1.4×10^{17}	10^{12}–10^{17}	2×10^{15}	2×10^{11}–10^{15}
Surface resistivity (23°C, 50% humidity), ohms	ASTM D 257-61 (1-in.^{2} mercury electrodes)	10^{14}	10^{13}			
Dielectric constant						
60 hz..............	ASTM D 150-65T	3.15	2.65	3.5–5.0	2.75–3.05	4–7.5
10^3 hz.............	(1-in.^{2} mercury	3.10	2.65	3.5–4.5	...	4–7.5
10^6 hz.............	electrodes)	2.95	2.65	3.3–4.0	2.6–2.7	6.5–7.1
Dissipation factor						
60 hz..............	ASTM D 150-65T	0.020	0.0002	0.002–0.01	0.007–0.001	0.015–0.017
10^3 hz.............	(1-in.^{2} mercury	0.019	0.0002	0.002–0.02	...	0.05–0.06
10^6 hz.............	electrodes)	0.013	0.0006	0.03–0.05	0.001–0.002	

* Properties measured on parylene films 0.001-in. thick.

† Properties and methods as reported in "Modern Plastics Encyclopedia," issue for 1968, vol. 45, no. 1A, McGraw-Hill, New York, 1967.

By permission of Union Carbide Corporation.

reactions or by-products are generated that can contaminate the coating. Exact controls must be maintained over deposition, however, since some dimer may be left unreacted and become entrapped in the coating.

Electrical Properties. The electrical properties of the parylenes are excellent. The dielectric strength and volume and surface resistivities are all quite high (see Table 2-18). The dielectric constants and dissipation factors are very low—better than for most epoxies, silicones, or polyurethanes, but not as good as for Teflon and some other fluorocarbons. These electrical properties and the low deposition temperature cited above render the parylenes particularly suitable as conformal coatings for high-frequency applications and as primary insulation for conductors and metal substrates. The best overall electrical properties are exhibited by parylene N, the unsubstituted polymer. Parylene C, the chlorine-substituted analog, has a higher dielectric constant and dissipation factor than parylene N, as is usually the case with chlorinated polymers, where a higher dipole moment is created (see Chap. 4). However, parylene C is superior to parylene N with respect to the dc dielectric-breakdown voltage for films under 5 microns (Figure 2-12). The differences in breakdown voltage between the two types are shown more strikingly in Figure 2-13, where the dc breakdown voltage per micron thickness is plotted against total film thickness in microns. Whereas the breakdown voltage per micron continues to increase, at least down to thicknesses of 0.7 microns for parylene C, it levels off for parylene N at about 2 microns and then decreases. However, when parylene N is deposited onto smoother surfaces, these decreases are minimized, and values of 15 volts (220 volts/micron) have been measured on parylene N films as thin as 0.067 microns (670Å).

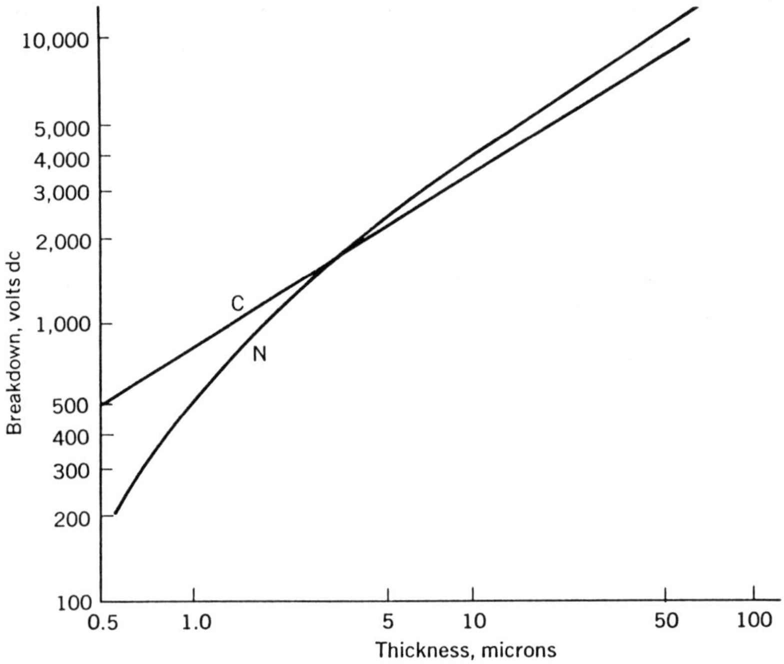

Figure 2-12: Breakdown voltage of parylenes C and N versus polymer film thickness.[51]

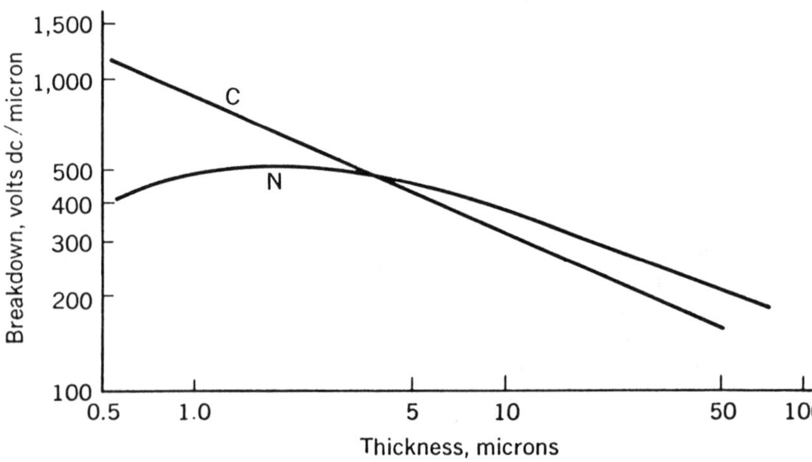

Figure 2-13: Breakdown voltage per micron of parylenes C and N versus film thickness.[51]

The electrical insulation resistance of Y-conductor-pattern test boards[52,53] coated with 0.5, 1.0, and 2.0 mils of parylene C were measured initially and during 10-day exposure to humidity-temperature cycling. Measurements were made (using a General Radio megohm bridge, Model 1644 A) at 500 volts and at a constant polarization of 100 volts. Initial measurements at 50 percent and 90 to 95 percent relative humidity at 25°C were consistently about 10^{14} ohms for all specimens. At 90 to 95 percent relative humidity and 65°C, the electrical insulation resistance decreased to a minimum of 10^{11} ohms for the 2-mil films and to 10^{10} ohms for the 0.5- and 1.0-mil films. At the end of the test period resistance values, again measured at 25° C and 90 to 95 percent relative humidity, had increased to 10^{11} to 10^{12} ohms. Results are given in Figure 2-14.

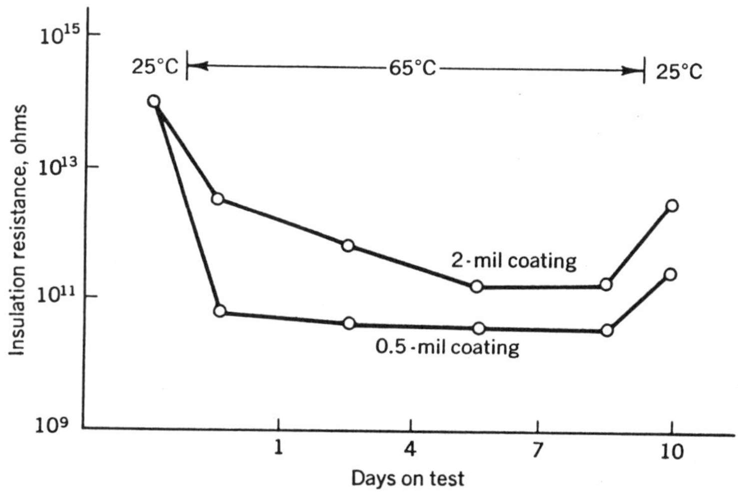

Figure 2-14: Insulation resistance of parylene C-coated Y-Pattern boards, NEMA grade G-10, during 10-day humidity tests, per MIL-STD-202, Method 302B.[51]

The *Q* factors of the test patterns were measured (using Boonton Radio Corp. *Q*-meters, Types 190A and 260A) before coating, after coating, and after immersion in distilled water for 24 hr.[43] Coating with 1- and 2-mil films of parylene C resulted in a decrease of about 5 *Q* units at both 50 and 100 Mhz. Epoxy, silicone, and urethane coatings resulted in similar decreases in *Q* values. Immersion of parylene-coated test panels in water caused an increase of about 30 *Q* units at both 50 and 100 Mhz. Results are given in Figure 2-15.

Other Properties. Among the other properties which render the polyxylylenes excellent candidates for electronic or electrical equipment are:

1. Their low coefficients of friction (slightly higher than Teflon, but about the same as nylon)
2. Their low moisture-vapor permeability (very low for parylene C, about the same as for Teflon and polyvinylidene chloride)
3. Their very high abrasion resistance (see Table 2-19)

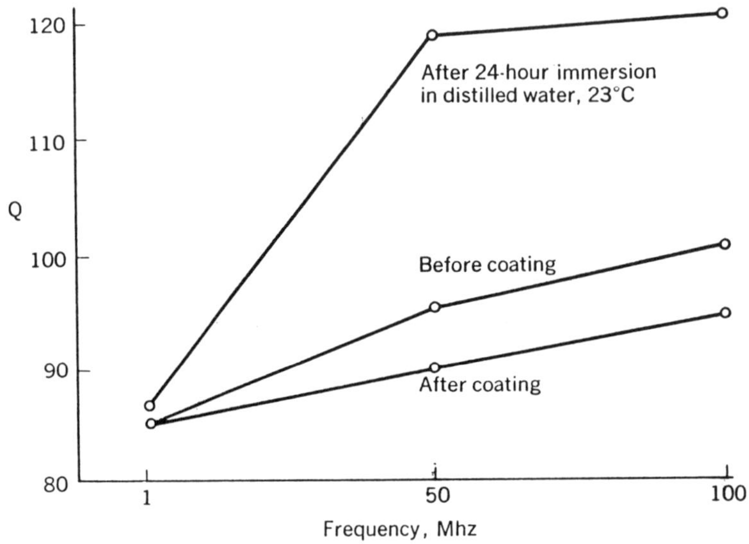

Figure 2-15: Changes in Q upon coating of Y-pattern G-10 boards, per MIL-I-46058B, with 1- and 2-mil films of parylene C.[51]

One possible disadvantage of parylene coatings is their poor temperature stability in air. Because of the presence of the aliphatic ethylene group adjoining the benzene rings, oxidation and thermal cleavage can readily occur at these sites. This apparently has no adverse effect on electrical properties but does affect physical characteristics such as tensile strength. Data for thermal stability, both in air and in an inert ambient, are given in Table 2-20.

The thermal stability of parylene N has also been established by thermogravimetric analysis and differential thermal analysis (Figures 2-16 and 2-17). The rate of weight loss in air is higher than in nitrogen, but in both cases deterioration and weight loss begin to occur at about 250°C. The DTA curve shows initial signs of deterioration at about 190°C, as indicated by an oxidation peak at this temperature.

Another limitation of parylene coatings is the difficulty in removing them. They are insoluble and inert in most commercially available solvents and chemicals and best removed by mechanical means[54] or by a reactive plasma. This makes it difficult and expensive to replace faulty components from printed circuit boards or hybrid circuits. However, there is one derivative, Parylene E, which is solvent strippable and may prove useful as a conformal coating.[55] This parylene has an ethyl group attached to the benzene ring. Parylenes E and C can be copolymerized due to their similar threshold condensation temperatures, producing a random copolymer. These copolymers have electrical properties close to those of Parylene C but the coating can be removed by solvents such as methylene chloride, toluene and xylene.

Table 2-19: Physical and Thermal Properties of Parylenes

Property	Parylene N	Parylene C
Coefficient of friction	0.25	0.29
Abrasion index, mg loss/100 cycles	2.6	4.4
Water absorption (2-mil films), %/24 hr	0.01	0.06
Thermal conductivity cal/(sec)(cm)(°C)	3×10^{-4}	
Coefficient of thermal expansion in./(in.)(°C)	6.9×10^{-5}	3.5×10^{-5}
Nitrogen gas permeability*	15	0.6
Oxygen gas permeability	55	5
Carbon dioxide gas permeability	420	14
Moisture-vapor transmission, g-mil/(100 in.^{2})(24 hr)	15	0.6

* This and other gas permeability values are in cm^3 mils/(100 in.^{2})(24 hr)(atm).
By permission from Union Carbide Corporation.

Table 2-20: Thermal-Stability Data for Parylenes

Material	Temperature, °C: In Air	Temperature, °C: In Inert Environment
Parylene N		
Long-term*	60	220
Short-term**	95	270
Melting temperature***	400 (decomposes)	—
Parylene C		
Long-term	80	230
Short-term	115	265
Melting temperature	280–300	280

*About 10 years.
**About 1,000 hr.
***Criterion for failure was 50% loss of tensile strength.

Note: By permission from Union Carbide Corporation.

ELECTRONIC APPLICATIONS

Parylene is being used by some manufacturers as a uniform conformal coating for the protection of printed-circuit boards. For complex circuit boards having a high density of electronic components, vapor-deposited parylene has a distinct advantage over spray-type coatings in its ability to penetrate beneath and around closely spaced components. Metallographic examination of coated boards has shown that parylene uniformly coats spaces 2 mils or less between wires or components and the substrate board. This is of considerable importance in preventing the entrapment of contaminants which cause corrosion or deterioration of electrical properties.[56] Another factor which makes parylene coatings suitable for use in the

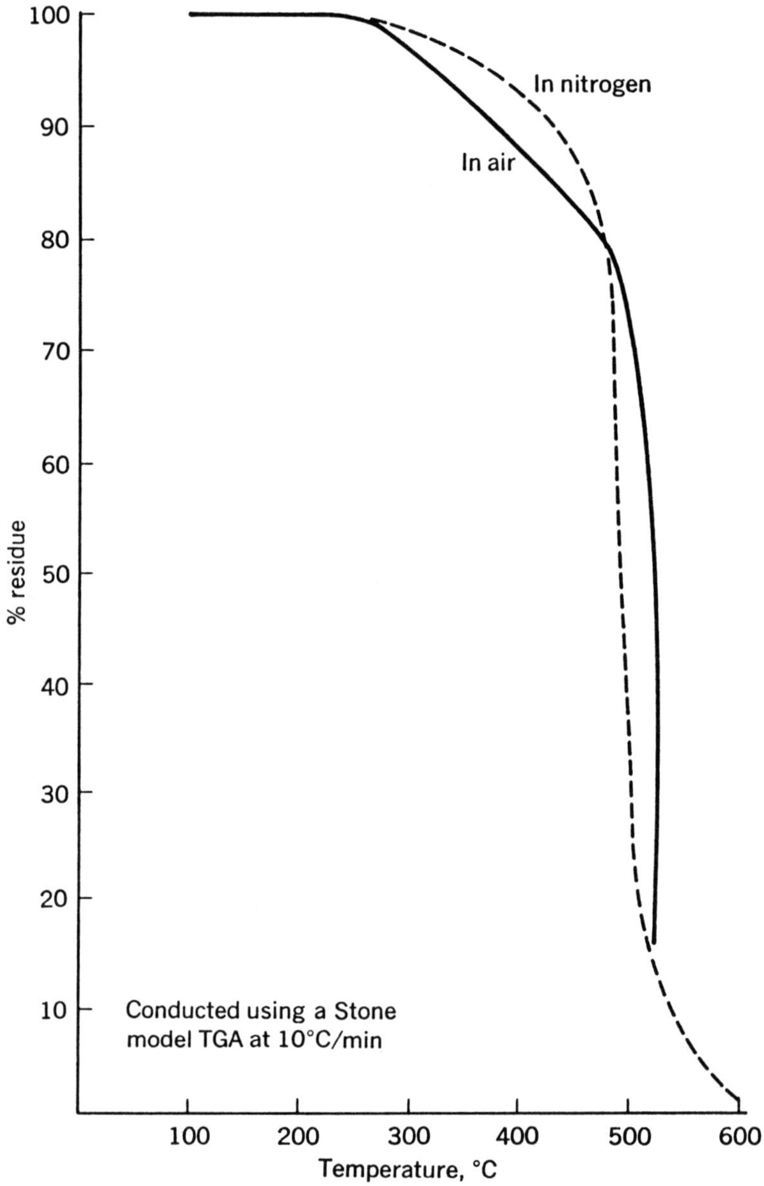

Figure 2-16: TGA curve for parylene N.

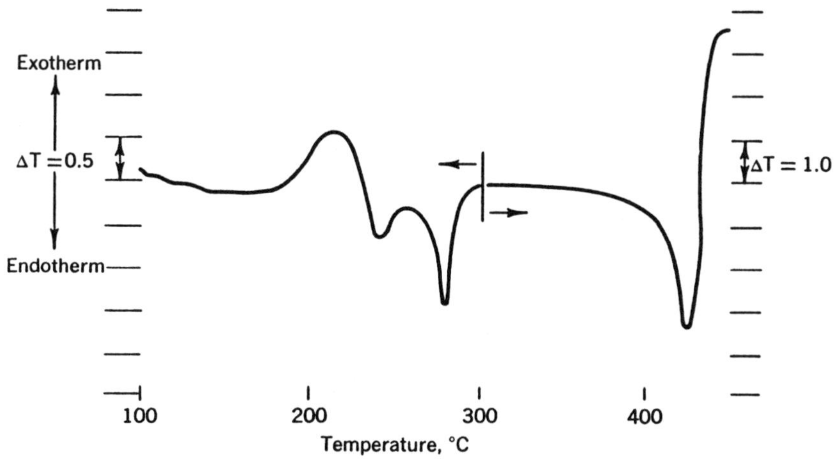

Figure 2-17: DTA curve for parylene N.

fabrication of printed-circuit boards is their good electrical properties. Their initial electrical insulation resistance is high, and they exhibit little deterioration after humidity and temperature cycling. Table 2-21 shows typical values for parylene C.

Table 2-21: Parylene C Circuit-Board Screening Tested per MIL-STD-202, Method 302

Sample thickness in.	Insulation resistance, ohms					
	Initial measurement (23°C, 50% humidity)	Precycle (23°C, 90% humidity)	Step 5, cycle 3 (65°C, 90% humidity)	Step 5, cycle 7 (65°C, 90% humidity)	Step 5, cycle 10 (65°C, 90% humidity)	Step 7, cycle 10 (25°C, 90% humidity)
0.002	2.0×10^{14}	1.8×10^{13}	2.3×10^{12}	2.5×10^{11}	1.4×10^{11}	7.5×10^{12}
0.0015	5.0×10^{14}	2.4×10^{13}	8.6×10^{11}	1.9×10^{11}	1.1×10^{11}	5.2×10^{12}
0.001	2.0×10^{14}	9.2×10^{12}	8.1×10^{11}	3.4×10^{11}	1.3×10^{11}	6.3×10^{12}
0.0005	5.0×10^{14}	2.3×10^{13}	4.1×10^{12}	2.4×10^{11}	1.1×10^{11}	4.7×10^{12}
0.0003	5.0×10^{14}	2.7×10^{13}	4.4×10^{12}	9.0×10^{10}	4.7×10^{10}	2.9×10^{12}
0.0001	5.0×10^{14}	3.2×10^{10}	1.3×10^{11}	1.1×10^{11}	6.4×10^{10}	2.3×10^{12}

In hybrid microcircuits parylene is used as both a protective coating for moisture resistance and as a coating to immobilize loose particles. Small fragments of conductive material can cause electrical shorts if they bridge wire bonds or conductor lines. These very fine particles are difficult to remove by conventional cleaning because of their strong electrostatic attraction to the circuit. However, a thin coating of parylene (0.10 to 0.25 mils) will immobilize particles, preventing future electrical failures.

Another interesting application for parylene is as an insulation and barrier coating applied over the inorganic passivation layers of semiconductor devices or over thin-film circuits. Parylene coatings as thin as 3 to 7 microns have been

deposited on metal-oxide semiconductor transistors, diodes, resistors, and capacitors. After being coated, the devices were subjected to a variety of stresses, including temperature stresses of 125 to 175°C for periods up to 128 hr and voltage stresses of 10 to 20 volts. The results indicated that changes in electrical parameters such as threshold voltage and leakage current were negligible for most devices. The average threshold voltage shifted less than 0.1 volt, the breakdown voltage changed less than 1 volt, and the leakage currents drifted within a few picoamperes for all groups.[57,58]

REFERENCES

1. Conference on High Temperature Polymer and Fluid Research ASD-TDR-62-372, August, 1962, *Chem. Eng. News,* April 13, 1964 (compiled by Lt. R. J. McHenry, Wright Patterson Air Force Base, Dayton, Ohio).
2. Wrasidlo, W., et al: Kinetics and Mechanism of Polyimides, *Proc. ACS Natl. Meeting,* April, 1964.
3. Chalmers, J. R.: Coating Compositions Comprising a Polyamic Acid Composition and 2,4-Diamino-6-Substituted S-Triazine, U.S. Patent 3342768, September, 1967.
4. Meinhard, J. E.: Final Report, NASA Contract No. NAS 8-21477; 6, 1971.
5. Air Force Contracts F33615-69-C-1463; F33615-71-C-1458; F33615-71-1228.
6. St. Clair, A. K. and T. L. St. Clair: "Polyimides, Vol. 2," K. L. Mittal, Ed., Plenum Press, New York, 1984.
7. Bilow, N.: "Resins for Aerospace," ACS Symposium Series, M. J. Comstock, Ed., American Chemical Society, 1980.
8. Landis, A. L. and A. B. Naselow: "Polyimides, Vol. 1," K. L. Mittal, Ed., Plenum Press, New York, 1984.
9. St. Clair, A. K., and T. L. St. Clair: Soluble Aromatic Polyimides for Film and Coating Applications, *Proceedings of the ACS Division of Polymeric Materials: Science and Engineering,* 1986.
10. *Ciba-Geigy Product Bulletin* CR 10321M87, 1987.
11. Rothman, L. B.: Properties of Thin Polyimide Films, *Solid-State Science and Technology,* October 1980.
12. Economy, J.: High Temperature Polymers for Electronic Devices, *Science Engineering,* **10**, 243, June 1974.
13. Bergeron, D. L., et al: Polyimide Interlevel Insulation Process Design Limitations, Presented at IRPS Conference, pp. 1–5.
14. Herndon, T. O. and R. L. Burke: Inter-Metal Polyimide Insulation for VLSI, *Kodak Interface Conference,* October 1979.
15. Wilson, A. M.: Polyimide Insulators for Multilayer Interconnections, *IEEE Thin Solid Films,* **83**, 1981.
16. Herndon, T. O., et al: "Polyimides for High Resolution Ion Implantation Masking," MIT Lincoln Lab.
17. Ray, D. R., et al: "Polyimide Planarization in Integrated Circuits," MIT, June 1984.
18. Rothman, L. B.: Process for Forming Passivated Metal Interconnection System with a Planar Surface, *Journal of Electrochemical Society; Solid-State Science and Technology,* May 1983.
19. Kitade, et al: Effectiveness of Prevention of Soft Errors with Polyimide Resin Coating, *Hitachi, Ltd, Product Literature Bulletin* 63, 1982.
20. Samuelson, G.: Polyimide for Multilevel VLSI and Alpha Protection, "Polyimide Coatings for Microelectronic Applications," August, 1981.
21. *Ciba-Geigy, Microelectronic Chemicals Technical Bulletin,* 1986.
22. Bower, G. M., and L. W. Frost: Aromatic Polyimides, *J. Polymer Sci.,* part A, vol. 1, no. 10 (1963).

23. Milek, J. T.: Polyimide Plastics: A State of the Art Report, *Hughes Aircraft Rept.* S-8, October, 1965.
24. Experimental Polyimide Insulating Varnishes, RC-B-24951 and RC-5060, *Du Pont Bull.* H65-4, January, 1965.
25. Sterling Amide-Imide Varnishes, *Sterling Varnish Co. Eng. Data Sheet,* April 1965.
26. Schweitzer, F. E.: Polyimide Binder Solutions, *Machine Des.,* April 23, 1964.
27. Pyre-M.L. Wire Enamel, *Du Pont Tech. Bull.* 19, August, 1967.
28. Freeman, J. H.: New Organic Resins and Reinforced Laminates for Service to 300°, *SPE Trans.,* **2:** 216–221 (1962).
29. New Class of Resins for High Service Temperature Applications, *Shawinigan Resins Corp. New Prod. Bull.,* November, 1964.
30. Dunnavent, W. R., and S. Palinchak: Trends in the Development of Thermally Stable Polymers, *Battelle Tech. Rev.,* **13:** 15–21 (September, 1964).
31. Pyre-M.L. Coated Glass Fabrics, *Du Pont Bull.* 6, rev. 2.
32. New 220 C All Purpose Magnetic Wire, *Insulation,* July, 1964.
33. Burton, A.: *Proceedings of the Technical Conference 3rd Annual International Electronics Packaging Conference,* October 1983.
34. Harper, A.: An Improved Materials System for Maintaining High Resistivity of Electrical Insulation in Humid Environments, *3M.*
35. Blome, R.: *Semiconductor International,* Nov 1979.
36. Kitchen, N. McK. and C. A. Russell: *IEEE Transactions on Parts, Hybrids, and Packaging,* Vol. PHP-12, **1**, March 1976.
37. *Du Pont Bull.* X-50 f.
38. *Du Pont Bull.* X-90 a.
39. Brenner, W., D. Lum, and M. W. Riley: "High Temperature Plastics," Reinhold, New York, 1962.
40. Renfrew, M. M., and E. E. Lewis: *Ind. Eng. Chem.,* **38:** 870 (1946)
41. Lontz, J. F., and W. B. Happoldt: Teflon Tetrafluoroethylene Resin Dispersion, *Ind. Eng. Chem.,* vol. 44 (August, 1952).
42. Sorenson, W. R., and T. W. Campbell: "Preparative Methods of Polymer Chemistry," Interscience, New York, 1961.
43. Szwarc, M.: *J. Polymer Sci.,* **6:** 319 (1951).
44. Szwarc, M.: *J. Chem. Phys.,* **16:** 128 (1948).
45. Gorham, W. F.: A New General Synthetic Method for the Preparation of Linear Poly-*p*-xylylene, *J. Polymer Sci.,* **4:** 3027-3039 (1966).
46. Parylene, *Union Carbide Prod. Data Bulls,* 1987.
47. Gorham, W. F.: Halogenated Di-paraxylylenes, U.S. Patent 3221068, Nov. 30, 1965.
48. Loeb, W. E., F. R. Tittmann, and J. H. Bowen: Apparatus for Vapor Deposition, U.S. Patent 3246627, April 19, 1966.
49. Gorham, W. F.: Para-xylylene Copolymers, U.S. Patent 3288728, Nov. 29, 1966.
50. Gorham, W. F.: Para-xylylene Polymers, U.S. Patent 3342754, Sept. 19, 1967.
51. Loeb, W. E., and C. E. White: Parylene for Conformal Insulation, *Proc. Natl. Electron. Packaging Convention,* Long Beach, Calif., Jan. 30, 1968.
52. Beccasio, A. J., L. S. Keefer, and A. Z. Orlowski: *Insulation,* November, 1965.
53. Military specification MIL-I-46058B.
54. Dunaetz, R. A., et al: Repair of Conformally Coated Printed Wiring Board Assemblies, Vol. 1, *Technical Report AFWAL-TR-80-4086,* 1980.
55. Corley, T. J. and D. A. Krcmaric, Solvent Strippable Parylene Conformal Coatings, *SAMPE Quarterly,* Vol. 17, (1), 1985.
56. "Parylene for Electronics," Union Carbide Corp.; Bound Brook, N.J., 1968.
57. Lee, S. M., J. J. Licari, and I. Litant: Reliability of Parylene Films, *Proc. Metallurgical Soc. Tech. Conf. Defects Electron, Matls. Devices,* Boston, Mass., March 1970.
58. Lee, S. M.: Polymeric Films for Semiconductor Passivation, Interim Rept. NASA Contract NAS 12-2011, March, 1969.

3

Chemistry and Properties of Coatings

Polyesters— Polyvinyls— Polystyrenes— Acrylics— Diallylphthalates— Polyamides— Phenolics— Polysulfides

POLYESTERS

The term "polyester" is a very broad one including any polycondensation product of a dicarboxylic acid with a dihydroxy alcohol. Polyesters are characterized by the ester group

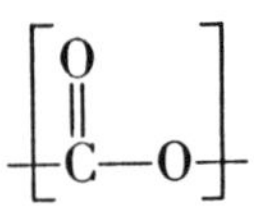

a repeating unit along the polymer chain. More specifically, polyesters may be divided into two main types: the unsaturated polyesters and the alkyd polyesters.

The Chemistry of Unsaturated Polyesters

In addition to the characteristic ester groups, unsaturated polyesters also contain double bonds (ethylenic groups) along the polymer chain. Because of these double bonds, they are often classified as vinyl types. Unsaturated polyesters are cured by cross-linking the long linear chains by means of vinyl monomers such as styrene, diallylphthalate, methylmethacrylate, triallylcyanurate, vinyl toluene, divinylbenzene, α-methylstyrene, dichlorostyrene, or combinations of these. Coating formulations generally contain the polyester resin dissolved in the vinyl monomer, and it is cross-linked or cured by stirring in a catalyst (initiator) just before use (Figure 3-1). The polymerization mechanism involves free-radical addition across the double bonds, and hence no volatile by-products are evolved. This is a distinct advantage over condensation polymerization, where water or other by-products that degrade electrical properties are evolved during the cure. Free radicals may be

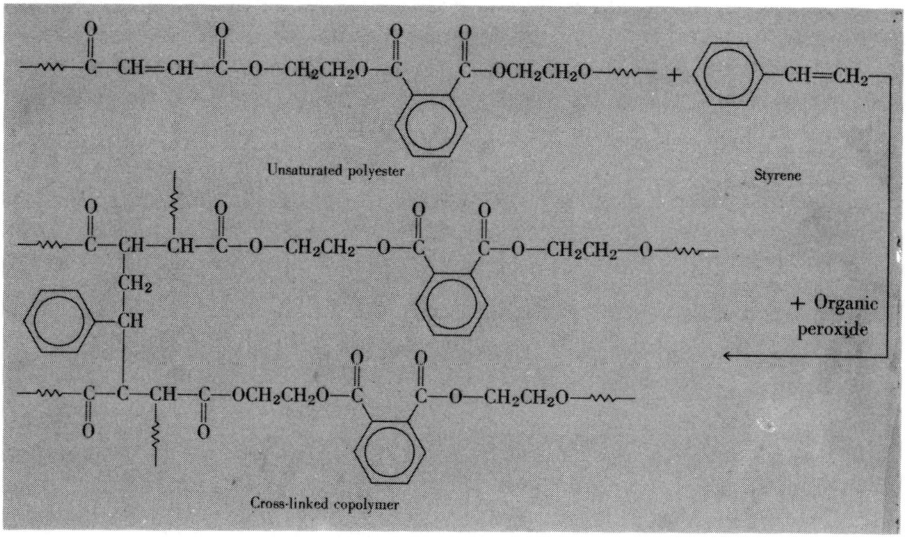

Figure 3-1: Mechanism for curing of unsaturated polyester resins.

initiated by thermal energy, ultraviolet light, or chemicals, notably the organic peroxides.

The use of unsaturated polyesters as coatings or encapsulants for electronic hardware may present several problems among which are:

1. Hydrolytic instability of the ester groups, resulting in deterioration and "whitening" of the coating
2. High shrinkage values, often as high as 10 percent, creating high stresses and separation from glass or metal substrates as the resin polymerizes
3. In the case of the peroxide cures, the possibility of corrosion of copper owing to the interaction of peroxide catalysts with copper

The Chemistry of Alkyd Polyesters

Alkyd resins are polyesters produced by the condensation of saturated dicarboxylic acids with polyhydric alcohols. The general reaction is represented by

$$\underset{\text{Dicarboxylic acid}}{HOOC{-}R{-}COOH} + \underset{\text{Dihydric alcohol}}{HO{-}R'{-}OH} \xrightarrow{-HOH} \underset{\text{Alkyd polyester}}{\left[O{-}\overset{\overset{\displaystyle O}{\|}}{C}{-}R{-}\overset{\overset{\displaystyle O}{\|}}{C}{-}O{-}R' \right]_n}$$

The most commonly used alkyd resins are those prepared from phthalic anhydride (source of phthalic acid) and glycerol. These resins are also referred to as *glyptals.* They are seldom used as such, but are modified with unsaturated oils or fatty acids, such as oleic or linoleic acids, in order to achieve rapid air-drying compositions. The first step in their preparation is the partial esterification of glycerol with the fatty acid:

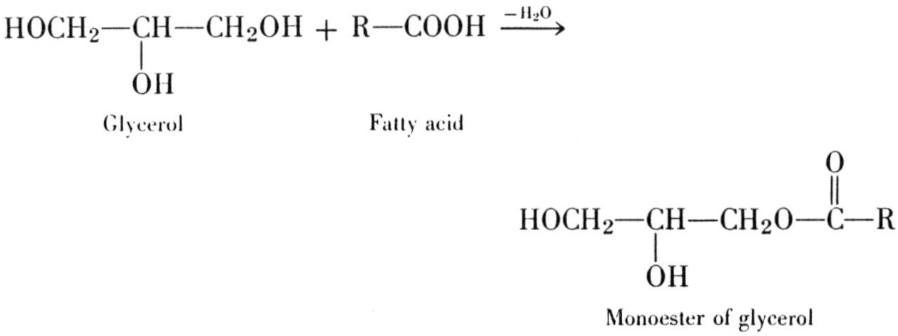

In the second step, phthalic anhydride reacts with one or more of the remaining hydroxyl groups of the partially esterified glycerol, resulting in the alkyd polymer:

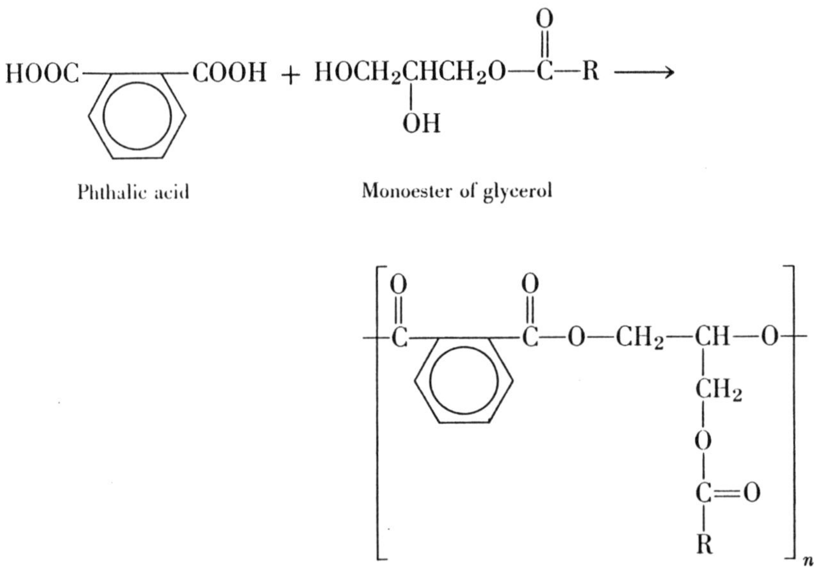

Modified alkyd polymer

Although glycerol and phthalic acid are the most commonly used starting materials, other polyols and dibasic acids can be used. Other polyols include pentaerythritol, polyglycerol, and the mono- and diglycerides of fatty acids of linseed oil. Other dibasic acids include isophthalic, terephthalic, hexahydrophthalic, maleic, tetrachlorophthalic, and adipic acids.

Alkyds may be further modified with rosin, styrene, phenolic resins, or silicones, depending on the characteristics desired. A more detailed discussion of the chemistry and formulation of alkyd polyesters appears in the literature.[1–3]

Properties of Alkyd Coatings

The alkyds and modified glyptals have been employed for many years as coatings, paints, and varnishes.[4] They are used primarily as protective and decorative coatings for metal surfaces, and hence they are of interest for electronic applications more as finishing systems for metal enclosures, housing for electronic assemblies, and computer chasses than as primary insulation coatings. As corrosion-protective coatings for metals, many proven and well-defined alkyds are available and covered by military specifications, among which are TT-E-508 and TT-E-489.

Alkyd coatings may be either air-drying or baking types. The air-drying types, as previously mentioned, are those which have been modified chemically by the incorporation of unsaturated fatty acid groups. Experimental studies have shown that the greater the percentage of these fatty acids in the formulation of the polymer, the shorter will be the drying time. Drying times may also be shortened by use of metal oxides such as those of lead, cobalt, or calcium, by the use of branched polyhydric alcohols such as pentaerythritol, or by the use of isophthalic acid instead of phthalic acid. The baking-type alkyds are modified with hardeners such as melamine or urea resins.

POLYVINYLS

Vinyl polymers, in the broad sense, are those prepared by the free-radical-addition polymerization of monomers containing a double bond. Monomers are caused to join head to tail to form long-chain linear polymers of high molecular weight without the release of any by-products. Ultraviolet light, heat, or catalysts are used to initiate the free-radical reaction. The best accelerators are peroxide catalysts such as benzoyl peroxide or dicumyl hydroperoxide. For example, the polymerization mechanism for styrene (a vinyl benzene) using a catalyst is as follows:

$$\text{Catalyst} \longrightarrow R^{\bullet} \quad \text{(free radical)}$$

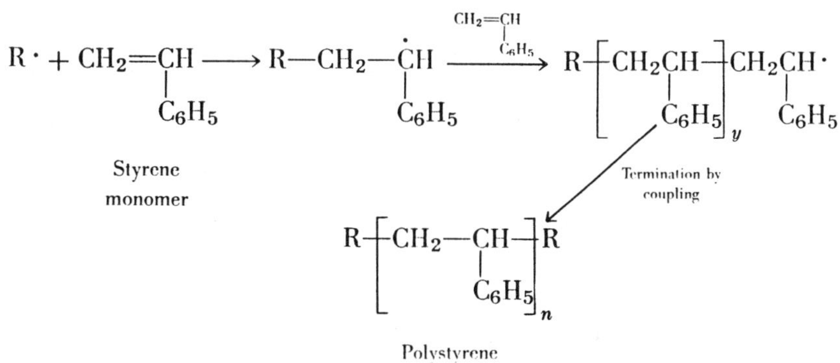

Included in the vinyl classifications are the well-known polymers listed in Table 3-1. Most of these are formed directly from the polymerization of vinyl monomer without further alteration and are referred to as *primary vinyls.* Others are formed

Table 3-1: Vinyl Polymers and Their Structures

Monomer	Monomer structure	Polymer	Polymer structure
		Primary types	
Vinyl chloride	$CH_2{=}CH{-}Cl$	Polyvinyl chloride	$\left[-CH_2-CH(Cl)-\right]_n$
Vinyl fluoride	$CH_2{=}CH{-}F$	Polyvinyl fluoride	$\left[-CH_2-CH(F)-\right]_n$
Vinyl acetate	$CH_2{=}CH{-}C({=}O){-}CH_3$	Polyvinyl acetate	$\left[-CH_2-CH(O-C(=O)-CH_3)-\right]_n$
Vinylidene chloride	$CH_2{=}C(Cl){-}Cl$	Polyvinylidene chloride	$\left[-CH_2-C(Cl)(Cl)-\right]_n$
Vinylidene fluoride	$CH_2{=}C(F){-}F$	Polyvinylidene fluoride	$\left[-CH_2-C(F)(F)-\right]_n$
Vinyl carbazole	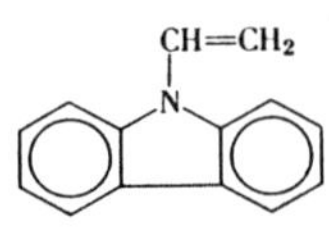	Polyvinyl carbazole	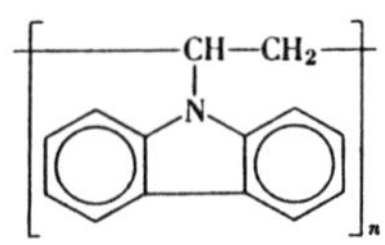
Ethylene	$CH_2{=}CH_2$	Polyethylene	$\left(-CH_2-CH_2-\right)_n$
Propylene	$CH_2{=}CH{-}CH_3$	Polypropylene	$\left[-CH_2-CH(CH_3)-\right]_n$
Styrene	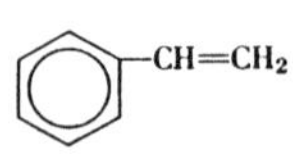	Polystyrene	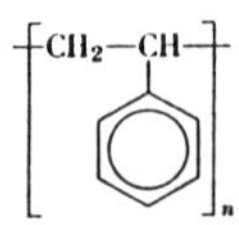
Methylmethacrylate	$CH_2{=}C(CH_3){-}COOCH_3$	Polymethylmethacrylate	$\left[-CH_2-C(CH_3)(COOCH_3)-\right]_n$

(continued)

Table 3-1: (continued)

		Secondary types	
Polyvinyl acetate	$\left[-CH_2-CH(-O-C(=O)-CH_3)-\right]_n$	Polyvinyl alcohol	$\left[-CH_2-CH(OH)-\right]_n$
Hydrolyzed polyvinyl acetate	$\left[-CH_2-CH(-O-C(=O)-CH_3)-CH_2-CH(OH)-\right]_n$	Polyvinyl formal	$\left[-CH_2-CH(-O-C(=O)-CH_3)-CH_2-CH(-O-CH_2OH)-\right]_n$
Hydrolyzed polyvinyl acetate	$\left[-CH_2-CH(-O-C(=O)-CH_3)-CH_2-CH(OH)-\right]_n$	Polyvinyl butyral	$\left[-CH_2-CH(-O-C(=O)-CH_3)-CH_2-CH(-O-HCOH-C_5H_7)-\right]_n$

by subsequently reacting the primary vinyl with reagents which alter the groups along the polymer chain. These are called *secondary vinyl resins.* For example, some of the acetate groups of polyvinyl acetate may be hydrolyzed, and the resulting hydroxyl groups may then co-react with aldehydes to yield acetal derivatives. There are three such derivatives of importance to the electronics industry: polyvinyl formal (Formvar), polyvinyl acetal, and polyvinyl butyral. Polyvinyl formals and polyvinyl acetal are widely used as tough high-temperature-resistant wire enamels. Polyvinyl butyral is useful as a surface coating and as the resin system for many wash primers (see Chapter 4).

Acrylics, polyethylenes, and polypropylenes also belong in the vinyl classification but are often considered and discussed separately because of their great commercial importance. The polymers listed in Table 3-2 represent a small sampling of the numerous vinyl modifications possible. In addition, many copolymers prepared by the reaction of two or more different monomers are available. They will usually be more soluble and possess properties intermediate between those of the respective homopolymers. An example of a copolymer reaction is:

$$\underset{\text{Vinylidene chloride}}{CH_2=CCl_2} + \underset{\text{Acrylonitrile}}{H_2C=CH-CN} \longrightarrow \underset{\text{Vinylidene chloride–acrylonitrile copolymer}}{\left[-CH_2-CCl_2-CH_2-CH(CN)-\right]_n}$$

Properties of Polyvinyls

Of the many vinyl types discussed, those used most widely as protective and electrical insulating coatings are the polyvinyl chlorides, polyvinyl fluorides, poly-

vinylidene chlorides and fluorides, acrylics, vinylaldehydes, and polystyrene. They are used as moisture barriers, primary-wire insulation, corrosion-protective coatings, dielectric impregnants, baking enamels, and as capacitor dielectrics. Some of the best electrical-grade and moisture-barrier coatings are formulated from vinyls.

Polyvinyl Formals. The electrical properties of polyvinyl formals are extremely good and render them especially useful for wire enamel coatings. Dielectric strengths range from 860 to 1,000 volts/mil and dielectric constants from 3.3 to 3.6. Coatings are applied to wire from a solution containing small amounts of phenolic resin, which functions as a cross-linking agent. They are generally insoluble in alcohols and hydrocarbons but soluble in acetic acid, dioxane, nitrobenzene, and tetrachloroethane.

Polyvinyl Alcohols and Acetates. Polyvinyl alcohols and acetates are seldom used for electrical applications because of their high dielectric constants and dissipation factors and the rapid and extensive deterioration of their electrical

Table 3-2: Vinyl Resins and Coatings[6]

Types and trade names	*Manufacturer or supplier*
Vinyl chloride	
polyvinyl chloride	Vista Chemical
PLIOVIC DR, M and WO	Goodyear
Geon	Goodrich
Polyvinylidene chloride	
76 RES	Unocal
DARAN	W.R. Grace
Vinyl chloride/acetate copolymer	
UCAR VYLF (low-viscosity)	Union Carbide
UCAR VYHH (medium-viscosity)	Union Carbide
UCAR VYNS (high-viscosity)	Union Carbide
Vinyl chloride/acetate/maleic acid copolymer	
UCAR VMCA (low-viscosity)	Union Carbide
UCAR VMCC (medium-viscosity)	Union Carbide
UCAR VMCA (high-viscosity)	Union Carbide
Vinyl chloride/acetate/alcohol copolymer	
UCAR VAGH	Union Carbide
Polyvinyl butyral	
UCAR XYSG	Union Carbide
UCAR XYHL	Union Carbide
Butvar (several grades)	Monsanto
Acrylic and acrylic esters	
Flexcryl	Air Products and Chemicals
Haloflex	ICI Americas
Poly-tex	Interez
Acryloid	Rohm and Haas
Lucite	DuPont
G-Cure	Henkel
Hycar	Goodrich
Styrene	
Styrene monomers	Dow Chemical

properties under humid conditions.[5] In fact, under normal atmospheric conditions sufficient water is absorbed to render electrical measurements useless.[7] They also have poor thermal stability in air at temperatures above 95°C. They are sometimes used as temporary coatings, serving as maskants or acid indicators (when formulated with pH-indicator crystals), because they can be removed easily by dissolving in warm or hot water. On the whole, however, polyvinyl alcohol and acetate resins find greater use in nonelectrical applications such as paper coatings, packaging films, and house paints.

Polyvinyl Chlorides and Fluorides. One of the widest uses for polyvinyl chloride (PVC) is as wire and cable insulation. For these applications the material is extruded from a dry blend or from hot fluxing. Polyvinyl chloride is well suited to these applications because of its good electrical-insulating properties, resistance to moisture and chemicals, and inherent flame-retardant properties.

The molecular weights of commercially available polyvinyl chlorides range from 50,000 to 150,000. As with other polymers, toughness, viscosity, and chemical and solvent resistance increase with an increase in chain length. The average molecular weight of the resin will largely determine its application. High molecular weight resins are used for wire insulation and extruded flexible tubing, whereas the intermediate and lower molecular weight resins are used in the production of films. Polyvinyl chloride is seldom used in its pure form. It is compounded with plasticizers for flexibility, with stabilizers to avoid thermal decomposition, and with lubricants and other additives to facilitate processing and increase impact strength. When it is used as an electrical-grade resin, vinyl suspension coatings are repeatedly washed to remove impurities. Compounding with large amounts of plasticizers, however, will degrade electrical properties, increase moisture absorption, or render the coating a fungus nutrient. There are hundreds of plasticizers on the market, and a judicious choice must be made to minimize these effects. Fillers such as clay or calcium carbonate are added at about 15 percent loading to reduce the price of the plastic and improve the electrical resistivity. Amounts larger than 15 percent, however, decrease mechanical properties such as abrasion resistance, tensile strength, elongation, and flexibility. A formulation typical of those used for primary wire insulation is given in Table 3-3.

Stabilizers are employed in these formulations because of the tendency of polyvinyl chlorides to decompose during high-temperature processing, cure, or operational use. Decomposition occurs by the release of hydrogen chloride, which

Table 3-3: Vinyl Wire-Coating Formulation[8]

Ingredient	*Proportion, pbw*
Polyvinylchloride homopolymer (medium mol. wt.)	100
Primary plasticizer	30
Secondary plasticizer	15
Pigment 33 (clay filler)	10
Calcium carbonate filler	8
Mark OHM (stabilizer)	2
Total	165

subsequently ionizes to hydrochloric acid owing to trace amounts of water contained in the plastic. Hydrogen chloride can cause corrosion of metals or deterioration of other plastics in an electronic enclosure and can also act as a catalyst to cause further decomposition of the polymer with the release of more hydrogen chloride. This process is reported to be accelerated at elevated temperature, under high-humidity conditions, in the presence of certain metals such as iron or zinc at 150°C, and under combinations of these conditions. Stabilizers consist of compounds which react with and remove trace amounts of hydrogen chloride as soon as it forms, preventing catalytic "unzipping" of the polymer chain. Examples of stabilizers include lead, epoxide compounds, and plasticized urea formaldehyde resins.

POLYSTYRENES

Except where very-high-temperature properties are required, polystyrenes are ideal for many specialized electronic applications. They have extremely low dielectric constants and power factors (Table 3-4), placing them in a class with polyethylenes, fluorocarbons, and silicones. Polystyrenes also have high dielectric-strength values (comparable to those of mica) and volume resistivities which do not vary much with high frequencies or in the presence of moisture. It is because of these electrical properties that they are especially suited for use as capacitor dielectrics, especially in the commercial markets, and for insulating high-frequency circuits. A second outstanding characteristic of polystyrenes is their transparency. Unmodified resin coatings are clear, colorless, and capable of transmitting 88 to 92 percent of light. Their refractive index is 1.59 to 1.60. They are therefore used in coating lenses and other optical equipment. Polystyrenes have very good tinting and coloring properties, giving them a wide range of transparent, translucent, and opaque colors for decorative uses.

Among the more serious limitations of polystyrenes are their inherent flammability and their upper temperature limit of 85°C, which is too low for most military electronic applications.

Table 3-4: Electrical Properties of Polystyrenes, Tested by ASTM Methods

Property Tested by ASTM Methods	*Value*
Dielectric strength, volts/mil	500–700
Dielectric constant (60 hz)	2.45–2.65
Power factor (60 hz)	1×10^{-5}–3×10^{-4}
Volume resistivity, ohm-cm	10^{16}
Arc resistance, sec	60–80

ACRYLICS

Acrylics are employed as general-purpose coatings, impregnants for the protection of printed-circuit boards, anticorrosive coatings for metals, wire enamel, and as wire-coil dope. They are commonly one-component low-viscosity solution coatings, formulated either as spray or dip types and as either air-drying or baking types.

Acrylics are available as either clear or pigmented protective coatings and are compatible with a large variety of surfaces. Adhesion to glass, ceramics, epoxy laminates and other plastics, and chrome plate, aluminum, magnesium, zinc, steel, and other metals is very good. Adhesion to metals may be improved by the use of primers of nonoxidizing alkyds.

Acrylic enamels are thermally very stable and retain their initial clarity and color characteristics at intermittent temperatures as high as 260°C. For electrical insulating applications, maximum continuous operating temperatures depend on the formulation and may range from 95 to 155°C. Some acrylic coatings are functional up to 150°C, but above this temperature the electrical parameters degrade rapidly. Intermittent-use temperatures for the better grades of acrylics may be as high as 180°C.

The electrical properties of acrylics are, in general, very good at room temperatures. However, they are not as good as those of epoxies or silicones, which are stable over a wider temperature range. The electrical properties of a number of acrylic films and of some electrical-grade coatings are given in Table 3-5 and 3-6.

Table 3-5: Electrical Properties of Acryloid Films[9]

Property	A-101	B-72	B-82	F-10
Dielectric constant at 25°C				
60 hz	3.6	4.4	4.7	3.2
10^3 hz	3.2	3.7	4.1	2.8
10^6 hz	3.1	3.1	3.3	2.6
3×10^7 hz	2.3	2.6	2.9	2.3
Dissipation factor at 25°C				
60 hz	0.07	0.096	0.088	0.09
10^3 hz	0.035	0.078	0.08	0.056
10^6 hz	0.03	0.046	0.057	0.023
3×10^6 hz	0.275	0.068	0.098	0.095

By permission of Rohm and Haas Co.

In summary, the chief properties that distinguish acrylics from other coatings are their transparency, color retention, dimensional stability, and excellent outdoor weatherability. Other important characteristics include:

1. Good adhesion to a wide variety of substrates
2. Good electrical properties at room temperature and at moderately elevated temperatures
3. Resistance to water, alcohol, alkalis, and acids

4. High resistance to mineral oil, vegetable oils, greases, and chemical fumes.
5. Low water absorption (less than 0.2 percent in 24 hr)
6. Excellent flexibility
7. Solderability

Table 3-6: Electrical Properties of HumiSeal Acrylic Coatings

Property	1B12	1B15	1B31	1B73
Dielectric constant, 1MHz at 25 C	2.8	2.3	2.5	2.6
Dissipation factor 1MHz at 25 C	0.010	0.010	0.010	0.010
Dielectric withstand voltage, volts	6000	8000	7500	6300
Insulation resistance, ohms	2.5×10^{14}	3.5×10^{14}	8.0×10^{14}	5.5×10^{14}
Insulation resistance, 10th cycle. 65°C, 98% RH, ohms	3.0×10^{10}	4.0×10^{10}	6.0×10^{10}	7.0×10^{10}

Courtesy of Columbia Chase Corporation

DIALLYLPHTHALATE AND OTHER ALLYLIC POLYMERS

Polymers based on diallylphthalate or those derived from monomers containing the functional allylic group —CH_2—CH=CH_2 are referred to as *allylic polymers.* They may also be regarded as a specific class of polyesters or as a class of vinyls. Diallylphthalate, the simplest and most commonly used monomer, is an ester of orthophthalic acid and allyl alcohol. In its synthesis the orthophthalic anhydride may be employed instead of the acid:

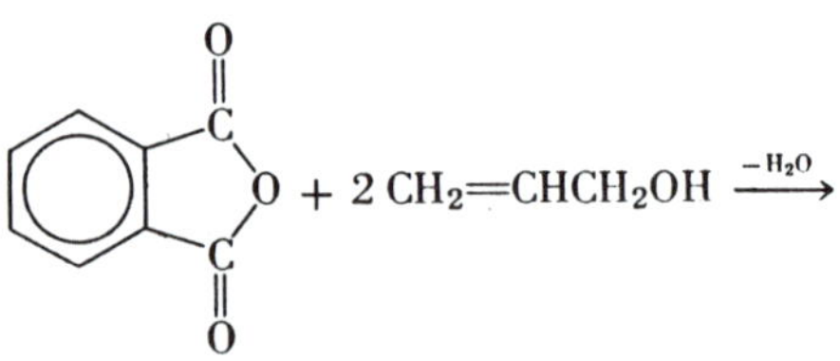

Orthophthalic anhydride   Allyl alcohol

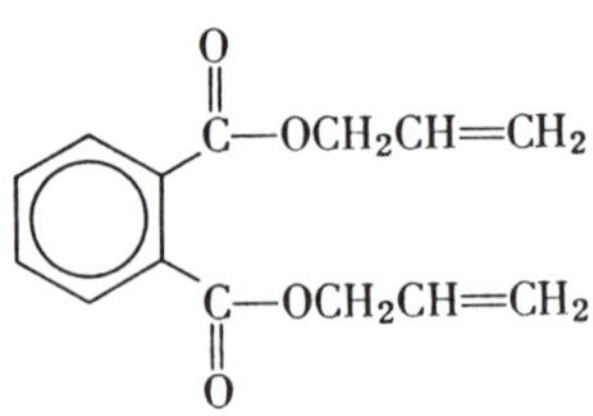

Diallylphthalate

Other commonly used ester monomers include diallylisophthalate, triallylcyanurate, diallylmaleate, and diallylchlorendate.

All allylic polymers are formed by a free-radical-addition type of polymerization through the vinyl groups in a manner similar to the polymerization of styrene and other vinyl monomers. Therefore no by-products are evolved during polymerization. A large variety of polymers and copolymers is possible, depending on the monomer used. However, the diallylphthalates and diallylisophthalates are the ones most commonly used as coatings, varnishes and impregnants. Shrinkage of monomers during polymerization can be as high a 11.8 percent, but this is considered low in comparison to styrene, where shrinkages of up to 30 percent have been encountered. Diallylphthalates possess excellent electrical properties, chemical resistance, and thermal stability. The triallylcyanurate polymer is stable up to approximately 300°C.[10]

POLYAMIDES

The polyamide resins are characterized by the presence of the repeating amide group

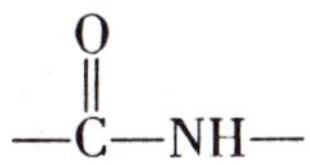

along the polymer chain. They are formed by the condensation of di- or polyamines with dibasic or polybasic acids and may be classified according to three main types: thermosetting fatty acid polyamines, thermoplastic fatty acid diamines, and nylons. Of these, the thermosetting polyamides are most widely used as coatings. They are derived from dimerized fatty acids and trifunctional amines. For instance, diethylenetriamine ($H_2NCH_2CH_2NHCH_2CH_2NH_2$) reacts with dimer acid (a mixture of dicarboxylic and polycarboxylic acids derived from linoleic acid) to give a series of liquid resins. These resins may be used as coatings as such, but more often they are co-reacted with epoxies as described in the section on epoxies in Chapter 1. The high concentration of amino groups along the polyamide chain readily opens the epoxy ring and cross-links numerous molecules. Epoxy-polyamide coatings thus formed have greater adhesion, higher flexibility, greater ability to wet surfaces, and better corrosion and chemical resistance than either epoxies or polyamides.[11, 12]

Polyamide resins, in addition to their high reactivity with epoxies, co-react with phenolics and other resins to yield highly cross-linked thermosetting coatings.[13]

Thermoplastic polyamides result from the condensation of dimerized fatty acids with short-chain diamines, often ethylenediamine ($H_2NCH_2CH_2NH_2$). They are quite soluble in many solvents and are compatible with plasticizers and resins. As such, they may be formulated as solution coatings, as aqueous dispersions, or as hot-melt coatings. Both the thermosetting and thermoplastic polyamides may be applied by conventional spray, dip, or brush techniques. The thermoplastic types may also be applied by fluidized-bed, electrostatic-spray, or hot-melt methods.

Nylons also belong to the polyamide family of polymers, but their structure, aside from the fact that they contain amide groups, is rather different from the other polyamides just described. They consist of linear molecules derived from the

condensation of aliphatic diamines with aliphatic dicarboxylic acids. The most common nylon is nylon 6,6, which is prepared from hexamethylenediamine and adipic acid as follows:

$$\underset{\text{Hexamethylenediamine}}{H_2NCH_2(CH_2)_4CH_2NH_2} + \underset{\text{Adipic acid}}{HOOC(CH_2)_4COOH} \xrightarrow{-H_2O}$$

$$\underset{\text{Nylon polyamide}}{\left[\overset{O}{\overset{\|}{C}}-(CH_2)_4-\overset{O}{\overset{\|}{C}}-NH(CH_2)_6-NH \right]}$$

The nylons have only limited use as coatings because of difficulties in formulating and applying them. Because of their low solubility and high softening temperatures, solution coatings are difficult to prepare. Of the various nylons, however, nylon 6 has been formulated both as a solution coating and as a dispersion coating using aqueous isopropanol or higher alcohols. A major use is for wire coatings.[13]

Properties of Polyamides

Some polyamide resins have softening temperatures of 110 to 200°C and hot-melt temperatures of 190 to 250°C.[14] The abrasion resistance of coatings derived from these resins is somewhat better than that of polyethylene coatings but not as good as that of many of the polyurethane and epoxy coatings (Table A-9).

Polyamide coatings formulated with zinc yellow, red lead, or other pigments have excellent corrosion-inhibiting properties. Extensive studies on the corrosion-inhibiting efficiency of polyamide primers and epoxy-polyamide coatings have been conducted, and detailed data on the protection of steel from salt water, hydrochloric acid, and hydrogen sulfide are given.[15] Alkyd- or oil-based coatings were least resistant to long-term exposure (2 years) to salt water. Polyvinyl chloride coatings were better, but epoxy polyamides were most resistant.

Polyamide coatings, as such, are not widely used as insulation for electronic or electrical equipment because of their marginal electrical properties, especially in the presence of moisture. In comparison to other polymer types, polyamides absorb above-average amounts of water. In a 24-hr period they may absorb 1.5 to 2 percent, and as much as 6 percent in an 8-day period at 100 percent relative humidity (Figure 3-2). Polyamide-epoxy combinations (see Chap. 1) are widely used as coatings for electronics and as adhesives, filleting compounds for component vibration damping, and encapsulants.

PHENOLICS

Phenolic resins, developed in 1909 by L. H. Baekeland, are the oldest completely synthetic polymers. They are formed by the chemical reaction of phenols with aldehydes, with either an acid or a base catalyst. The simplest and most widely used resin is based on the condensation of phenol and formaldehyde:

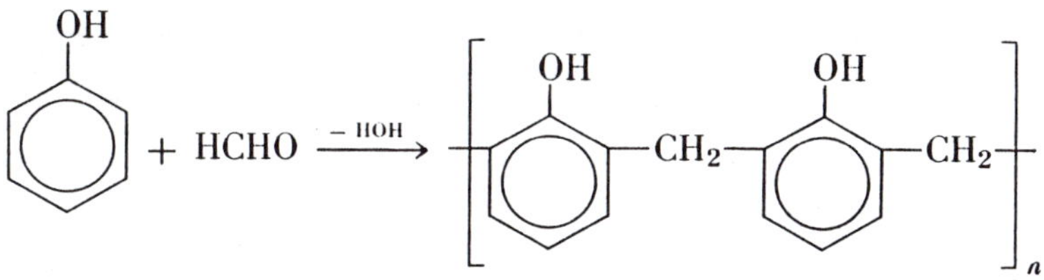

Phenol Formaldehyde Phenol-formaldehyde resin

Phenolic resins thus formed may be liquid or solid, depending on their structure. Hardening or curing is achieved by heating either alone or in the presence of hexamethylenetetramine, which acts both as a catalyst and as a source of methylene groups.[17, 18] In this way cross-linking of the resin molecules occurs opposite the hydroxyl groups:

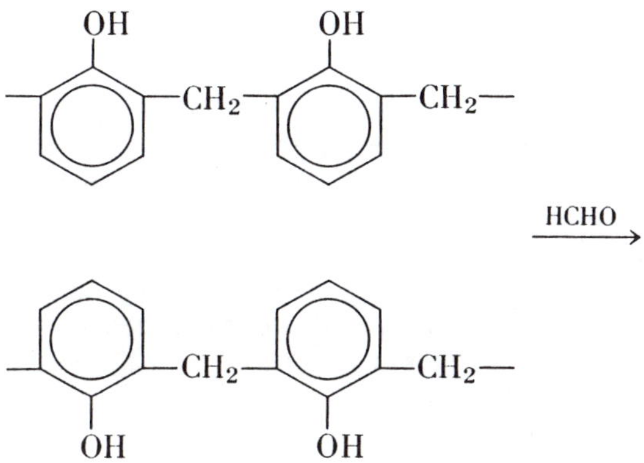

Phenol-formaldehyde resin molecules

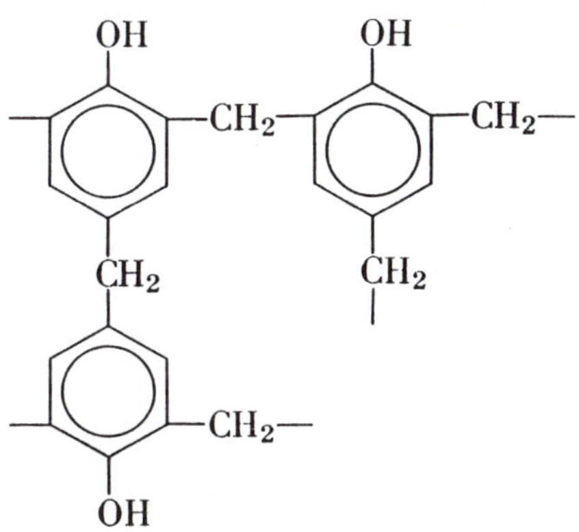

Cross-linked phenolic plastic

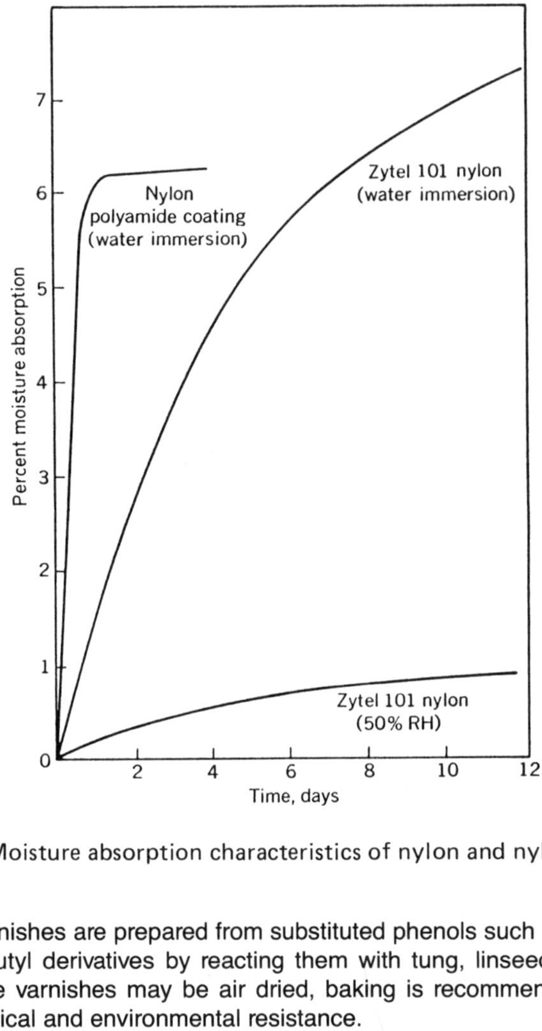

Figure 3-2: Moisture absorption characteristics of nylon and nylon coatings.[16]

Phenolic varnishes are prepared from substituted phenols such as the *p*-phenyl or *p*-tertiary butyl derivatives by reacting them with tung, linseed, or other oils. Although these varnishes may be air dried, baking is recommended to achieve optimum chemical and environmental resistance.

Properties of Phenolics

The biggest markets for phenolic resins are for molding compounds and laminates. They are used for electrical insulation, impregnating and surface varnishes, and as chemical-resistant coatings. They are available as solution coatings and may be applied by spraying, roller coating, or dipping. After a short air-drying period to allow evaporation of the solvents, they are baked at temperatures as high as 205°C to complete the polymerization and attain optimum properties.

Phenolic coatings have four outstanding properties: high thermal stability, high chemical resistance, good adhesion, and low cost. Some formulations are reported to withstand 205 to 260°C continuously with little or no loss in mechanical or electrical properties. Mineral fillers are normally incorporated to achieve this high-temperature performance. They are also employed to improve electrical properties;

for instance, they increase arc resistance and dielectric strength and lower dissipation factors.

Nevertheless, phenolic coatings are losing ground to epoxies for high-reliability electronic applications. Phenolic resins have several limitations not found in epoxies. For example, there are many reports that phenolic plastics have caused corrosion of copper and other metals because of the acidic or alkaline nature of the catalysts used in their cure. Shrinkage is also above average in phenolics and can cause high stresses. Moisture released during cure may be entrapped in the coating and degrade its electrical properties. The electrical properties of phenolics are generally poorer than those of epoxies because of their more highly polar nature. Dielectric constants and dissipation factors are higher (Table 3-7). Finally, the very high temperatures required to cure phenolics militate against their use on temperature-sensitive electronic components. Some of these inherent limitations can be overcome by coreacting phenolics with other resins. In this form they are more useful for electronic and electrical applications. Epoxy-phenol and polyvinyl formal-phenol coatings are two notable examples.

Table 3-7: Properties of Cured Phenolic Resins[19]

Property	*Value*
Volume resistivity (23°C, 50% humidity), ohm-cm	10^{12}–10^{13}
Dielectric strength (⅛-in. thick sample), volts/mil	
Short time	350–400
Step by step	250–300
Dielectric constant	
60 hz	6.5–7.5
10^3 hz	5.5–6.0
10^6 hz	4.0–5.5
Dissipation factor	
60 hz	0.10–0.15
10^3 hz	0.01–0.05
10^6 hz	0.04–0.05
Arc resistance, sec	200–250
Water absorption after 24 hr (⅛-in. thick, sample), %	0.3–0.4
Thermal conductivity, cal/(sec)(cm²)(°C/cm)	$3–5 \times 10^4$
Coefficient of thermal expansion in./(in.)(°C)	$60–80 \times 10^{-6}$

POLYSULFIDES

Polysulfides are organic polymers containing sulfur in the form of disulfide (—S—S—) linkages and mercaptan or thiol groups (—SH). The generalized structure may be represented as follows:

$$HS—[CH_2CH_2OCH_2OCH_2CH_2S—S]_n— CH_2OCH_2CH_2SH$$

Liquid polysulfide resins of molecular weight 3,000 to 4,000 may be polymerized further to yield solid rubbery coatings by oxidative free-radical mechanism via the

terminal mercaptan groups. Consequently they require the admixture of a second component or curing agent. Curing agents commonly used are dicumene hydroperoxide, lead dioxide, manganese dioxide, calcium dioxide, zinc dioxide, *p*-quinonedioxime, and organic peroxides in general.[20]

Polysulfide resins can also co-cure with various polymers to yield elastomeric and/or plastic properties based on the ratio of ingredients used. For example, epoxy resins in conjunction with aliphatic or aromatic amines are commonly co-cured with polysulfides.[21]

Properties of Polysulfides

Polysulfides provide flexible, rubbery coatings which are excellent water and moisture sealers, have excellent adhesion to many surfaces, and are resistant to solvent, oil, fuel, ozone and oxidation. Their electrical properties are good but not outstanding (Table 3-8). Dielectric constants are high (about 7 or greater), and electrical insulation resistances, though initially high, can drop rapidly and irreversibly upon exposure to humidity and applied voltage. The choice of curing agent may improve electrical properties. For instance, better electrical properties are obtained by employing lead dioxide rather than cumene hydroperoxide as the curing agent.

Table 3-8: Typical Electrical Properties of Polysulfides (Morton Thiokol)

Volume resistivity (ohm-cm)	1×10^{11} to 7×10^{12}
Surface resistivity (ohms)	1×10^{12} to 2×10^{14}
Dielectric constant (1 KHz)	5.5 to 8
Dissipation factor (1 KHz)	0.001 to 0.010

The intermediate electrical properties of polysulfides are valuable for many low-cost electrical applications, principally for transformer coatings and for connector potting and sealing. Polysulfides are seldom used for circuit-board protection or for microelectronic insulation. The application of a voltage potential across polysulfide-coated Y conductor patterns under humid conditions invariably lowers insulation resistance values to less than 10^8 ohms and results in decomposition of the polysulfide coating. The electrical properties of polysulfides may be altered by the addition of fillers. In general, the electrical and solvent-resistance properties of polysulfides are poor compared to epoxies.

REFERENCES

1. Korshak, V. V., and S. V. Vinogradova, "Polyesters," translated from the Russian by B. J. Hazzard, Pergamon, New York, 1965.
2. Lawrence, J. R.: "Polyester Resins," Plastics Application Series, Reinhold, New York, 1960.

3. Bjorksten, J.: "Polyesters and Their Applications," Reinhold, New York, 1956.
4. Beavers, E. M.: Applications for Alkyds and Polyesters, *Ind. Eng. Chem.,* **41**:738 (1949).
5. Smith, W. M.: "Vinyl Resins," Plastics Application Series, Reinhold, New York, 1961.
6. "Chemcyclopedia 87", American Chemical Society, 1987.
7. *Matls. Des. Eng.,* Materials Selector Issue, October, 1963.
8. Struber, V. R., "Theory and Practice of Vinyl Compounding." Argus Chemical Corp., New York, 1968.
9. Acryloid Acrylic Resins, *Rohm and Haas Product Bulletin,* 82A4, September 1985.
10. Raech H.: "Allylic Resins and Monomers," Plastics Application Series, Reinhold, New York, 1965.
11. Glaser, D., D. Floyd, and H. Wittcoff: Coatings Based on Blends of Polyamide and Epoxy Resins, *Official Dig.,* **29:** 159–169 (1957).
12. Floyd, D. E., D. E. Peerman, and H. Wittcoff: Characteristics of the Polyamide-Epoxy Resin System, *J. Appl. Chem.,* **7:** 250–260 (1957).
13. Roberts, A. G.: Organic Coatings, Properties, Selection, and Use, *Building Sci. Ser.* 7, February, 1968.
14. *Gen. Mills Bull.* 4–63, November, 1963.
15. Anderson, D. L., D. E. Floyd, D. W. Glaser, and H. Wittcoff: Corrosion Control Advantages of Polyamide-Epoxy Coatings, *Corrosion,* December, 1961.
16. *Du Pont Tech. Bull.,* Designing with Zytel Nylon Resin, 1969.
17. (a) Martino, C. F., and W. A. Keutgen: Phenolics, "Modern Plastics Encyclopedia," vol. 45, McGraw-Hill, New York, 1968. (b) ibid, vol 64, 1988.
18. (a) Jost, H. D.: Phenolic Resins, "Modern Plastics Encyclopedia," vol. 45, McGraw-Hill, New York, 1968. (b) ibid, vol 64, 1988.
19. (a) Plastics Properties Chart, "Modern Plastics Encyclopedia," Plastics Catalogue Corp., 1959. (b) ibid, vol 64, 1988.
20. Hockenberger, L.: Preparation and Application of Solid and Liquid Organic Polysulfide Polymers, *Chemie-Ing. Technol.,* vol. 36 (1964).
21. *Morton Thiokol Product Bulletin* LP 2/85.

4

Functions of Coatings

Coatings may serve numerous functions, but in general, the two most important ones are to provide high electrical insulation between conductors, interconnections, components, and other electrical parts and to provide protection from environmental damage. Severe environmental conditions necessitating protection include moisture, heat, salt spray, oxygen, radiation, microorganisms, chemicals, and solvents. In addition, fragile microelectronic assemblies need coatings for protection against abrasion, handling, shock, and vibration. Specially synthesized or formulated coatings may also serve a number of functional requirements, such as to provide electrical conductivity for radio-frequency interference, for ohmic contacts, or to provide either thermal insulation or thermal conductance. Coatings play a vital part in the successful operation of electronic assemblies, and if superior electrical coatings had not been developed, many of the highly dense and intricate windings, coils, and microelectronic assemblies would not now be possible.

ELECTRICAL PROPERTIES

One of the most important functions of organic coatings is to provide electrical insulation and dielectric isolation for active and passive electronic components. Their effectiveness in this respect is expressed in terms of insulation resistance, volume resistivity, surface resistivity, and dielectric strength. Other functions include the storage of electric current and the conductance of current. These are expressed in terms of dielectric constant, capacitance, dissipation factor, and conductivity. A knowledge of exact electrical values and how these parameters vary with changes in composition, purity, structure, or environment is important in selecting the most reliable coatings for electronic equipment. In the following sections these electrical parameters will be defined, and their dependence on temperature, frequency, and other variables will be presented. More theoretical discussions of electrical parameters may be found elsewhere.[1, 2]

Resistance and Resistivity

Insulation resistance is the ohmic relationship of the ratio of applied voltage to the total current between two electrodes in contact with a specific material. It is directly proportional to the length and inversely proportional to the area of the specimen according to the equation

$$R = \frac{\rho l}{A}$$

where R = insulation resistance, ohms
l = length, cm
A = area, cm^2
ρ = a proportionality constant called the *specific resistance or resistivity*, ohm-cm

It is convenient to compare materials in terms of their resistivity values because these values reduce resistance measurements to a common denominator. The frequently used *volume resistivity,* for example, is the ohmic resistance of a cube of bulk dielectric material 1 cm or 1 in. per side and is expressed in ohm-centimeters or ohm-inches, respectively. *Surface resistivity* is the resistance between two electrodes on the surface of an insulating material, expressed in ohms per square centimeter.

All materials may be roughly classified according to their ability to conduct or impede the flow of electrons. They range from metals, which are extremely good conductors, to plastics, which are very good insulators. Materials intermediate between these two types are referred to as *semiconductors,* but this does not necessarily mean that they have properties suitable for semiconductor devices. A select few inorganic materials, such as doped silicon or gallium arsenide are, of course, the basis for the fabrication of almost all semiconductor electronic devices. Organic semiconductors, as a rule, have not been useful for fabricating such devices. It is difficult to define a material as either completely insulating or completely conducting, since all solid materials conduct electricity to some degree, and we are therefore dealing with relative degrees of conduction (or resistance). In general, materials may be classified according to volume resistivity as follows:

Good conductors	10^{-6}–10^{0} ohm-cm
Semiconductors	10^{0}–10^{8} ohm-cm
Good insulators (poor conductors)	10^{8}–$>10^{12}$ ohm-cm

Organic coatings are, in general, good electrical insulators and as such are useful for electronic hardware. Although the resistivity of plastics is quite high (usually greater than 10^{12} ohm-cm), and the current which they conduct is considered negligible for most power equipment, in some electronic assemblies even this small current flow can be a critical factor. Therefore, a knowledge of the electrical tolerances that a circuit or device must meet is important before a coating can be selected. Volume resistivities of some of the more familiar materials are compared in Figure 4-1. Note the wide difference (up to 24 orders of magnitude) between the most conductive materials, such as copper or silver, and the least conductive, such as polytetrafluoroethylene or polystyrene.

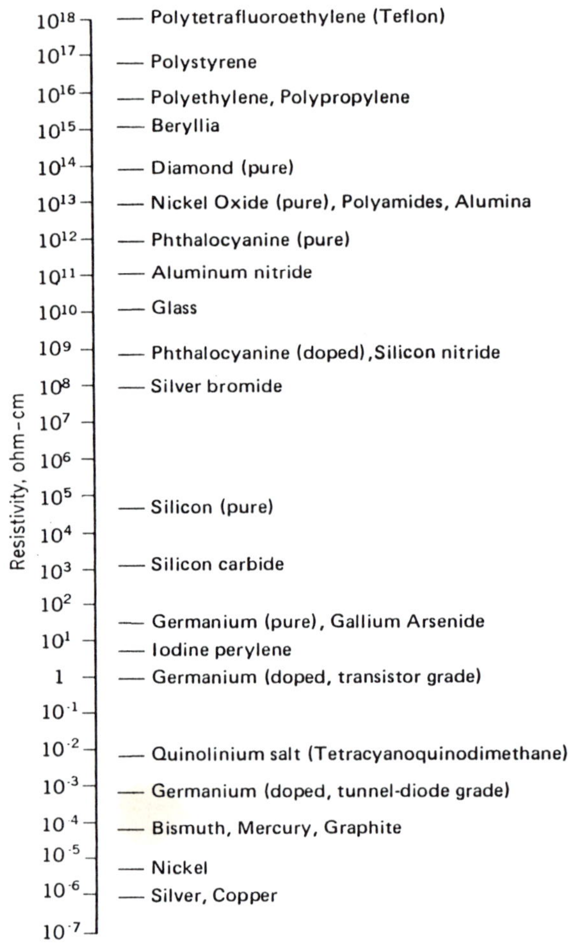

Figure 4-1: Comparative electrical resistivities of materials.

The Effect of Variables on Resistivity. Resistivity is a constant characteristic of a material. There are, however, many variables which can affect it. First, changes in chemical composition or formulation can have pronounced effects. Note in Figure 4-2 the effect of blending a flexibilizing resin such as polyamide into an epoxy resin. The excellent electrical properties of the epoxy decrease rapidly after the amount of polyamide added exceeds 40 percent. It may be assumed that reduction of electrical properties occurs after the stoichiometric quantity of polyamide has been reacted with epoxy.

Besides the effects of gross changes in composition, minor amounts of impurities also affect resistivity values. Large resistivity variations due to trace amounts of impurities have been successfully utilized in the design and fabrication of transistors and other semiconductor devices. For example, the addition of only 1 ppb of gallium or arsenic to pure germanium increases its conductivity by two

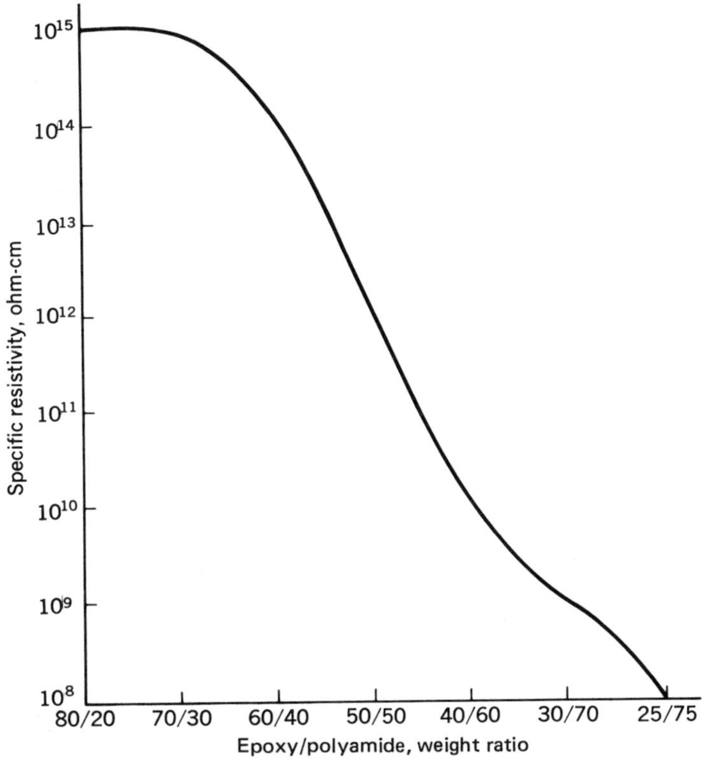

Figure 4-2: Variation of resistivity with change in epoxy-polyamide ratios.[3]

orders of magnitude, rendering it suitable as a transistor material. The variation of resistance with impurities, however, may also be undesirable, especially as it affects the performance of organic insulating coatings. Ionic impurities in plastics, coupled with the presence of moisture, are known to lower resistivity values by as much as 6 to 11 orders of magnitude.

Resistivity values also depend on the degree of cure or advancement in the state of polymerization of the coating resin. As the cure advances, electrical resistivity increases. A small decrease may be noted when the material first reaches its peak exothermic temperature. Two competing phenomena are occurring: a decrease in resistance as the temperature rises to reaction exotherm and an increase in resistance as the resin polymerizes and becomes fully cured. This is exemplified for an amine-cured epoxy system in Figures 4-3 and 4-4. It is apparent that these curves may also be useful in determining when resin polymerization is essentially completed.

Variation of resistivity with temperature is also important. Plastics, like inorganic semiconductors but unlike metals, possess negative temperature coefficients of resistivity, and large decreases in values can occur at elevated temperatures. Hence a coating which is an adequate insulator at room temperature may be completely unsatisfactory when the equipment is operating at higher temperatures.

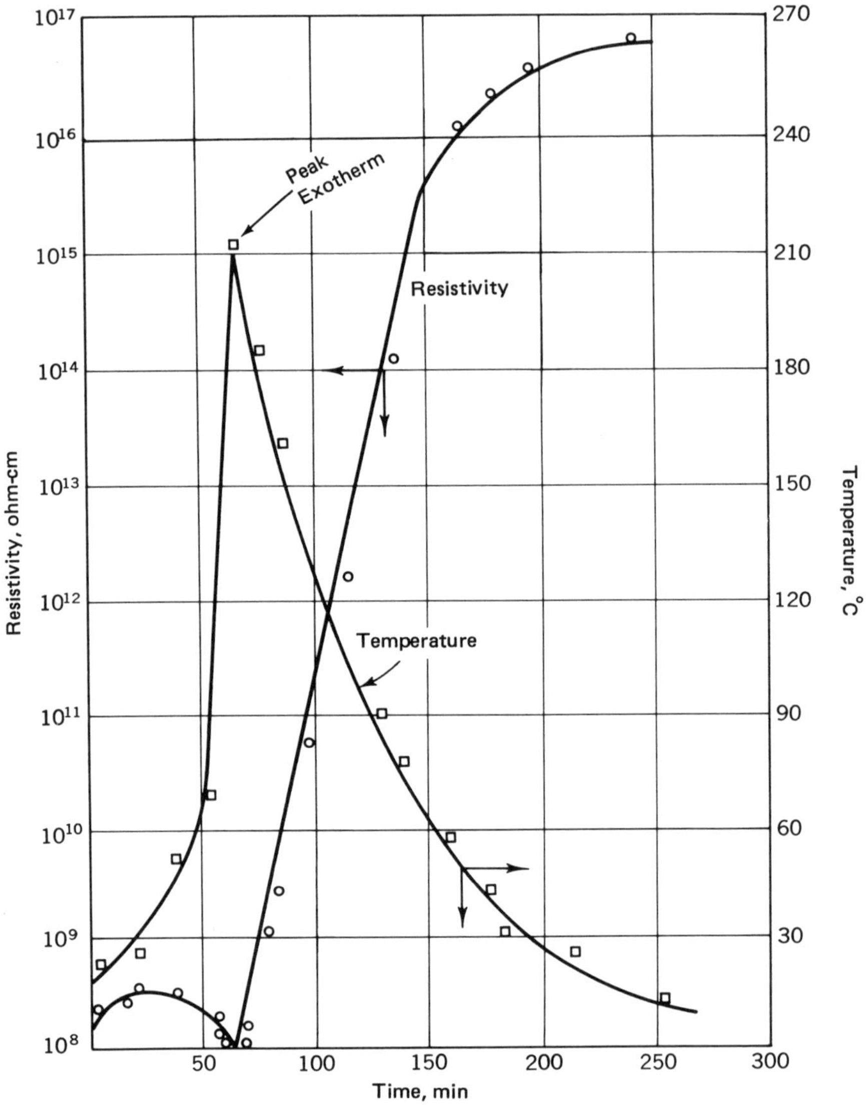

Figure 4-3: Electrical resistivity as a function of cure conditions for an amine-cured epoxy.[4,5]

Examples of temperature -vs.-resistivity curves for some polymer coating types are given in Figure 4-5. The temperature dependence of resistivity is given by the equation.

$$\rho = \rho_0 e^{\Delta E/2KT}$$

where ρ = resistivity at temperature T
ρ_0 = is a constant
K = Boltzmann's constant
ΔE = energy
T = absolute temperature

Finally, volume- and surface-resistivity values are also dependent on frequency, voltage, pressure, light (photoconductivity of some organic crystals and polymers), and the conditioning of the sample.

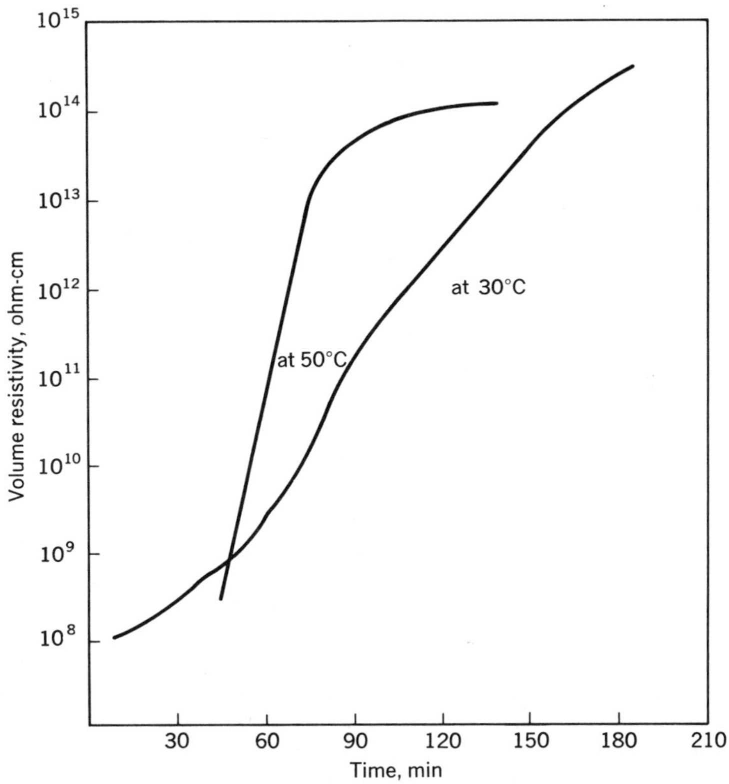

Figure 4-4: Isothermal polymerization of an amine-cured epoxy as a function of volume resistivity.[5]

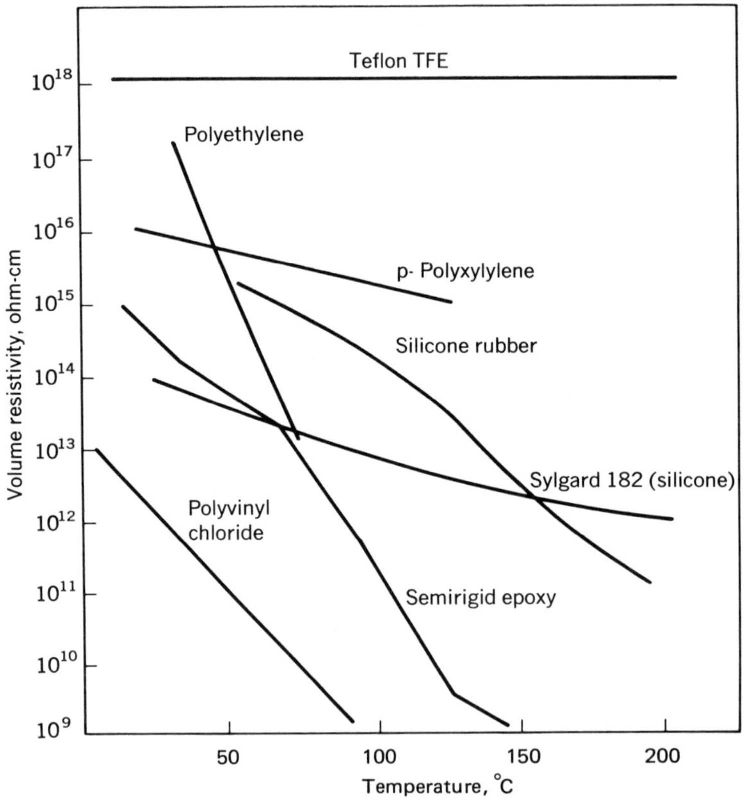

Figure 4-5: Electrical resistivity-temperature curves of polymer coating types.

Surface Resistivity. Moisture and contaminants have a more pronounced effect on surface resistivity than on volume resistivity. Whereas it may take several days or weeks for volume resistivity to change in humid or contaminating environments, surface effects are almost immediate. Drastic reductions in surface resistivity occur through contamination even for coatings which have very high initial values. For example, Mathes[6] reports reductions of 4 to 11 orders of magnitude in surface resistivities of polymers such as polyethylene and polystyrene which were contaminated with fingerprints and subsequently exposed to 96 percent relative humidity (Table 4-1).

Buchoff and Botjer[7] studied the effects of humidity on the surface resistivity of epoxies cured with various hardeners. As can be seen from Figure 4-6, electrical stability as a function of relative humidity is hightly dependent on the hardener used. Aromatic amine-cured epoxies were found to be stable at higher relative-humidity levels than anhydride- or aliphatic amine-cured types. However, it should be pointed out that in the anhydride-cured sample resistivity leveled off to about 5×10^{12} ohms/square, a value that is still considered adequate for most electrical applications. Furthermore, resistivities recover after removal from the humid environment, and again the aromatic amine-cured epoxies displayed the fastest rate of recovery (Figure 4-7).

Table 4-1: Effects of Contaminants on Surface Resistivity[6]

Material	Surface resistivity, ohms/square	
	At 50% humidity	At 96% humidity
Polyethylene		
As received	$>10^{18}$	2.4×10^{9}
Contaminated with fingerprints	...	4.6×10^{7}
Freshly shaved surface	...	2.2×10^{11}
Polystyrene		
As received	$\sim10^{18}$	1.8×10^{11}
Contaminated with fingerprints	...	5.5×10^{9}
Polymethylmethacrylate		
As received	$\sim10^{18}$	2.0×10^{14}
Contaminated with fingerprints	...	1.2×10^{12}
Silica glass, clean	10^{16}	5.0×10^{8}

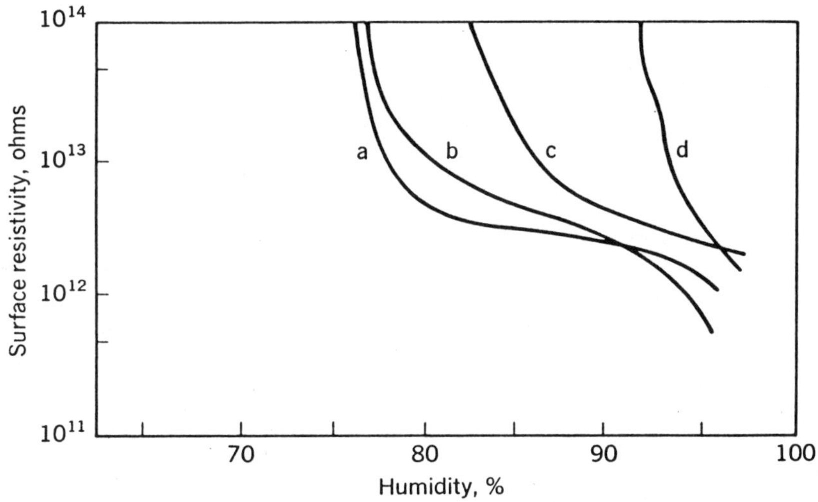

Figure 4-6: Effect of humidity on surface resistivity of cured epoxy resins at 35°C: (a) epoxy resin cured with methyl nadic anhydride, (b) epoxy resin cured with diethylenetriamine, (c) epoxy resin cured with aromatic amine, (d) Novolac-epoxy resin cured with aromatic amine.[7]

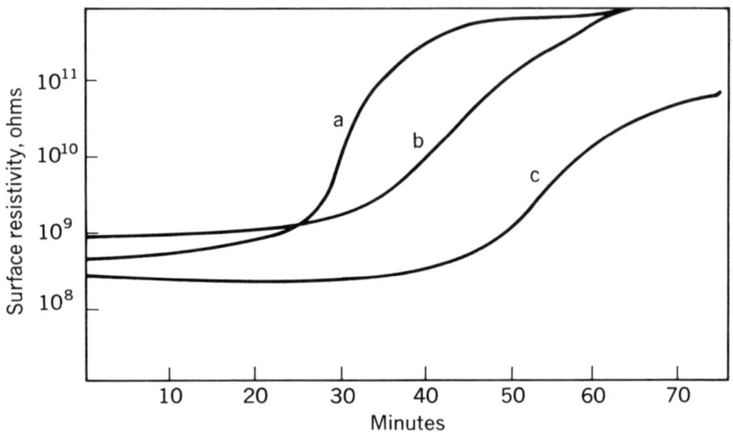

Figure 4-7: Recovery of surface resistivity for cured epoxy resins at 25°C after 80% humidity: (a) epoxy cured with aromatic amine, (b) epoxy cured with methyl nadic anhydride, (c) epoxy cured with diethylenetriamine.[7]

Conductance and Conductivity

The reciprocal of resistance is called *conductance,* expressed in units of $ohms^{-1}$, or mhos, and the reciprocal of resistivity is the *specific conductance,* expressed in units of $ohms^{-1}$ cm^{-1} or mhos per centimeter.

Metal-Filled Polymers. Coatings may be formulated with metal or other conductive fillers to render them electrically semiconducting or conducting. Conductive coatings are useful in applications which require ohmic contact for circuits and bridging of conductors, radio-frequency interference, and the dissipation of electrostatic charges. Silver, gold, copper, and carbon black are the most commonly used fillers, and epoxies, polyurethanes, silicones, vinyls, and acrylics are typical resin binders. Standard compositions are available or may be formulated in all viscosities, ranging from thick pastes to sprayable liquids.[8]

Conductivity values vary widely with the conductivity of the filler chosen, the technique for cleaning and processing the filler particles, and the percentage (loading) of the filler. Except for the noble metals (gold, platinum, etc.), meticulous cleaning of the metal is necessary to remove high-resistance surface oxide layers. Even after cleaning, metal oxides may slowly form again, and over long periods of time or under accelerated conditions, conductivity will decrease. The degree of cure of the resin binder also affects conductivity, and in general, better values are obtained with the baking-type formulations than with air-drying varieties (see Table 4-2). For very high conductance, silver or gold fillers are always employed; and with optimum formulation and cure, conductivity values of the order of 10^3 mhos/cm may be obtained.

Metal-filled polymers are used extensively for die-attach adhesives in the assembly of hybrid microcircuits and as coatings to protect electronic circuits from electromagnetic interference (EMI) and electrostatic discharge (ESD).

Table 4-2: Conductivity Data for Filled Epoxies

Composition	Conductivity (mhos/cm) Initial Value	After 40 Days Humidity Cycling	After 50 hr 20% Salt Spray
Silver-filled epoxy cured 6 hr at 180°C	3,000	1,250	1,400
Silver-filled epoxy air dried 36 hr	100	700	–
Gold-filled epoxy cured 6 hr at 50°C	200–700	–	–

Table 4-3: Comparison of Conductive Fillers

Filler (Use)	Advantages	Disadvantages
Silver (Electrical conductor)	Very high conductivity	High cost; silver migration can occur under certain conditions; tarnishes and corrodes
Gold (Electrical conductor)	Very high conductivity; very inert and stable	Higher cost than silver; subject to government controls and audit
Copper (Electrical conductor)	High conductivity; low cost	Requires extra steps for cleaning to remove oxides; conductivity decreases with age
Aluminum (EMI shielding)	High conductivity; low cost; good EMI shield	Can oxidize; shielding decreases with rising frequency
Steel (EMI shielding)	Low cost; good EMI shield at low volume percentage	Low conductivity
Graphite and carbon black (conductor and EMI shielding)	Low cost; good EMI shield	Low conductivity; shielding decreases with rising frequency; must be large volume percentage.

Molecularly Conductive Polymers. Most organic materials, as pointed out before, are insulators. However, some have recently been synthesized and others formulated that are semiconductive or even conductive. Polymers that are inherently conductive may be synthesized by doping a linear, conjugated polymer that has electrons in easily excited valence bands, with either an electron donor or an electron acceptor.[9] Some polymers under study include polyacetylene, polypyrrole, polyphenylene sulfide, polyparaphenylene, polythiophene and polyaniline (Figure 4-8).

In an undoped state these polymers have low conductivity, ranging from semiconductors (polyacetylene) to insulators (polyparaphenylene). The polymers may be doped by exposing thin films or the powder to the doping agent as a gas or liquid. Electron donating dopants include alkali metals such as lithium, sodium and potassium and sodium-naphthalide. Electron accepting dopants include arsenic pentafluoride, iodine and bromine.[10]

The final conductivity of a polymer is dependent primarily on the nature of and amount of the dopant and can vary between that of a semiconductor and a true metallic conductor. For example, polyparaphenylene doped with iodine is a semiconductor [4×10^{-15} $(\Omega\text{-cm})^{-1}$] but doped with arsenic pentafluoride converts to the metallic range [500 $(\Omega\text{-cm})^{-1}$].[10]

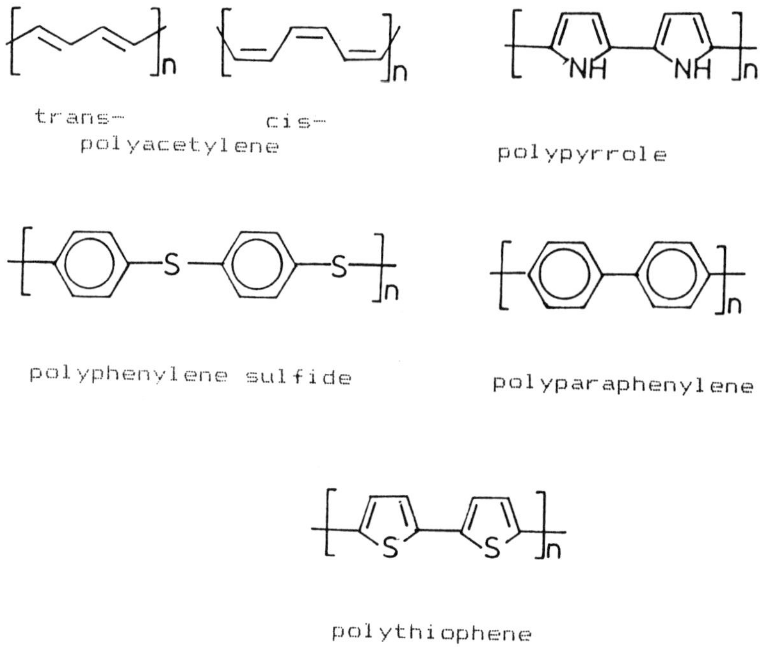

Figure 4-8: Examples of inherently conductive polymers.[9]

It has recently been reported that improved polymerization techniques can produce polymers that are more stable and conductive. In 1987 researchers at BASF synthesized a polyacetylene doped with iodine reported to be twice as conductive as copper on a weight to weight comparison, 147K (Ω -cm)$^{-1}$.[11] The electrical conductivity of this material is reported to remain stable for weeks in air, whereas that for polyacetylene-iodine polymers synthesized by standard methods degrades rapidly after a few days.[12]

Inherently conductive polymers are still in the research stage and not yet commercially available. These materials are still difficult to handle and shape and are insoluble in most organic and aqueous solutions. There are also problems in long-term stability since many of them are susceptible to oxidation.[12]

Capacitance

Consider a parallel-plate capacitor with the charge on one plate designated as $+\sigma$ and the charge on the other plate as $-\sigma$ per cm^2. If the space between the plates consists of a vacuum, the electrical field E_v within the capacitor is given by

$$E_v = 4\pi\sigma$$

Capacitance, by definition, is the total charge divided by the potential difference between plates.[13] Thus

$$C_v = \frac{Q}{V} = \frac{\sigma A}{4\pi d\sigma} = \frac{A}{4\pi d}$$

where Q = the total charge, coul
A = area of the plates
d = distance between plates
V = potential difference between plates, volts
C_v = capacitance, farads

For high-frequency, linear, high-speed digital circuits, the capacitance effects arising from plastic substrates, adhesives, or coatings are critical to the reliable functioning of the circuit. High capacitance can result in delays in switching times, changes in component values. Present computer operations are limited by the coupling capacitance between circuit paths and integrated circuits on multilayer boards. The computing speed between integrated circuits is reduced by this capacitance, and the power required to operate them is increased. Also, with the likelihood that future systems will employ many metal-oxide semiconductor devices, these problems will be compounded, since the coupling capacitance has a greater effect on computer speed.

Reductions in such parasitic capacitance can be achieved in a number of ways through proper selection of materials and design of circuit geometry. Capacitance C is directly proportional to the dielectric constant of the insulator separating the conductors, directly proportional to the area of the conductors, and inversely proportional to the distance between conductors; thus

$$C = \frac{kA}{d}$$

Hence low capacitance may be achieved by keeping A low, k low, and d high.

With recent trends toward microminiaturization and the utilization of very thin conductor lines, close spacings, and very thin insulation (5 mils or less), greater demands are being placed upon the insulating layer. Insulating materials must possess very low dielectric constants and at the same time must retain other required engineering and manufacturing properties. For high-frequency linear circuits, such as those used in radar assemblies, the dielectric constant of insulators again becomes important, especially since it may vary with changes in frequency. Graphs showing the variation of the dielectric constant as a function of frequency for some commonly used polymers are given in Figure 4-9.

Dielectric Constant

If the space between the plates of a parallel-plate capacitor is filled with a dielectric material, the capacitance will be increased by a factor which is a constant for a particular material. This constant k is referred to as the *dielectric constant* or *permittivity* and is represented by the equation

$$C_m = k\ C_v \qquad \text{or} \qquad k = \frac{C_m}{C_v}$$

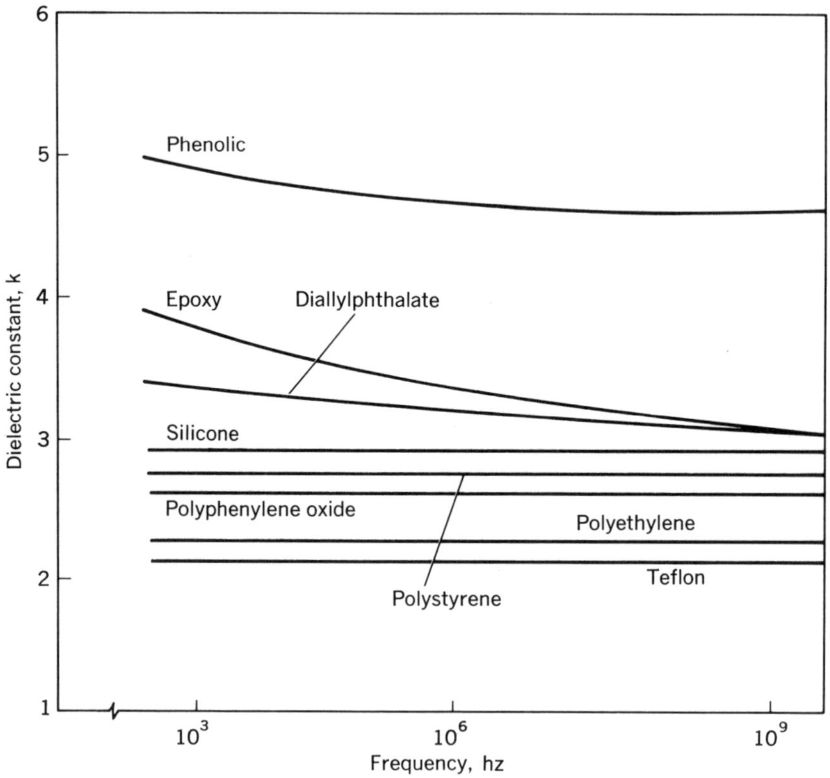

Figure 4-9: Variation of dielectric constant with frequency.[14]

where C_m is the capacitance of the dielectric material and C_v is the capacitance of a vacuum. The dielectric constant of a material may therefore be defined as the ratio of the parallel electrical capacitance with the material between the plates to the capacitance when a vacuum separates the plates. Since the dielectric material affects the force with which two oppositely charged plates attract each other, it may also be defined as the relative effect of the medium on this force of attraction, according to *Coulomb's equation,*

$$F = \frac{QQ'}{kd^2}$$

where F = force of attraction between the two plates
Q = charge on one plate
Q' = charge on the second plate
k = dielectric constant
d = distance between plates

Therefore the higher the dielectric constant of the material between the plates, the more attenuated will be the force of attraction between plates. It should be pointed out that the dielectric constant of vacuum is 1; and since the dielectric constant of air is just slightly above 1, for all practical purposes it is also taken as 1, which simplifies the measurement somewhat. Details of sample preparation, measuring methods, and equipment for measuring dielectric constants are given in ASTM D 150-59T.

As can be seen from the tables in the Appendix, the dielectric constants of gases are only slightly greater than 1, but values up to 100 are common for many organic compounds, especially the highly polar types, and values up to 1,000 are known for some inorganic materials such as barium titanate. Dielectric constants of materials arise from their electronic polarizability. Materials with polar groups (that is, those having permanent dipole moments, such as CO or COOH) will have large dielectric constants because of the orientation of the dipoles in an applied field.[15] Polar polymers also tend to absorb more water from the atmosphere, which again will impair their electrical properties. It is for these reasons that the nonpolar polymer types such as polyethylene, polystyrene, and fluorocarbon (Teflon) have better electrical properties. The more polar structures, such as polymethylmethacrylate, polyamides, and polyvinyl chlorides are, as a rule, inferior in electrical properties.

Dielectric constants for the insulating or protective polymer coatings discussed in this book will range from 2 to 8. Those coatings which have low dielectric constants and low dissipation factors and are able to maintain these low values over a wide temperature and humidity range are preferred as electrical insulating materials. Those combining high dielectric constants and low dissipation factors are useful as capacitors, because they can store and hold large amounts of electrical energy.

Dielectric constants of 4.5 maximum at 1,000 hz and 25°C are normally considered satisfactory as insulating materials for electronic and electrical assemblies. This is the minimum requirement specified in MIL-I-16923. However, as pointed out earlier, for microelectronic and miniaturized circuitry operating at high frequencies capacitance effects must be minimized, and materials with low dielectric constants are needed. A knowledge of dielectric constants at high frequencies is therefore important to the engineer designing such circuits. Values of dielectric constants from 60 hz to 10^6 hz are given in the Appendix.

The dielectric constant of a coating may increase or decrease, depending on variations in its composition. For example, the addition of glass or ceramic fillers, which have fairly high dielectric constants, will usually increase the *k* value of the coating. The blending of resins with high *k* factors with those of low *k* factors will increase the *k* value of the lower one in proportion to the amount and type of resin added. A typical curve for the blending of polysulfide with epoxy is given in Figure 4-10.

When experimental data are not available, a rough calculation of the dielectric constant of a composite system consisting of two components of different *k* values (resin and filler) may be made by means of the equation [16]

$$\log k_x = v_1 \log k_1 + v_2 \log k_2$$

where k_x = dielectric constant of the composite

v_1 = volume fraction of the first component

k_1 = dielectric constant of the first component
v_2 = volume fraction of the second component
k_2 = dielectric constant of the second component

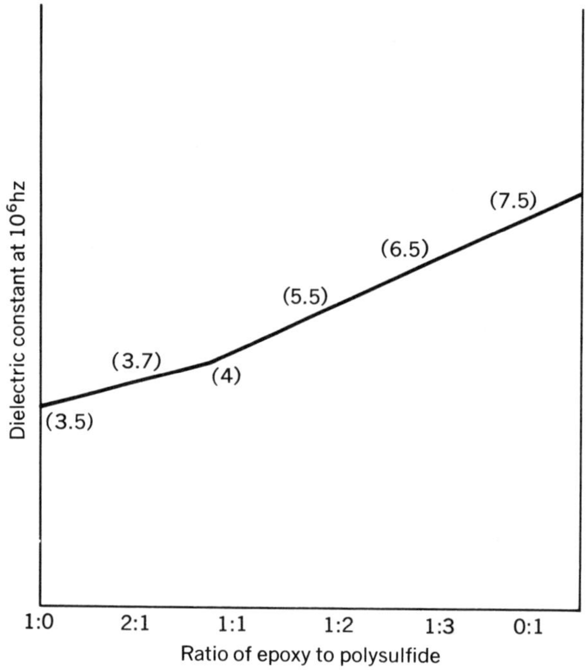

Figure 4-10: Dielectric constant variation of epoxy polysulfide blends.[17,18]

Dissipation Factor and Power Factor

The *dissipation factor D* is the ratio of the resistive (loss) component of current I_r to the capacitive component of current I_c and equals the tangent of the dielectric loss angle δ as follows:

$$D = \frac{I_r}{I_c} = \tan\delta$$

The *power factor PF* is a ratio of power dissipated to the current (volts × amperes) and is a measure of the dielectric loss in the insulation acting as a capacitor. It is related to dissipation factor by the equation

$$D = \frac{PF}{\sqrt{1 - (PF)^2}}$$

Since values for dielectric materials are low, dissipation factors are essentially equal to power factors. Consequently these terms are often used interchangeably in the literature. Military specifications such as MIL-I-16923 call for dissipation values no greater than 0.020 at 1,000 hz and 25°C.

The *loss factor* is the product of the power factor and dielectric constant and is a measure of signal absorption according to the equation

$$\text{Loss factor} \simeq \text{watts loss} \simeq k \tan \delta \simeq kD$$

Low values are desirable for all these parameters, especially for high-speed, high-frequency circuitry operating in the 10^7-to-10^{10}-hz range. A low value indicates low conversion of electrical energy to heat energy and reduced power loss for the network. The dissipation factor, as well as other electrical parameters, depends on frequency, temperature, humidity, and purity of the sample. A compilation of dissipation factors is given in the Appendix.

Variation of Dielectric Constant and Dissipation Factor with Temperature

At constant frequency, both the dielectric constant and dissipation factor for insulating coatings will, in general, increase with increasing temperature. Because coating formulations are not homogeneous, and because they contain constituents which become volatile or change on heating, electrical values may be quite erratic, and no simple linear relationship with temperature exists. Figures 4-11 and 4-12 show the variation of dielectric constant and dissipation factor with temperature for an

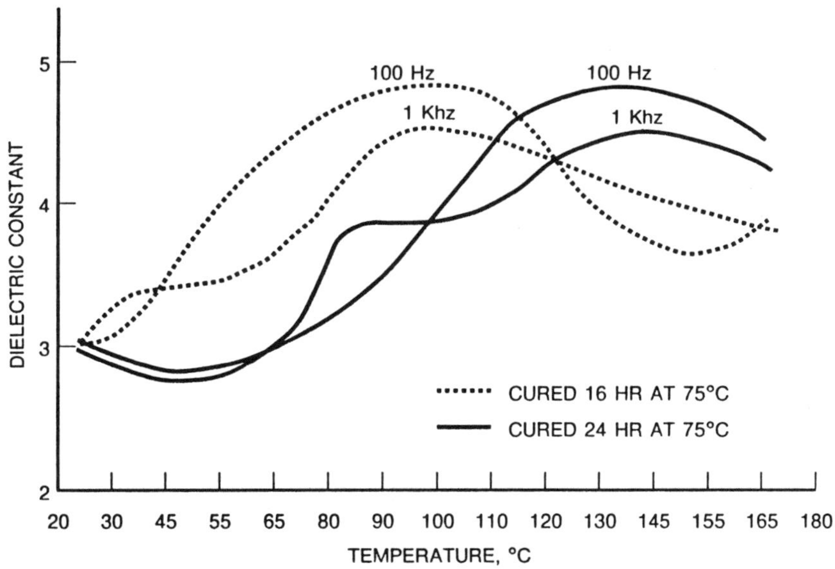

Figure 4-11: Variation of dielectric constant with temperature, degree of cure, and frequency for an epoxy coating cured with anhydride-castor oil adduct.

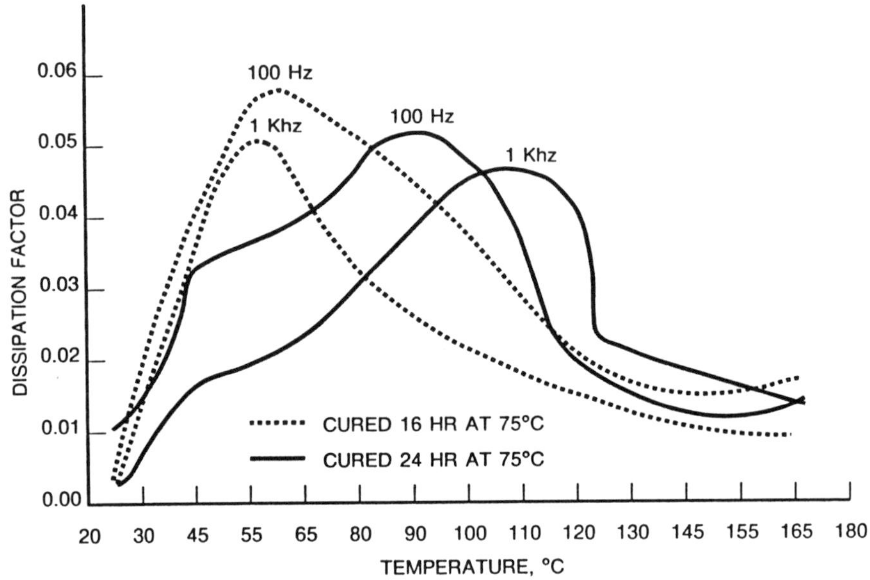

Figure 4-12: Variation of dissipation factor with temperature, degree of cure, and frequency for an epoxy coating cured with anhydride-castor oil adduct.

epoxy coating cured with an anhydride-castor oil adduct. The sample temperature was allowed to stabilize for about 25 min prior to the measurement of the electrical properties at each temperature. Therefore part of the change attributed to temperature may be masked by further curing of the polymer at each temperature.

Variation of Dielectric Constant and Dissipation Factor with Cure

The rate of change in the dielectric constant and the dissipation factor with increasing temperature may be used as an indication of the degree of cure of a polymer. Electrical properties of a fully cured polymer change only slowly with increasing temperature, in comparison with the rapid changes in a partially cured polymer. This is apparent from Figures 4-11 and 4-12, where an additional 8 hr of cure at 75°C resulted in a lower rate of change for both the dielectric constant and the dissipation factor. At room temperature the electrical properties of the two samples are almost identical, and therefore at this temperature one cannot differentiate between the two cure conditions. Further data on the change of dielectric constant and dissipation factor with the degree of cure are given in Table 4-4. In these studies the temperature was held constant at 23°C. At all frequencies dielectric constants and dissipation factors decreased progressively as the cure schedule was increased. The coating tested was Epon 828 cured with an anhydride-castor oil adduct.[19]

The optimum curing time and temperature obtained from a plot of the dielectric constant or dissipation factor versus time at constant temperature are even more apparent for the epoxy system shown in Figures 4-13 and 4-14. Initial values

Table 4-4: Effect* of Cure on Electrical Properties of Epoxy Cured with Anhydride—Castor Oil Adduct

Sample	After Cure of 16 hr at 74°±2°C	After Postcure of 5 hr at 74°±2°C	After Additional Postcure of 16 hr at 120°±2°C
		Dielectric Constant	
At 100 hz			
1	4.31	3.77	3.08
2	4.30	3.73	3.04
3	4.32	–	3.07
At 1 khz			
1	3.91	3.52	3.07
2	3.87	3.49	3.02
3	3.93	–	3.07
At 10 khz			
1	3.63	3.39	3.05
2	3.57	3.35	3.00
3	3.65	–	3.04
At 100 khz			
1	3.45	3.28	3.03
2	3.40	3.24	2.96
3	3.49	–	3.01
		Dissipation Factor	
At 100 hz			
1	0.0844	0.0664	0.0039
2	0.0912	0.0690	0.0040
3	0.0841	–	0.0036
At 1 khz			
1	0.0626	0.0385	0.0037
2	0.0668	0.0396	0.0038
3	0.0586	–	0.0035
At 10 khz			
1	0.0433	0.0257	0.0051
2	0.0452	0.0266	0.0053
3	0.0398	–	0.0048
At 100 khz			
1	0.0371	0.0254	0.0091
2	0.0380	0.0259	0.0094
3	0.0343	–	0.0083

*Measured at 23°C.

immediately after mixing a two-component system or upon activating a one-component coating are high, but they decrease as polymerization and hardening progress. Ultimately, at some optimum time, these values will level off and stabilize. It can be seen that the cure of this epoxy resin is essentially complete in 3.5 hr at 120°C but requires 12 hr at 66°C[20] to achieve the same degree of cure.

Other investigators[21] have also shown that the degree of hardening and cross-linking of an epoxy system may be followed by dielectric measurements over a frequency range of 30 to 10^{10} hz. A correlation of mechanical properties with electrical changes may also be performed, and once the exact correlation has been established and standard curves have been obtained, they may be used for rapid quality-control testing and inspection.

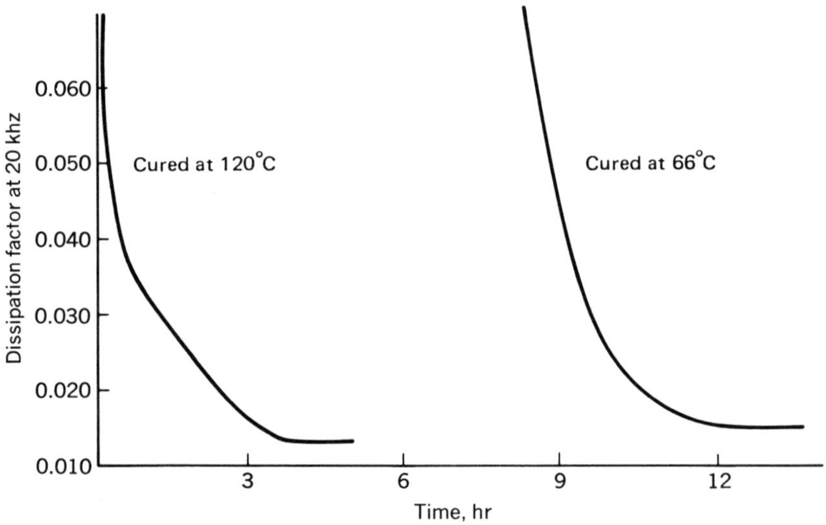

Figure 4-13: Establishment of epoxy cure schedule from dissipation-factor data.[20]

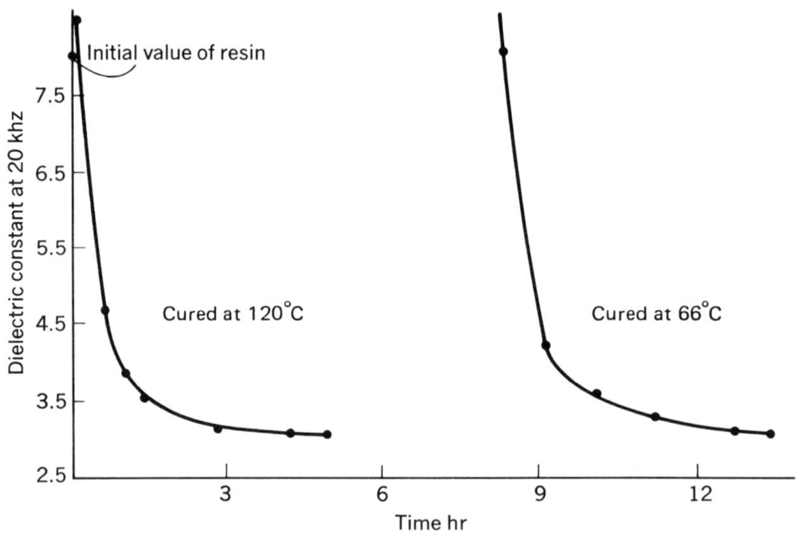

Figure 4-14: Establishment of epoxy cure schedule from dielectric-constant data.[20]

Dielectric Strength and Breakdown Voltage

Still another important electrical property is the dielectric strength or the ability of a coating to withstand applied voltage without breakdown. *Dielectric strength* is defined as the minimum voltage (usually reported in volts per mil thickness) at or below which no breakdown occurs. Others prefer to speak about *dielectric breakdown,* that is, the voltage threshold at or above which actual failure occurs. This is also reported in volts per mil thickness. The two terms are often used interchangeably. The physical breakdown of coatings at a critical voltage is believed to be due to a large increase in electrons (electron "avalanche" effect) within the insulation.[22] Even the best insulation materials contain a small number of ions and free electrons owing to thermal agitation or molecular imperfections. In such cases very high voltages will be necessary to cause breakdown. Other, less perfect insulation materials, containing impurities, voids, and moisture, will have lower breakdown voltages.[23]

Accurate dielectric-strength data are important in designing reliable electrical parts of high component density intended for operation at high voltages. For both low- and high-voltage operating parts, organic coatings are widely used because of their high dielectric strengths. In fact, dielectric strengths of organic coatings are, in general, much higher than those of inorganic or ceramic materials.

Dielectric-strength values should be measured carefully under specified conditions to obtain reproducible results and a reliable comparison between materials or samples. A large number of test conditions are known to affect dielectric-strength values. For example, it is always important to standardize the electrode configuration and the thickness of the specimen. Although all values are ultimately reduced to volts per mil thickness, the thickness of the sample has a marked effect on the values obtained (Figure 4-15). In general, thin specimens result in higher values than thicker ones. In most cases the 17-mil thickness specified in the ASTM procedure is too thick and is difficult to achieve with most coatings. The same test procedure, however, may be followed for samples of 2-5 mil thickness. Other variables which affect readings are the manner in which the voltage is applied (continuously or stepwise), the rate of voltage increase; the frequency of the applied power, the integrity of the coating film (lack of pinholes and air voids), and the purity of the sample. In general, the dielectric strength of a coating decreases with an increase in sample imperfections and impurities.

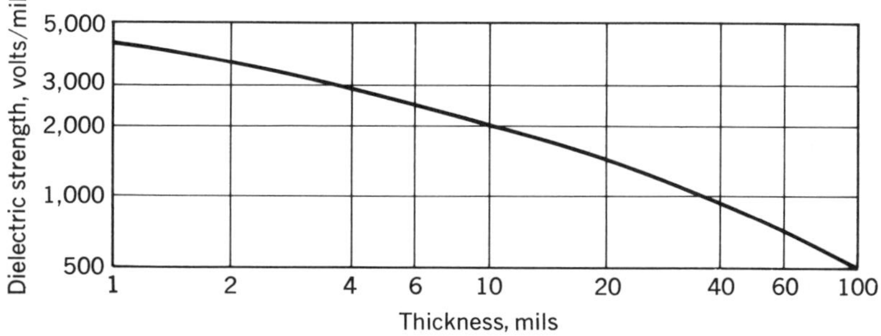

Figure 4-15: Effect of coating thickness on dielectric strengh of Teflon TFE.

The atmospheric conditions, such as temperature and humidity, under which the test is performed are also important variables. A plot of the variation of dielectric strength with temperature shows the same general shape of curve for all polymers. There is less change at low temperature but a rapid decline above a critical temperature which is characteristic of the coating being tested[24] (Figure 4-16).

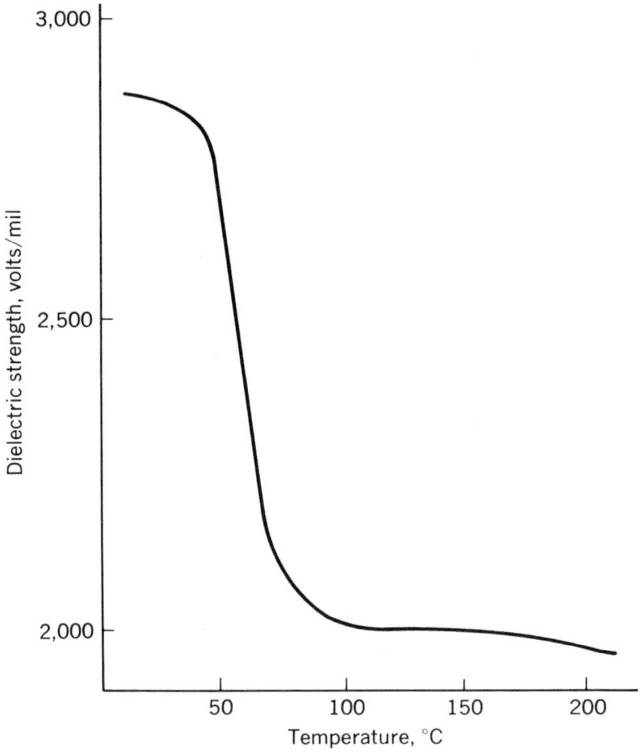

Figure 4-16: Temperature dependence of the dielectric strength of polytetrafluoroethylene (5-mil film, ¼ inch electrode).[24]

Although elevated temperature and moisture cause a pronounced reduction in dielectric strength, in some cases final dielectric-strength values may still be sufficiently high to be of little concern. As an example, the dielectric-breakdown voltage of an 8-mil silicone coating (Dow Corning DC-630) with an initial value of 1,000 volt/mil dropped to 400 volts/mil after 24 hr of immersion in distilled water and to 500 volts/mil after 24 hr of exposure at 260°C.[25]

Measuring the dielectric strength under ambient conditions may result in readings which are difficult to interpret because of electrical discharges and surface arcing around the sample. In these cases samples may be tested while immersed in a highly stable fluid having a minimum breakdown voltage of 26,000 volts, with 0.1-in. separation between the electrodes.[26] Examples of high-dielectric fluids used for this purpose are silicones and various hydrocarbon and transformer oils.

There are several standardized procedures for measuring dielectric strength, such as ASTM D 149. Dielectric-strength values for the commonly used organic coatings and for some ceramic and nonorganic materials are given in the Appendix.

Arc Resistance

Arc resistance is defined as the time, in seconds, that an arc may play across the surface of a material before electrical breakdown of the surface occurs. A high-voltage, low-current arc is used to simulate service conditions such as those existing in ac circuits operating at high voltages and currents limited to the low-milliampere region. The commonly employed testing method is that described in ASTM D 495. Several arc testers complying with this test method are commercially available. According to ASTM D 495, the specimens are placed between the electrodes, and an arc is generated at scheduled intervals and at specified current densities.

There are three modes by which failures due to arcing can occur:

1. By tracking; that is, by the formation of a thin wirelike line between the electrodes.
2. By carbonization of the surface due to heating. A conductive path is formed because of the formation of carbonaceous material which has less resistance to the flow of current than the original surface.
3. By self-ignition of the coating without the formation of any visible conducting path.

Arc resistances of different coatings vary widely, depending on the molecular structure of the polymer resin (see the Appendix). Constituents of the coating formulation, such as the curing agent and the nature and amount of filler, also affect arc resistance. In the case of epoxies, aromatic amine and anhydride curing agents result in higher values than aliphatic amines. Fillers also have pronounced effect, generally improving the arc resistance of a coating. A study of the effects of fillers such as mica, alumina, silica, and gypsum on the arc resistance of an epoxy resin was made by Delmonte[27] and by others.[28] It was found that the addition of fillers could readily double or even triple the arc resistance of an unfilled epoxy system.

Not all coatings may be defined in terms of precise arc-resistance values. Some may melt, gum up, or crack. When precise values are obtained, they are a function of the inherent stability of the material, its purity, and its suface cleanliness. With regard to surface cleanliness, arc resistance may be improved by keeping the surface dry and free from contaminants. Even touching the surface of the sample will reduce the arc resistance because of moisture, salts, and grease transferred from the fingers.

In addition to its usefulness as a design criterion, arc-resistance testing may be useful as a quality-control method and as a means of determining the changes a surface can undergo on aging, irradiation, and exposure to temperature or humidity.

ENVIRONMENTAL PROTECTION

Protection Against Moisture

Probably the most important function of coatings for printed circuits, and indeed for all electronic assemblies, is that they act as moisture and gas barriers to prevent corrosion and breakdown of electrical insulation. It should be understood that all present coatings absorb water and are permeable to moisture and other gases to a greater or lesser degree and cannot be considered true hermetic seals, as is achieved with welded metal packages. However, it is possible to select plastic coatings which will reduce moisture penetration sufficiently that assemblies can pass long-term storage and operational requirements.

Failures due to penetration of moisture may be of two types. First, moisture mobilizes ionic contaminants in the plastic or on the surface of the device, resulting in the normally observed deterioration of electrical insulation or the formation of inversion layers in semiconductor devices. Second, moisture, again in the presence of impurity ions, permits electrolytic corrosion to take place. For these reasons both active and passive devices such as transistors, diodes, and resistors are hermetically sealed in metal packages, and entire electronic assemblies are sealed in dry inert-gas atmospheres. Hermetic sealing is expensive and is not practical for most commercial applications. In such cases organic coatings or plastic encapsulants may provide sufficient protection.

Because all organic coatings are permeable to moisture in varying degrees, quantitative values for their resistance to water are important. Water resistance may be defined in terms of three factors: (1) the percent absorption of water by the coating at constant temperature over a period of time, (2) the rate of water-vapor transmission through the coating, and (3) functioning of a coated electronic assembly after simulated or accelerated humidity cycling tests.

Water Absorption. Coatings absorb water to varying degrees, the amount being a function of the molecular structure of the coating and the degree of cure. As a rule, coating resins containing hydrophilic polar groups such as hydroxyl or amide or resins containing low molecular weight water-soluble additives have high water-absorption values. Also, semirigid coatings tend to absorb more water than highly cross-linked systems. The percentage of water absorption eventually becomes constant at constant temperature for each material. In many cases, however, equilibrium is achieved only after 1 or more weeks of immersion. Unfortunately, most of the reported data are for a 24-hr immersion period and do not give the true equilibrium values, which are often considerably higher. Furthermore, some coatings (the polyurethane in Figures 4-17) reach a maximum value and then proceed to lose weight and reach equilibrium at a value much lower than the maximum. This is believed to be caused by the leaching out of water-soluble additives or of trapped solvents. Twenty-four-hour data are meaningful for practical applications only where the test conditions are equal to or more severe than those to which the part will be subjected in actual operation.

For the most part, the percentage of water absorption is not in itself a very meaningful value. It should be correlated to deleterious effects on the coating or on the part which the coating protects. Examples of these effects include corrosion, insulation breakdown, hydrolysis of the coating, dimensional change, or delamination and loss of adhesion. Absolute judgments concerning these effects cannot be

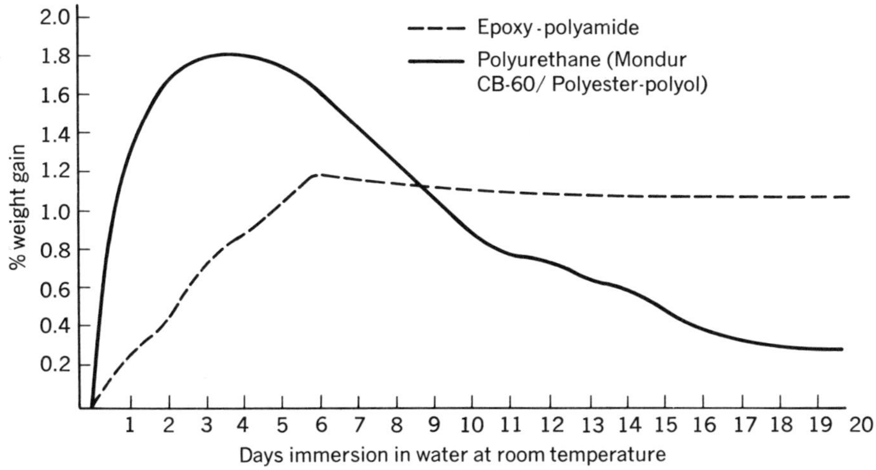

Figure 4-17: Water-Absorption curves for circuit-board coatings.

drawn from water-absorption data alone. Although water absorption is a critical factor in the decrease in electrical insulating properties, it is not the only factor. Ionic impurities and the polar nature of the coating also play a vital role. In some coatings, electrical properties are adversely affected when only a small amount of water is absorbed, and conversely, others with very large water absorptions may be quite stable electrically.[29] In general, however, surface resistivity, volume resistivity, and dielectric constant decrease while dissipation factors increase according to the increase in the percentage of water absorbed.[30] It is in this light that comparative water-absorption values should be studied and used (see Table 4-5 and Appendix).

Permeability. Moisture may seep through a coating by one or both of two mechanisms. One entails passage through microcracks, channels, or pinholes arising from physical imperfections in the coating. Pinholes and pores in the coating can be reduced considerably by applying multiple coats of the same material (Figures 4-18). The second moisture-penetration mechanism is one inherent in the molecular structure of polymers and entails permeation of the water-vapor molecules through the space between polymer molecules and within polymer structures.

This permeability or moisture-vapor transmission rate (MVTR), expressed in grams per hour per square centimeter per centimeter, is a more significant criterion than water absorption in defining water resistance. Under specified conditions of humidity and temperature, MVTR is expressed by the equation

$$\text{MVTR} = \frac{Ql}{at}$$

where Q = water vapor permeating the film, g

l = film thickness, cm

a = film area, cm^2

t = time, hr

Table 4-5: Water-Absorption Values for Epoxies

Resin and curing agent	Curing schedule	Water absorption, % 1 hour, 100 C	10 days, 20 C
Araldite 502 20 phr HY 956	7 days at 25 C	1.1	0.6
Araldite 6010 85 phr HT 907 + 3 phr XU 183	gel at 100 C + 1 hr at 150 C	0.3 to 0.8	0.2 to 0.3
85 phr HT 907 + 3 phr XU 183 + 400 phr quartz	gel at 100 C + 1 hr at 150 C	0.1 to 0.2	0.1 to 0.2
Araldite CY 179 105 phr HT 107 + 12 phr XU 183	gel at 120 C + 3 hr at 120 C	0.35	---
105 phr HT 107 + 12 phr XU 183	gel 2 to 4 hr at 80 C + 6 to 10 hr at 120 C	0.45 to 0.50	0.55 to 0.60
105 phr HT 107 + 12 phr XU 183 + 40 phr PPG 425 + 450 phr 325 mesh silica	gel 2 to 4 hr at 80 C + 6 to 10 hr at 120 C	0.33 to 0.37	0.29 to 0.33
105 phr HT 107 + 12 phr XU 183 + 40 phr PPG 425 + 450 phr hydrated alumina	gel 2 to 4 hr at 80 C + 6 to 10 hr at 120 C	0.25 to 0.29	0.31 to 0.35

Courtesy of Ciba-Geigy

MVTR values are generally reported in grams in 24 hr per 100 in.^{2} per mil, by multiplying by $24 \times 645.2/0.00254$ or 6.1×10^6.

Organic polymers differ widely in permeability because of differences in their molecular structures. A highly cross-linked compact structure should be less permeable than a loose, only partially cross-linked polymeric network. The geometry and size of a vapor molecule also determine its permeability. The rate of permeation is inversely proportional to the molecular or atomic size of the permeating gas. In addition to these geometrical considerations, the chemical natures of both the plastic and the vapor are important factors. For instance, hydrophobic polar groups available in a polymer chain provide great cohesive strength and will result in a film that is more resistant to moisture. The excellent moisture-barrier properties of polyvinylidene chloride (Saran) are thus attributable to the many polar chlorine atoms in its structure.

Permeability is also a function of the nature and amount of pigment or filler used. Pigmented coatings have lower permeabilities than unpigmented versions because of the longer path the water molecules must travel to penetrate the thickness of the film. The nature of the pigment, its geometry (for example, spherical or flake), and the packing density all affect permeability to moisture.[32] Pigments in flake

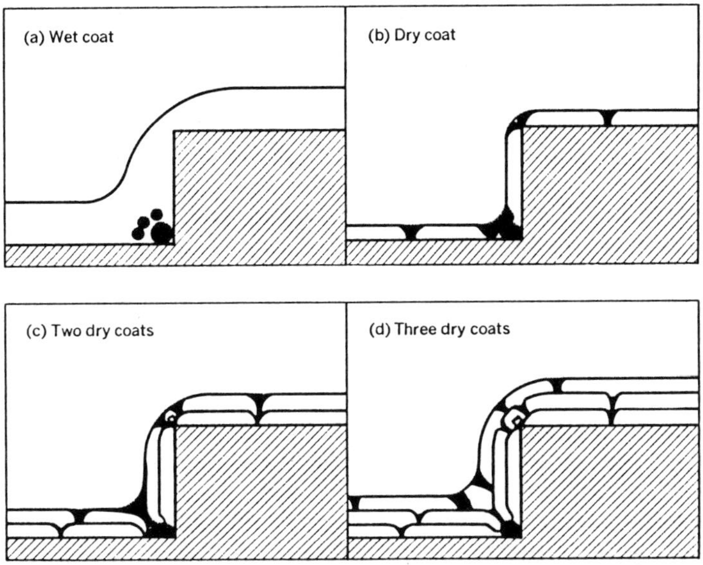

Figure 4-18: Reduction of pinholes by the use of multiple coats: (a) and (b) illustrate, on an exaggerated scale, the formation of pinholes and the stretching of the coating over sharp edges and protrusions after a single coating application; (c) illustrates a substantial reduction in the number of contiguous pinholes after a second application; (d) illustrates the same component with three coats, showing practically pinhole-free insulation.[31]

form, such as aluminum, will considerably improve resistance to moisture penetration. Permeability values drop with an increase in the concentration of pigment up to a critical pigment-volume ratio and then rise sharply. This critical ratio differs for different pigments and usually ranges from 40 to 70 percent. Several other factors which affect permeability are shown in Table 4-6. The moisture permeability properties of polymer coatings are normally presented as moisture vapor transmission rates. Values for some typical plastics are given in the Appendix.

Humidity Testing. Both water-absorption and permeability values may serve as preliminary data for selecting coatings for further evaluation. As pointed out above, to ensure long-term reliability, electronic assemblies should be subjected to long-term or accelerated environmental testing. Humidity tests such as those stipulated in specifications MIL-E-5272 and MIL-STD-202 are very severe. Generally, the test requires a 10-day exposure to moisture in both the vapor and condensate form, with temperature cycling from -10 to 65°C. A 100-volt direct current applied stress is also required by MIL-STD-202. The test is an accelerated one, but it does give an indication of the long-term reliability of the assembly under outdoor and operating conditions. The relatively inexpensive Blue-M humidity chamber (produced by the Blue M Electric Co.) may be used for laboratory testing. Because of the variety of substrates, metal combinations, and sensitive components, very few pieces of uncoated hardware will pass this test without serious physical or electrical impairment.

Table 4-6: Factors Affecting Moisture Permeability

Factor	*Usual effect on permeability*
Solvent entrapment	Increase
Plasticizers	Increase
Structure of film	
Polar, hydrophobic	Decrease
Polar, hydrophilic	Increase
Nonpolar	Decrease
Degree of cross-linking	Decrease
Crystallinity	Decrease

In selecting a protective coating, care should be taken to ensure that it will not only shield the assembly, but will not itself deteriorate by blistering, lifting, whitening, softening, or absorbing excessive amounts of water. Unfortunately, since coating formulations vary widely, it is impossible to give a list of good coatings and poor ones from which to make selections. It is imperative that humidity testing of some sort be performed in each case prior to use or recommendation.

Protection Against Corrosion

Coatings are applied to metal surfaces for protection from the corrosive effects of the surrounding medium. The most commonly encountered corrosive environments are listed in Table 4-7. As a rule, elevated temperatures accelerate the effects of these environments.

Table 4-7: Environments and Their Corrosive Constituents

Environment	*Corrosive constituents*
Normal air ambient	Moisture, oxygen, sulfur dioxide, carbon dioxide
Water	Calcium salts and other metal salts, ions
Salt water or salt spray	Sodium and chloride ions, marine organisms
Chemicals and solvents	Numerous acids, bases, oxidizing or reducing agents, solvents, strippers, etc., which may come into contact with the part during processing or operation
Soil	Moisture, fungus, other microorganisms

Metals corrode in different ways, depending on the nature of the metal and the environment to which it is exposed. Some of the various modes of failure for metals and alloys commonly employed in electronic assemblies are given in Table 4-8. Corrosion occurs either by a chemical or electro-chemical process, and in almost every case, moisture must be present. Hence the moisture-barrier and moisture-resistance properties of a coating become key factors in corrosion prevention. With many metals, inorganic oxide films will normally form on storage in air; these films

are quite passive and will slow down or prevent further deterioration of the metal. In other cases artificially produced inorganic coatings may be formed, and these will also provide the same protection. These films, called *chemical-conversion coatings,* are formed by the reaction of the metal surface with oxygen, phosphates, chromates, or other chemicals. Other passive films, such as the anodized films, are formed electrochemically. For very harsh environments or for very long-term exposure, organic coatings are applied over these conversion coatings and provide a very effective barrier system. Lacquers, baking enamels, and highly resistant epoxies, vinyls, polyurethanes, or silicones are often used.

The effectiveness of conversion coatings may vary widely, depending on differences in surface preparation, cleanliness of the substrate, and thickness of the coating. Thicker coatings reduce the probability of imperfections in the coating, increase the protection from abrasion and mechanical damage during use, and increase the moisture resistance of the film. In general, the thicker the coating, the better its anticorrosive properties.

Magnesium. Because of its high strength-to-weight ratio, good fatigue strength, and dimensional stability, magnesium has become a very popular metal for electronic applications, especially those intended for flight hardware. Heat sinks, housings, coolant tubes, and frames are a few examples of the use of magnesium. In a benign environment such as indoor exposure, or within a container containing dry inert gas, a protective finishing system for magnesium is normally not necessary. Even for normal weathering or short-term intermittent outdoor exposure, an electrochemical conversion coating alone or a single coating of clear lacquer may provide sufficient protection.

For military uses anodic conversion coatings are preferred. Some commonly used anodizing treatments are galvanic anodize (Dow 9 and Dow 17), fluoride anodizing plus post-treatment (usually dichromate), and stannate immersion (Dow 23).[33]

Other chemical treatments useful as protective coatings, either in themselves or as bases for primers and topcoats, are Dow 1 (Chrome Pickle), 19 (Dilute Chromic Acid), and 21 (Ferric Nitrate Bright Pickle).[34] The treatments most commonly used for electronic hardware are Dow 17, an excellent paint base for long-term exposure to humid and salt environments; Dow 9, used for optical and electronic components where a nonreflective coating is required or as a paint base; and Dow 19, applied as a touch-up by brush, spray, or swab for rework on Dow 9 or Dow 17 surfaces. The various treatments are compared in Table 4-9.

These surface treatments are all acid types which either neutralize or acidify the normally alkaline magnesium surface. Neutralization of alkalinity is essential to improve the adhesion and durability of paints and coatings. To be effective, primers for magnesium should be based on alkali-resistant vehicles such as polyvinyl butyral (contained in wash primers), acrylic, polyurethane, vinyl epoxy, or baked phenolic. Zinc chromate or titanium dioxide pigments are often employed as further corrosion inhibitors.[34] Because of their greater insolubility, strontium chromate and basic zinc chromates are reported to provide sufficient corrosion inhibition and longer service life.[35]

For severe or extensive exposure, as might be encountered in ground support, ground operating equipment, or marine equipment, the best finish system usually consists of a conversion coating, a primer, and a final organic topcoat. Vinyl,

Table 4-8: Corrosion Modes for Metals and Alloys Commonly Used in Electronic Assemblies

Metal or alloy	*Type of Corrosion*
Copper	Thermal-air exposure results in black oxidation product; forms green copper carbonate (verdigris) which inhibits further corrosion except as accelerated by galvanic coupling; exposure to salt environment produces high electrical resistance due to formation of copper salts: tarnishes in sulfur containing ambients.
Copper-nickel alloys and copper-beryllium alloys	Tarnishes in sulfur containing ambients and in moist air.
Aluminum, pure or 5052, 6061, 1100, 3003	Forms white oxide, which is usually superficial and entails no structural damage.
Aluminum high-strength alloys 2024, 2014	Susceptible to tunneling, exfoliation, and stress corrosion.
Aluminum high-strength alloys 7075, 7079	Susceptible to stress corrosion.
Aluminum alloys, general	Galvanic corrosion from dissimilar metal couples.
Magnesium AZ31B	Rapid local dissolving at breaks in metallic coatings or at points of coupling with nobler metals; oxidizes slowly in moist air.
Magnesium-lithium alloy	Interaction of moisture with the lithium portion of the alloy results in rapid evolution of hydrogen and formation of white lithium hydroxide; organic protective coatings used are permeable to moisture and blister and lift due to hydrogen generation; also reacts with carbon dioxide from the ambient to give white lithium carbonate.
Magnesium-thorium alloy	Very susceptible to moist ambient.
Beryllium	Stable in air; forms a metal-oxide protective layer on heating in air; will react rapidly with methyl alcohol; attacked by alkalis with evolution of hydrogen.
Nickel	Little corrosion problem except for galvanic couples; very stable in air and water owing to presence of nickel oxide layer.
Silver	Black silver sulfide "tarnish" on exposure to ambient owing to interaction with SO_2 and H_2S contaminants in air.
Gold	Extremely inert, no corrosion products.

Table 4-9: Dow Chemical Treatments Used in Finishing of Magnesium[33]

Name	Military spec	Appearance	Alloys used on	Uses	Remarks
Dow 1 Chrome Pickle	MIL-M-3171 Type 1	Matte gray to yellow red	All	For protection during shipment and storage.	Simple dip treatment, slight dimensional loss, good paint base.
Dow 7 Dichromate	MIL-M-3171 Type 3	Brassy to dark brown	All except Th-containing alloys	Good combination of paint base and protective qualities.	Does not materially affect dimensions.
Dow 9 Galvanic Anodize	MIL-M-3171 Type 4	Dark brown to black	All	Used on alloys which will not take No. 7 or where a black coat is needed.	Requires galvanic couple between work and steel cathode; no external current required.
Dow 17 Anodize					
Thin	MIL-M-45202	Light gray to light green	All	Excellent protective and decorative finish.	Thin, 0.0001 to 0.0003 in.; heavy, 0.001 to 0.0015 in. Coating is stable above melting point of metal; best abrasion resistance, best paint base, and most consistent of all Dow treatments for magnesium; license required.
Heavy	MIL-M-45202	Dark green			
Dow 19 Dilute Chromic acid	MIL-M-3171	Brassy to brown	All	Touch-up treatment; used as paint base when refinishing.	Less critical than Dow No. 1 with brush application; solution not harmful when trapped in joints.
Dow 23 Stannate Immersion	None	Medium to dark gray	All	Good paint base and protective finish; used with dissimilar metal inserts (except aluminum or in assemblies).	Must be neutralized if painted; retards galvanic corrosion; license required.

acrylic, epoxy, or polyurethane topcoats proved to give excellent results. Another popular system consists of a Dow treatment, an epoxy primer, and a final epoxy-polyamide topcoat; for example, MIL-M-45202 surface treatment, MIL-P-23377 epoxy primer, and MIL-C-27251 epoxy topcoat. Other formulations recommended for magnesium protection include MIL-C-15328, MIL-P-15930, MIL-P-15936, MIL-L-14486 (ORD), and MIL-C-27337 (USAF).

Tin-Lead Solder. For maximum protection of solder joints and solder-plated surfaces, conformal coatings should be used.[36] However, as with other metals, corrosion can be prevented or minimized by following certain design principles. For example, bimetallic combinations of tin-lead solder with either copper or aluminum should be avoided in high-humidity or salt-water environments, where corrosion is known to occur.[37] The coating of such bimetallic combinations will provide only temporary protection unless the two dissimilar metals are isolated from each other. Other difficulties associated with the corrosion protection of solder joints or solder plate are caused by the incomplete removal of corrosive fluxes during cleaning operations. Conformal coatings will not provide protection unless these contaminants are adequately removed. Cleaning procedures for flux residues are discussed in Chapter 5.

Aluminum. Like magnesium, aluminum is widely used because of its low density. It, too, must be protected from corrosion, although the normal oxide film which forms on its surface on exposure to air is highly protective and prevents deterioration of the bulk of the metal. A number of oxide coatings (anodize) may purposely be formed as protective coatings and as surface treatments to improve the adhesion of subsequent organic coatings (MIL-A-8625). Primers such as wash primer, phenolic, and epoxy are useful in serving as a bridge between the anodized surface and a topcoat paint. As with magnesium, epoxy primers and epoxy or acrylic topcoats offer outstanding long-term corrosion protection. A proven finish system for outdoor use consists of MIL-C-5541 chemical surface treatment, MIL-P-23377 epoxy polyamide primer, and MIL-L-81352 acrylic topcoat. This system has been successful for the protection of many types of aluminum alloys, both clad and nonclad. For interior use, this finishing system may be employed without the topcoat.[38]

Magnesium-Lithium Alloys. Magnesium-lithium alloys such as LA 141 are attractive to designers of airborne electronic equipment because of their stiffness and high strength-to-weight ratio. These alloys afford a 20 to 25 percent weight saving over magnesium or other magnesium alloys. However, the high reactivity of magnesium lithium with moisture and carbon dioxide limits its use to hermetically sealed boxes with dry inert-gas ambients. Often, even in an enclosed box an incompatible situation can arise owing to the evolution of moisture from plastics or from electronic components within the box. Moisture reacts with the lithium portion of the alloy, forming hydrogen gas and lithium hydroxide. In the presence of carbon dioxide, white lithium carbonate is formed on the alloy surface. In addition to the formation of these corrosive products and deterioration of the alloy, a secondary, perhaps even more harmful, effect may result from reaction of evolved hydrogen gas with active devices, chemically sensitive thick- or thin-film resistors, and other components.

Although there are reported to be several coatings which will protect magnesium lithium in normal environments, no organic coating or coating system is known to provide sufficient protection to pass military environmental tests such as tem-

perature and humidity cycling, or the salt-spray test. Brands and Kilgore[39] investigated 40 finishing systems chosen for their high resistance to alkali and low permeability. These included epoxies, polyurethanes, polyvinylidene chlorides and fluorides, waxes, and polymer-wax blends. No combination of inorganic pretreatments, primers, or topcoats was found to be satisfactory. In all cases the mode of failure was blistering. Although some of these systems provided excellent protection of pure magnesium metal from humidity and temperature exposure, they failed on magnesium-lithium alloy after only laboratory ambient storage conditions.

Wash Primers. Wash primers consist of rust-inhibiting pigments, chiefly basic zinc chromate or lead chromate, in an alcoholic solution of polyvinyl butyral resin, as one part of a two-part system.[40] The second part consists of a phosphoric acid-water-isopropyl alcohol solution. The first solution is acidified with the second just prior to use. This primer is commonly referred to as WP-1, and its precise composition is given in Table 4-10. Wash primers provide effective corrosion protection for many metal parts, such as aluminum, magnesium, steel, zinc, and tin. They may be used alone or, where superior protection against salt spray and moisture is

Table 4-10: WP-1 Composition[33]

Ingredients	*percent by weight*
Pigment base	
Polyvinyl butyral resin	9.2
Basic zinc chromate	8.8
Magnesium silicate	1.3
Lampblack	0.1
Butyl alcohol, normal	20.5
Isopropyl alcohol	57.7
Water	2.4
Total	100.0
Acid Diluent	
85% Phosphoric acid	18.5
Water (maximum)	16.5
Isopropyl alcohol	65.0
Total	100.0

required, in conjunction with a topcoat such as epoxy, vinyl, or alkyd. They are applied in thicknesses of 0.3 to 0.5 mil by spraying, brushing, or swabbing. The best corrosion protection is provided by using a wash primer followed by a topcoat. A wash-primer-epoxy topcoat system provides protection for either magnesium or aluminum for over 800 hr of salt-spray exposure, performed according to MIL-STD-881. Over 2,000 hr of salt-spray exposure without deterioration has been reported for a vinyl (VAGH) coating applied over either a wash primer or a red lead primer. Wash primers may be purchased from paint manufacturers according to military specifications, chiefly MIL-C-8507, MIL-C-8514, and MIL-C-15328.

Other Primers. Other types of primers are also employed to improve adhesion of topcoats and hence provide better corrosion protection. The performance of vinyl coatings, for instance, is greatly improved by priming metal surfaces. Primers

may consist of other specifically formulated vinyls, phenolics, epoxies, or butadiene acrylonitriles.[41]

Epoxy primers are very effective in improving the adhesion of polyurethanes. In the protection of either aluminum or steel, the use of an epoxy primer followed by two topcoats of moisture-curing polyurethane resulted in panels which passed 800 hr of 20 percent salt spray exposure, according to ASTM B 117-57.[42] Epoxy primers may consist of normal coating formulations which have been thinned further and sprayed on very thin (called *flash primers*) or of coatings formulated to contain a number of ingredients, such as formula MP (Table 4-11).

Table 4-11: Primer Formulation MP

Ingredient	*Proportion, pbw*
Red iron oxide, R. K. Williams R-4800	7.5
Basic lead silicochromate, National Lead M-50	23.6
Talc, Whitaker, Clark & Daniels No. 399	7.8
Lecithin (soya 50 percent in xylene)	0.4
Epoxy, Shell Epon 1004	15.5
Cellosolve acetate, toluene, xylene (2:1:1)	39.2
Mondur CB-75	3.3

Mechanical Protection

Abrasion Resistance. Applications in which there is excessive handling, rubbing, or contact with other objects require that surfaces have a high degree of abrasion resistance. Abrasion resistance may be determined by periodic examinations while the coating is in actual use. However, excessive time would be required to obtain meaningful data by this method, and results would be more or less qualitative. Fortunately, several accelerated and quantitative methods are available that may be correlated with actual service performance or used to compare the relative abrasion resistances of coatings. These include projection of a controlled stream of abrasive such as carborundum (ASTM D 658-44) or sand (ASTM D 968-51) against the sample, use of an abrasive tape (Armstrong test, ASTM D 1242), controlled scraping with a metal tool until the coating is penetrated,[43] or the use of rotating abrasive wheels. This last method, referred to as the *Taber abrasion test,* is the most widely used for plastic coatings and is described in Federal Standard 141, Method 6192, and ASTM D 412 and D 1044. According to the Taber method, two standardized abrading wheels rotate against a coating sample which consists of a 4-in.-square coated metal panel with a center spindle hole (Figure 4-19). The panel, preweighed to the nearest milligram, is placed on a turntable operating at a constant speed while two abrading wheels are fixed on each side of the spindle, in direct contact with the coating. A 1,000 g load may be placed on the wheels. A built-in tachometer records the number of turntable revolutions, and a vacuum attachment removes plastic particles as they are abraded from the surface. The instrument is operated for a specific number of cycles or until a portion of the

substrate becomes exposed. Results may be reported as either the number of cycles to failure per mil thickness of coating or the weight loss, in milligrams, for a specified number of cycles, ranging from 100 to 1,000, referred to as the *Taber wear index.* Quantitative data for a number of coatings may be found in the Appendix.

Figure 4-19: Taber abrasion tester.

Friction Resistance. When repeated sliding or insertions must take place as in electronic drawers or consoles, surfaces with a low coefficient of friction will facilitate operation and will minimize surface abrasion and deterioration of the coating. The best way to assess the friction resistance of materials is by comparing their coefficients of friction; the lower this value, the greater the resistance to friction (Table 4-12). Theoretical treatments of both friction and abrasion are reported elsewhere.[44]

With the exception of some of the fluorocarbon polymers, plastics have coefficients of friction of the same order of magnitude as metals. Values for plastics, however, depend more on test conditions, such as temperature, humidity, loading, and operating speed, and on material conditions, such as degree of cure and moisture content. As an example, the coefficient of friction of nylon varies from 0.91 to 1.19 with a variation of moisture content from 0.2 to 10 percent, respectively.[45] The surface against which the sample is tested is also an important variable. A highly polished steel surface is often employed, although polished glass or wood may also be used. In any event, the nature of this surface should be specified.

Teflon appears to be unique among both plastics and metals in possessing one of the lowest coefficients of friction. The coefficient of friction of nylon and of some specially formulated epoxies and polyurethanes containing Teflon as a filler, although not as low as that of Teflon, are still considered satisfactory for many sliding or insertion applications. The use of Teflon as a filler for epoxy coatings can reduce the coefficient of friction and the Taber wear index (milligram loss per 1,000 cycles) by as much as 40 percent.[46]

Hardness. The wear resistance of a coating may also be indicated by its hardness. A soft coating is more subject to abrasion, scratching, or other mechan-

Table 4-12: Coefficients of Friction of Typical Coatings

Material	Coefficient of friction	Source
Coatings		
Epoxy, air-dried coating with Teflon filler	0.15	a
Polyamide, nylon	0.3	b
Polyethylene	0.6–0.8	b
Polyimide (Pyre-M.L.)	0.17	c
Polymethylmethacrylate	0.4–0.5	b
Polystyrene	0.4–0.5	b
Polytetrafluoroethylene (Teflon TFE)	0.05–0.1	b
	0.04	d
	0.016	e
Polyvinyl chloride	0.4–0.5	b
Polyxylylene, parylene N	0.25	Union Carbide
Parylene C	0.29	Union Carbide
Parylene D	0.31–0.33	Union Carbide
Other materials for comparison		
Graphite	0.18	c
Molybdenum disulfide	0.12	f

[a] PT-401TE, *Product Techniques Inc. Bull.*, Oct. 17, 1961.
[b] Bowder, F. P.: *Endeavor,* **16** (61):5 (1957).
[c] Pyre-M.L. Wire Enamel, *Du Pont Bull.* 19, August, 1967.
[d] Shooter, K. V., and P. H. Thomas: *Research,* **2:**533 (1949).
[e] Thompson, G. B., G. C. Turell, and B. W. Sandt: *SPE J.*, April, 1955.
[f] Pillsbury, R. D.: *J. Teflon* 19, May, 1961.

ical abuse than a hard coating. One of the more popular methods of determining hardness is the *Sward hardness test.* The tester (Figure 4-20) automatically records the number of swings made by a standard rocker in contact with the test surface before it comes to a stop. The automatic counter is based on the photoelectric-eye principle; a shutter on the rocker interrupts the light beam once during each complete oscillation. The test is reported to be sensitive to ±1 or 2 units.[47] The coating is normally applied to a rigid, thin substrate, such as aluminum. Values are related to the value for polished plate glass, which is taken as 100. Although the Sward hardness test is nondestructive, it often cannot be used on an actual part because of the limited size and shape of samples that the tester will accept. Sward hardness values for some typical coatings and paints are presented in the Appendix.

Hardness values are also a means of establishing the approximate state of cure of a coating, because the coating hardens progressively as it cures or dries. Hardness values change rapidly at first but eventually level off to a constant value. Sward hardness values are useful in determining optimum cure schedules (time and temperature) or optimum catalyst concentrations. As an example, Figure 4-21 shows a plot of Sward hardness versus the percentage of amine catalyst used to cure a polyurethane coating. It is evident from this curve that hardness values have stabilized at 1 to 3 percent of the catalyst and that a higher percentage of catalyst

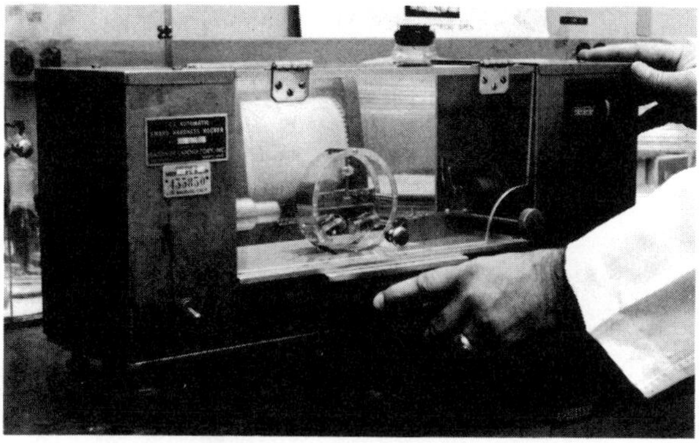

Figure 4-20: Sward hardness rocker.

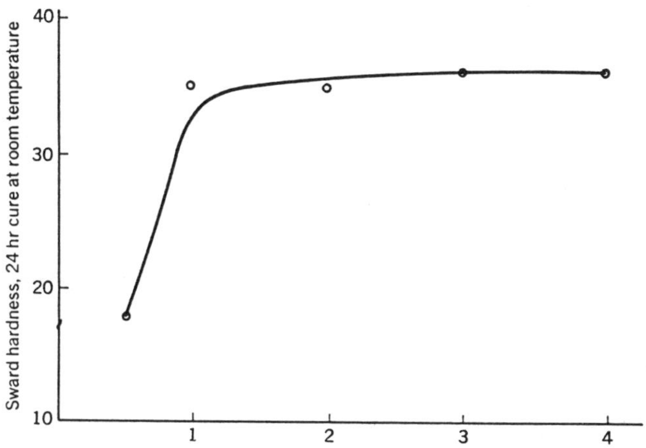

Figure 4-21: Sward hardness versus catalyst concentration for polyurethane coating.[49]

would provide no further benefit under the time and temperature conditions used. If the catalyst concentration is maintained constant at 0.5 percent for the same resin system, the Sward hardness values may be used to determine the optimum cure time, as shown in Figure 4-22. The use of hardness values to establish cure endpoints applies more to slow room-temperature-curing or air-drying coatings than to the bake- or high-temperature-curing types. Hardness values are not sensitive enough to detect the very small changes in cure occurring on postbaking a coating after it has had an initial high-temperature bake.

Another hardness test employs the use of a series of pencils of varying hardness to scratch through the coating to the substrate. The method, called the *pencil hardness test,* is relatively simple to perform but highly subjective and destructive in nature. Numerous other hardness and scratch tests have been developed and are occasionally used.[48]

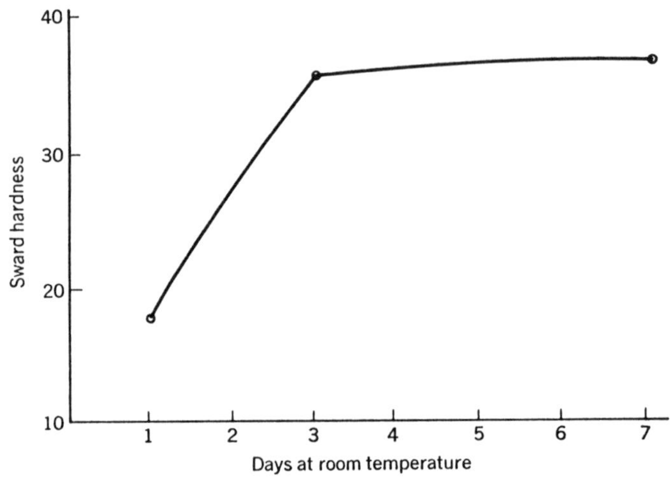

Figure 4-22: Sward hardness versus cure time for polyurethane coatings.[49]

Protection Against Electromagnetic and Radiofrequency Interference

Electromagnetic and radiofrequency interference (EMI and RFI) can result in electrical failure of sensitive electronics. They can cause electrical transients in electronics and jam computers and communication systems. EMI/RFI can be caused by electrically incompatible parts in the same system or by spectrum pollution from an outside source.[50] To prevent these problems, the parts must be shielded to isolate them from the interference. The best shielding is a conductive encasement such as a metal package or enclosure. However, not all parts can be packaged in metal and those that are may require openings for input and output leads or cables.

Plastics that contain a conductive material are commonly used as gaskets, coatings, sealants, and filters for EMI/RFI shielding. Silver is the most commonly used metal filler followed by nickel, copper, stainless steel, carbon fiber and carbon black, nickel-coated graphite, and aluminum flake and fiber.[51] Silicones, epoxies, and polyesters are used as resin bases for these materials. Examples of some of these materials are shown in Table 4-13.

Protection Against Electrostatic Discharge

Each year millions of dollars are lost by the electronics industry because of electrostatic damage to devices and circuits. This loss can come from damage to semiconductor devices causing a product to fail or, in a more extreme case, an explosion due to a static spark in a combustible atmosphere.

To minimize this problem electrostatic-sensitive devices such as integrated circuits, transistors, and diodes are stored and transported in specially treated plastic enclosures. These plastic enclosures are manufactured primarily from polyethylene (PE), polypropylene (PP), or Aclar. In order to dissipate the electrostatic charge that develops on the plastic surface, the plastic must be formulated or coated with electrically conductive or semiconductive fillers such as graphite or

ionic surfactants such as long-chain fatty amines or carboxylic acids (octanoic acid, octylamine or octanoic acid amines and ionic salts thereof). In other cases, the plastic surface may be coated with an ionic salt such as a quaternary ammonium chloride. Migration of ionic salts onto a circuit must be avoided to prevent shorting and/or corrosion. A few examples of such products are shown in Table 4-14.

Microbial Protection

Microorganisms also cause millions of dollars of loss a year through material deterioration. Microbial growth can cause corrosion of metals and bimetallic combinations, deterioration of mechanical or electrical properties of coatings or substrates, or an increase in surface roughness producing increased friction and drag. This last effect is of special significance for underwater vehicles. The resistance of

Table 4-13: Examples of Products Used for EMI/RFI Protection (Tecknit)

Product	Plastic/Filler	Shielding Performance (dB)* at Different Frequencies
Elastomet oriented wire in gasket material	Silicone or fluoro-silicone/Monel or phosphor bronze	100 kHz: 75 (Monel); 80 (Phr brz) 10 MHz: 130+ (Monel); 130+ (Phr brz) 1GHz: 110 (Monel); 100 (Phr brz) 10 GHz: 115 (Monel); 100 (Phr brz)
Consil-C custom gaskets	Silicone/silver-copper	100 kHz: 75; 10 MHz: 120+; 1 GHz: 115; 10 GHz: 110
Consil-N custom gaskets	Silicone/silver-nickel	100 kHz: 75; 10 MHz: 120; 1 GHz: 110; 10 GHz: 100
Consil-A custom gaskets	Silicone/silver-plated aluminum particles	100 kHz: 75; 10 MHz: 120+ 1 GHz: 110; 10 GHz: 100
SC-Consil custom gaskets	Fluorosilicone/carbon	100 kHz: 40; 10 MHz: 100; 1 GHz: 60, 10 GHz: 50
ECTC windows	Acrylic/thin-film metal	100 kHz: 20; 10 MHz: 90; 1 GHz: 30
Teckfilm transparent sheets	Polyester/thin-film metal	100 kHz: 20; 10 MHz: 90; 1 GHz: 30
Teck-bond RTV sealant	Silicone/silver-plated copper	–
Conductive caulking joints/seals	Polyolefin acetate/ silver-coated inert particles	–
Epoxy adhesive gasket bonding, sealing	Epoxy/silver	–
Conductive coatings coating plastic enclosures	Acrylic or latex/ silver, nickel, or graphite	–

*Test methods used either from MIL-STD-285 or MIL-G-83528.

Table 4-14: Examples of Products Used for ESD Protection of Electronic Parts (Scientific Products)

Product	Antistatic Material	Surface Resistivity (ohms/square)
Baxter AmeriStat Shield Bags	Conductive thin-film metal	1×10^{10}
Baxter AmeriStat Pink Poly Bags	Antistat	—
NesTier Modubox Antistatic Tote	Thin layer of water	10^{9} to 10^{13}
NesTier Modubox Conductive Tote	Carbon powder	10^{3} to 10^{5}

coatings to attack by microorganisms is particularly important for electronic hardware that must be stored or operated in humid, tropical, or semitropical environments. Contact of materials with moist soil is an ideal condition for microbial growth, since it is estimated that 1 cm^3 of soil contains as many as 50,000 fungi, 500 million bacteria, and 250 million actinomyciles (plant or animal biotics, a class of molds midway between fungi and bacteria).

Most microorganisms thrive in an environment of only 50 percent relative humidity and temperatures ranging from 20 to 40°C. Some may even adapt and grow well under less favorable conditions.[52-54] Fortunately, most plastics and plastic coatings are resistant to microorganisms by virtue of their inherent nonnutrient characteristics or because of the fungicides or bactericides used as additives in the formulation. In fact, because of their fungus-resistant properties, certain organic coatings are often used as barriers to protect otherwise nutrient substrates. Such protection, however, is only as good as the integrity of the coating; if the coating is scratched, abraded, or in any way detached from the surface, microbial growth will occur in the exposed areas.[55]

As a rule, the newer synthetic coatings, such as the epoxies, silicones, and polyurethanes, are inherently fungus resistant. Some formulations may not be as resistant as others because of the presence of plasticizers or other nutrient additives, and some may deteriorate with age or exposure to humidity, producing nutrient substances or hydrolysis products. The use of naturally occurring nutrient substances in polymer coatings does not necessarily render the coating fungus nutrient. There is still a wide misconception that coating resins based on vegetable oils are nutrients. The important factor is the form in which the nutrient material exists in the formulation. If it is co-reacted with the base resin and becomes an integral part of the polymer structure, the coating is not likely to be nutrient. One clear example is the use of castor oil in curing isocyanate resins. The resulting polyurethane coatings are generally fungus inert. However, if the vegetable oil is physically admixed in the formulation and does not become chemically bound with the polymer structure, it is probable that the formulation will support microorganisms to some degree. A knowledge of the composition and chemistry of a coating formulation is therefore helpful in predicting its fungus-resistant properties.

Plasticizers are probably the greatest cause for fungal and bacterial growth. On the basis of available information, the plasticizers which are good nutrients are

diesters of saturated aliphatic dibasic acids containing 12 or more carbon atoms. Some common nutrient plasticizers are di-*n*-hexyladipate, di-2-ethyl-hexylazelate, epoxidized soybean oil, triethylene glycol, derivatives of soybean or tall oil fatty acids, methyl acetylricinoleate, dioctylsebacate, and tetrahydrofurfuryloleate. However, the maleates are somewhat fungus resistant, as are the alkyl derivatives of phosphoric and phthalic acids. Polyhydric alcohols can be readily assimilated by fungi and bacteria if the hydroxyl groups are on the adjacent or terminal carbon atoms. Ether linkages in a polymer structure, as in epoxies, tend to reduce fungal growth. Much of the success of epoxy resins as replacements for naturally occurring resins and varnishes has been due to their fungus-resistant properties.

It has been observed that under suitable environmental conditions enzymatic action of microorganisms converts the plasticizers in vinyl resins to water-soluble salts, which can then be easily extracted. Stiffening, weight loss, and dimensional changes may follow. Air and water entering the tunnels left by extraction cause further erosion. This process continues until the tunnels penetrate the entire material and destroy its physical integrity. The electrolytic solutions which may collect in the crevices and tunnels as a result of the metabolic processes of the microorganisms destroy the insulating properties of the material. In addition, stains and odors can develop from the waste products of the microbial growth.

Considerable work has been carried out on the effects of microorganisms on plastics, but attention has focused mainly on the effects of fungi on vinyl polymers. Polyvinyl chlorides, polyolefins, polyesters, and many other polymer types are inherently resistant to fungus attack, but some of the formulation ingredients—plasticizers, fillers, pigments, stabilizers, and processing aids—support mildew formation and other microbial growth. Surface greases and other contaminants can also act as nutrients for microorganisms, including bacteria such as pseudomonas, salmonella, and bactillus. The genera of fungi which may possess the necessary enzymes to assimilate these formulation ingredients are Aspergillus, Paecilomyces, Trichoderma, Penicillium, Fusarium, Mucor, and Curvularia.

Nutrient coatings can be rendered nonnutrient by incorporating special chemicals which function as biostats or biocides. Salicylanilide and bis(8-quinolinolato)-copper are two examples of widely used biocides. Others, such as the organotin compound bis(tri-*n*-butyltin)-oxide, have been reported to control a wide variety of bacteria, fungi, algae, and even barnacles.[56] Examples of biocides or biostats are given in Table 4-15. Coating types and their relative nutrient properties are summarized in Table 4-16.

THERMAL CONDUCTIVITY

The total amount of heat a material can conduct is directly proportional to the surface area, time of contact, and temperature gradient and is inversely proportional to the thickness of the sample, according to the equation

$$Q \propto \frac{(T_2 - T_1)At}{d}$$

where Q = total heat flow
A = surface area, cm^2
d = thickness of sample, cm
T_2 = temperature of the hot surface, °C
T_1 = temperature of the cold surface, °C
t = time, sec

When a proportionality constant k is introduced, the thermal-conductivity equation becomes

$$Q = \frac{k(T_2 - T_1)At}{d}$$

The constant k is a material constant called the *coefficient of thermal conductivity.* It is the time rate of heat flow under steady-state conditions through a unit area and unit thickness per unit temperature gradient.

Table 4-15: Commonly Used Biocides

Salicylanilide
Brominated salicylanilides
Organotin compounds
Bis(8-quinolinolato)-copper or zinc
Phenylmercuric oleate, -salicylate, stearate, or phthalate
Di(phenylmercuric)dodecenylsuccinate
N-(trichloromethylthio) phthalimide
Phenylmercuric-*o*-benzoic sulfimide
Zinc dimethyldithiocarbamate
2-Mercaptobenzothiazole
Zinc pentachlorophenoxide
Pentachlorophenol
Quaternary ammonium carboxylates
Bis(tri-*n*-butyltin)-oxide
Sodium or zinc pyridinethione-*N* oxide
p-Toluenesulfonamide

Organic coatings, as a rule, are seldom used for the primary purpose of conducting heat. Unfilled coatings and plastics have rather low k values and are very good thermal barriers or insulators. Furthermore, there is very little difference in thermal conductivity among the different polymer types; the order of magnitude is about 10^{-4} cal/(sec.)(cm.)(°C). This value is thousands of times lower than for metals or metal oxides (see the Appendix).

From a design standpoint, the inherently low thermal conductivities of organic coatings may be desirable, as in the case of ablative and thermal protective materials, or undesirable, in the case of assemblies or modules with a high density of electronic components, which require the dissipation of large amounts of heat. Accordingly, plastics used in intimate contact with electronics such as spacers, filleting compounds, substrates, heavy coatings, and encapsulants require high

Table 4-16: Resistance of Coating Resins to Microorganisms[57]

Coating type	*Resistance to attack*
Acrylic	Good
Alkyd	Poor
Cellulosics	
Cellulose acetate	Good to poor
Cellulose acetate butyrate	Good
Cellulose nitrate	Poor
Ethyl cellulose	Good
Chlorinated rubber	Good
Chlorosulfonated polyethylene	Good
Epoxy amine	Good
Epoxy phenol formaldehyde	Good
Epoxy polyamide	Good
Epoxy urea formaldehyde	Good
Melamine formaldehyde	Good to poor
Phenol formaldehyde	Good to poor
Phenoxy	Good
Polyester	Good
Polyethylene, low molecular weight	Poor
High molecular weight	Good
Polychlorofluoroethylene	Good
Polyisobutylene	Good
Polystyrene	Good
Polyurethane	Good
Polyvinyl acetate	Poor
Polyvinyl butyrate	Poor
Polyvinyl chloride	Good*
Polyvinyl chloride acetate	Good
Polyvinylidene chloride	Good to poor
Polyvinylidene fluoride	Good
Polyxylylene (parylene)	Good
Silicone	Good
Silicone alkyd	Good to poor
TFE fluorocarbon	Good
Urea formaldehyde	Good

* Except when plasticized.

thermal conductance. Electrically insulated metal heat sinks or beryllium oxide spacers are commonly employed. No plastic yet exists which combines the high thermal conductivity and low electrical conductivity of beryllia or of aluminum nitride, materials that are unique in this respect.

If some degree of thermal conduction in organic coatings is required, either or both of two approaches may be used:

1. Employ a coating which is formulated with a thermally conductive filler, since filled coatings have substantially higher thermal conductivities than unfilled versions

2. Employ as thin a coating as possible, since thermal conduction is inversely related to thickness

In practice, it is found that any gain in filling a coating is offset by the fact that the coating becomes thixotropic, and very thin layers cannot be achieved. For most applications the fillers selected should be good electrical insulators in order to maintain the good insulating properties of the coating. Commonly used fillers which meet these requirements are aluminum oxide or alumina ($A1_2O_3$), silicon dioxide or silica (SiO_2), mica, and beryllium oxide or beryllia (BeO). The thermal conductivity of an Epon 828/diethylenetriamine formulation is improved three- to fourfold by silica, aluminum, or beryllium oxide fillers, but it is still far below the thermal conductivity of the filler alone. Thermal-conductivity values for several fillers and polymer materials are given in the Appendix. Little thermal-conductivity data for coatings are reported in the literature, probably owing to the difficulty in preparing samples thick enough for measuring thermal conductance. Many of the materials reported are molding or casting types, which may be easily formed in thick samples.

Some investigators[58,59] have attempted to increase the thermal conductivity of epoxies by further cross-linking the structure by postcuring or by exposure to gamma radiation, but they have only been able to achieve at most a 5 percent improvement over the initial value. The addition of selected fillers still remains the most effective way to increase the *k* factor of plastics and plastic coatings.[60,61]

Test Methods

Several methods of measuring thermal conductivity are available, of which the guarded-hot-plate and heat-flow meter methods are the most widely used. The most accurate method for plastic samples having thermal conductivities of 3.4×10^{-3} cal/(sec.)(cm.)(°C) or less (most filled or unfilled plastics are in this category) is the guarded-hot-plate method described in ASTM C 177. This and other methods are discussed elsewhere.[62,63]

The Guarded-hot-plate Method. Two different types of guarded-hot-plate apparatus can be used. They are similar in principle but differ somewhat in construction. One example, shown in Figure 4-23, consists of three sections: a central heating plate, guard heating plates, and cooling plates. The guard section ensures unidirectional heat flow from the central heater and eliminates any influences from the edges of the sample. Thermocouples are located at various positions in the guarded hot plate to measure the difference between the temperature of the guard ring and central heater and the temperature of the hot and cold surfaces. The central heating unit and the cooling units are adjusted until the temperature drop through the two sides does not differ by more than 1 percent and the fluctuation over a 1-hr period is not more than 0.5 percent of the temperature difference between hot and cold plates. When equilibrium is reached, measurements are made of the temperature differences across specimens, the hot-plate temperature T_2, the cold-plate temperature T_1, and the electrical power input Q to the central heater. Once the thickness and area of the sample are known, the thermal conductivity can be calculated by means of the thermal-conductivity equation given above.

The thickness of the sample can vary and the maximum specimen thickness is based on calculations listed in Table 1 of ASTM C 177.

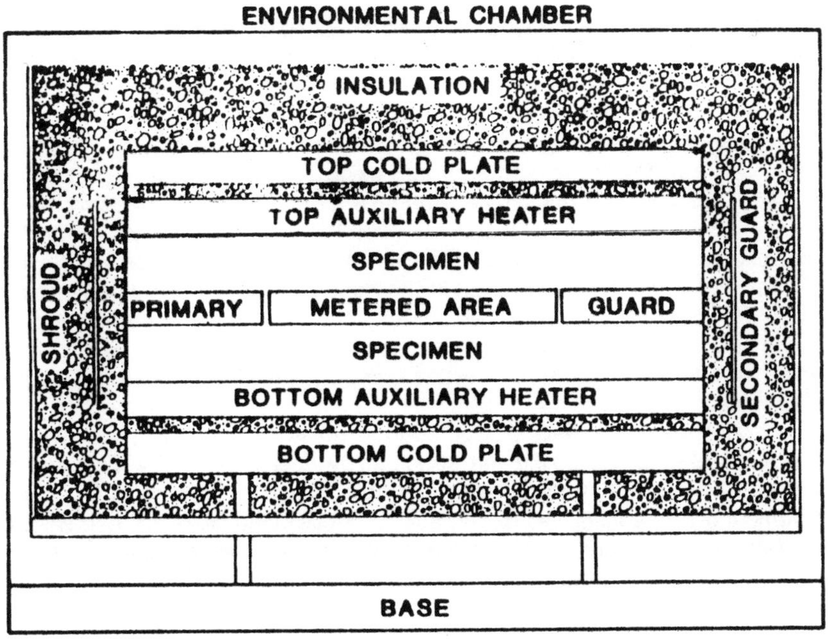

Figure 4-23: Guarded hot-plate apparatus.

Heat-Flow Meter Method. The steady-state thermal transmission properties of an insulator can be measured using a heat-flow meter. This is a secondary or comparative method for measuring thermal conductivity since only the ratio of the thermal resistance of the specimen to that of a standard specimen is measured.

In this method a sample is placed between a warm and a cold plate. The meter measures heat flow through the specimen by generating a temperature difference across a slab of material of known thermal resistance. There are commercially available units, one example is shown in Figure 4-24.

A calibration curve can be set up using samples of known thermal conductivity. These standards should have thermal conductivities close to that expected for the test sample to obtain accurate results. Sample thickness can range from 0.001 to 0.8 inches, but for accuracy the sample should have a C value (thermal conductivity divided by sample thickness) of approximately 400 W/m K or less.

In the heat-flow meter method the spacing between the warm and cold plates is based on the theoretical maximum thickness of the specimens in the conventional guarded-hot-plate apparatus. Calculations of this spacing are found in Table 1 for ASTM C 518.

Figure 4-24: C-Matic thermal conductivity tester used for measuring the thermal conductivity of materials by the heat-flow meter method. (Courtesy of Scientific, Inc.)

REFERENCES

1. Von Hippel, A. R. (ed.): "Dielectric Materials and Applications," M.I.T.-Wiley, New York, 1954.
2. Baer, E. (ed.): "Engineering Design for Plastics," Reinhold, New York, 1964.
3. Versamids, *Gen. Mills Chem. Div. Tech. Prod. Bull.,* 1960.
4. Harper, C. A.: "Electronic Packaging with Resins," McGraw-Hill, New York, 1961, p. 325.
5. Warfield, R. W.: Studying Electrical Properties of Casting Resins, *SPE J.,* vol. 14, no. 11 (November, 1958).
6. Mathes, K. N.: Electrical Properties of Insulating Materials, *Proc. Seventh Elec. Insulation Conf., Chicago,* Oct. 15–19, 1967.
7. Buchoff and Botjer: Effect of Humidity on Surface Resistance of Filled Epoxy Resins, *Proc. Ann. Tech. Conf. SPE, Chicago,* January, 1960.
8. Delmonte, J.: "Metal Filled Plastics," Reinhold, New York, 1961.
9. Wirsen, A.: Electroactive Polymer Materials, NTIS report FOA Report C 20596-F9, January 1986.
10. Gill, W. D., et al: Electrically Conducting Polymers, IBM Res Lab ADA-129-488, April 1983.
11. *Chemical and Engineering News,* pp. 20–21, June 22, 1987.
12. Will, F. G. and D. W. McKee: *J. Pol. Sci. Polym. Chem.,* **21,** p 3479 (1983).
13. Moore, Walter J.: "Physical Chemistry," Prentice-Hall, Englewood Cliffs, N.J., 1950.
14. Piser, J.: Pick Plastics for High Speed Circuits, *Electron. Des.,* Aug. 16, 1967.

15. LeFevre, R. J. W.: "Dipole Moments," Methuen, London, 1953.
16. Lechtenecher, K., and K. Rather: *Phip. Z.,* **32:** 255 (1931).
17. Thiokols, *Thiokol Chem. Corp. Tech. Prod. Bull.,* 1959.
18. Volk, M. C., J. W. Lefforge, and R. Stetson: "Electrical Encapsulation," Reinhold, New York, 1962.
19. Electrical Properties of Experimental Epoxy Resins, *Autonetics Internal Rept.* TR-61-549, December, 1961.
20. Knudsen, J. J.: Dielectric Measurements for Process and Product Evaluation, *Thirteenth Ann. Exhibit Conf. Soc. Plastics Ind., Reinforced Plastics Div., Chicago,* 1958.
21. Kobale, M., and H. Löbl: *Z. Elektrochem.,* **65:** 662–669 (1961); H. Löbl, Dielectric Studies of Epoxy Resins for Casting, *NTG Convention,* Karlsruhe, September, 1957.
22. O'Dwyer, J. J.: An Examination of Avalanche Theories of Dielectric Breakdown in Solids, Electrical Insulation Conference, *Natl. Acad. Sci./Natl. Res. Council Publ.* 1356, 1966.
23. Philofsky, H. M., and E. M. Fort: Degree of Correlation between Tan and Electric Strength of Insulation, *Proc. Seventh Elec. Insulation Conf., Chicago,* October, 1967.
24. Oakes, W. J.: *Proc. IEE,* **96** (I): 37 (1949).
25. *Dow Corning Bull.* 07-062, May, 1963.
26. O'Dwyer, J. J.: "The Theory of Dielectric Breakdown of Solids," Oxford, New York, 1964.
27. Delmonte, J.: Arc Resistance of Epoxies, *Plastics Technol.,* March, 1958.
28. Comparative Properties of Filled Epoxy Electrical Potting Compounds, *Union Carbide Rept.,* 1959.
29. Mathes, K. N.: Selection of an Insulation System in Product Design, *Insulation,* October, 1967.
30. Killam, D. L.: Effect of Humidity on the Dielectric Properties of Some Polymers, Electrical Insulation Conference, *Natl. Acad. Sci. Publ.* 1238, 1965.
31. Kirkpatrick, S. D., and J. R. Callaham: *Chem. Engr.,* December, 1952.
32. Drinberg, A. Ya, E. S. Gurevich, and A. V. Tikhomirov: "Technology of Non-metallic Coatings," Pergamon, New York, 1960.
33. "Metal Finishing Guidebook and Directory Issue," Metals and Plastics Publications, Hackensack, NJ. 1982.
34. "Magnesium Finishing," Dow Chemical Company, Midland, Mich., 1963.
35. Stevens, J. A.: Designing for Long Service: A Primer of Magnesium Finishing, *Magnesium Topics,* October–December, 1958.
36. Bailey, J. C., and J. A. Hirschfield: *Research,* **7:** 320 (1954).
37. LaQue, F. L., and H. R. Copson (eds.): "Corrosion Resistance of Metals and Alloys," ACS Monograph Series, Reinhold, New York, 1963.
38. Wheeler, E.: *Proc. AF Matls. Lab. Fiftieth Anniv. Tech. Conf. Corrosion Mil. Aerospace Equipt., Denver,* May, 1967.
39. Brands, E. R., and L. C. Kilgore: Finishing Systems for Magnesium-Lithium Alloy LA141, *Proc. AF Matls. Lab. Corrosion Mil. Aerospace Equipt., Denver,* May, 1967.
40. Whiting, L. R., and P. F. Wangner: Coating Compositions for Metals, U.S. Patent 2525107, Oct. 10, 1950.
41. Hayward, J. A.: Vinyl Coatings: Latest Aspects in Anti-corrosion, *Corrosion Technol.,* September, 1965.
42. *Hughson Chem. Co., Div. Lord Corp., Tech. Bull.* 7030A.
43. Smits, P.: *Metal Progress,* **80** (3): 75 (1961).
44. Ritchie, P. D. (ed.): "Physics of Plastics," Van Nostrand, 1965.
45. Shooter, K. V.: *Proc. Roy. Soc. (London),* **A212:** 490 (1952); Shooter, K. V.: *Plastics,* **26:** 281, 117 (1961).
46. *Autonetics Internal Rept.,* Evaluation of Teflon Modified Epoxy Coatings TR-62-424, 1962.
47. Moore, D. T.: Precision of the Sward Hardness Rocker, *Paint, Oil, Chem. Rev.,* **113** (21): 41 (1950).

48. Gardner, H. A., and G. G. Sward: "Paint Testing Manual, Paints, Varnishes, Lacquers, and Colors," Gardner Laboratory, Bethesda, Md., 1962.
49. *Spencer Kellogg, Div. Textron, Tech. Bull.* TS-6083.
50. White, D. R. J.: "Electromagnetic Interference and Compatibility," Vol. 1, Don White Consultants, Germantown, Maryland, 1971.
51. Toensmerea, P. A.: *Modern Plastics,* Dec. 1986.
52. Swatch, F. E.: Fungus Testing, *Environmental Quart., vol. I,* no. 1g (January, 1961).
53. Landfield, H., and W. S. Taylor: Protecting Vinyl Compounds from Microorganisms, *Plastics Technol.,* vol. 8, no. 8 (August, 1962).
54. Berk. S., H. Ebert, and L. Teitell: Utilization of Plasticizers and Related Organic Compounds by Fungi, *Ind. Eng. Chem.,* vol. 49 (1957).
55. Baseman, A. L.: "Antimicrobial Agents for Plastics," *Plastics Technol.,* September, 1966.
56. Zedler, R. J.: *Metal Finishing,* **60** (12): 57 (1962).
57. Licari, J. J., and E. R. Brands: Organic Coatings for Metal and Plastic Surfaces, *Machine Des.,* May 25, 1967.
58. Janssen, J. E., and R. H. Torborg: "ASME Thermodynamics and Transport Properties of Gases, Liquids, and Solids." McGraw-Hill, New York, 1959.
59. Evans, A. G., and M. Palanys: *Nature,* **152:** 738 (1943).
60. Kline, D. E.: *J. Polymer Sci.,* **50:** 441 (1961).
61. Schilling, H.: *Kautschuk Gummi,* **16:** 84 (1963).
62. ASTM C 177-63, part 14, 15–26, 1965.
63. Anderson, D. R.: Thermal Conductivity of Polymers, *Chem. Rev.,* vol. 66, 1966.

5

Manufacturing Technology

CLEANING

The cleaning of a substrate or assembly prior to coating is an extremely important step, but one which, unfortunately, is often neglected. Efficient cleaning is critical to the immediate or long-term performance of both the coating and the assembly. Contamination of surfaces with ionic (saltlike) residues, oily residues, or particulate contaminants can result in deleterious effects such as corrosion, electrical failure, and poor coating adhesion. A proven cleaning procedure should therefore always be called out on a drawing or process specification.

To establish a scientific basis for the selection of a cleaning method, the coatings specialist needs answers to the following questions:

1. What contaminants are on the surface?
2. How much damage can they cause?
3. How can they be removed?
4. What tests can be used to assure their complete removal?

A discussion of these questions and of the materials, procedures, and equipment used for cleaning interconnect substrates and electronic assemblies is given in this section.

The Nature and Effects of Contaminants

Contamination has always been a serious problem in the manufacture and operation of electronic equipment. With recent advancements in high-density microelectronic assemblies, where the total size of a component may be smaller than a speck of dirt, the effects of particulate contaminants become even more critical. There are hundreds of different types of contaminants; those most often found on electronic hardware have been categorized as shown in Table 5-1. Many

Table 5-1: Typical Contaminants and Their Sources

Contaminant	*Possible source*
Fibers (nylon, cellulose, etc.)	Clothing, paper towels, tissues, and other paper products
Silicates	Rocks, sand, soil, fly ash
Oxides and scale	Oxidation products from some metals
Oils and greases	Oils from machining, fingerprints, body greases, hair sprays, tonics, lotions, and ointments.
Silicones	Hair sprays, shaving cream, aftershave lotions, hand lotions, soap
Metals	Slivers and powder from grinding, machining, and fabricating of metal parts; particles from metal storage cans and other metal containers
Ionic residues	Perspiration, fingerprints (sodium chloride); residues from cleaning solutions containing ionic detergents; certain fluxes such as the glutamic acid–hydrochloride types; residues from previous chemical steps such as etching or plating
Nonionic residues	Rosin fluxes, nonionic detergents, organic processing materials
Solvent residues	Cleaning solvents and solutions

of these particles are not visible to the naked eye but are disclosed only by separation and magnification, as in the photomicrograph of particles removed from an electronic assembly (Figure 5-1).

The incomplete removal or the entrapment of these contaminants under a coating can result in electrical short circuits, corrosion, deterioration of the coating, and hence increased penetration of moisture and other contaminants. The severity of these effects will, of course, depend on the nature and amounts of contaminants and on the amount of moisture and applied voltage. Several dramatic examples of the seriousness of leaving contaminants on a surface may be cited. In thin film circuits, nichrome resistors, contaminated by salt residue from fingerprints, were observed to disappear when tested with applied voltage in a humid environment. Apparently an electrolytic cell was created in which the resistors ionized and went into solution. This phenomenon was not observed when the resistor surface had been kept completely clean and dry. In another case fingerprints left beneath the coating on a circuit board caused visible blistering of the coating over them after exposure to humidity (Figure 5-2). In other cases poor surface cleaning resulted in peeling, lifting, or blistering of the coating immediately after drying or after a period of high-humidity exposure.

Ionic contaminants may be detected by measuring the electrical resistivity of a 75/25 isopropyl alcohol/water solution used to extract a part suspected of being contaminated. The lower the resistivity value, the greater the concentration of ionic impurities, usually calculated and reported as micrograms of sodium chloride per square centimeter.

Printed circuit boards used in military applications are required to be tested for ionic contamination per MIL-P-28809. Two methods can be used: (1) solvent

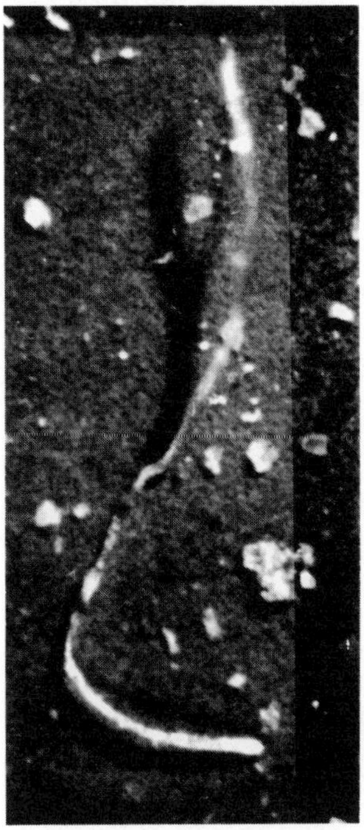

Figure 5-1: Photomicrograph of contaminants removed from electronic assembly, showing long-fiber and smaller metallic and nonmetallic particles. (X 100).

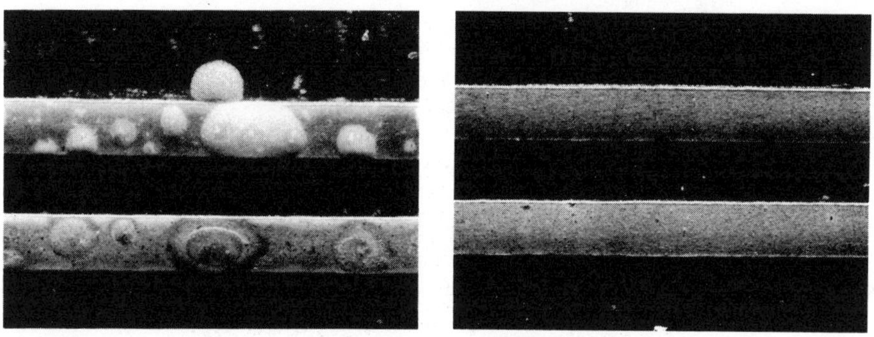

(a) (b)

Figure 5-2: The effect of aqueous cleaning in preparing circuit boards for coating: (a) test board prepared without aqueous cleaning; (b) test board prepared with aqueous cleaning.

extraction of ions from the part and then measuring the resistivity, or (2) extraction and resistivity done simultaneously in an Omega Meter (trade name of Kenco Industries, Inc.) (Figure 5-3). The Omega Meter contains a tank filled with a 75/25 isopropanol/water solution into which the part is submerged. The resistivity is measured and the sodium chloride equivalents are calculated by the Omega Meter. The extracted ions are then removed from the solution by circulating it through ion-exchange resin filters.

The presence of metal particles on electronic devices or circuits, even in trace quantities, is serious. For example, even a single metal particle wedged between two closely spaced conductors can result in catastrophic failure; an electrical short can occur immediately; or, in time, corrosion may result from the formation of ionic species. Metal particles detected in several commercial-grade cleaning solvents have probably been introduced from storage cans or during the transfer of solvent from one container to another. Other particles, such as fibers, greases, and soil minerals, have been isolated and identified and can also be detrimental, depending on the application.

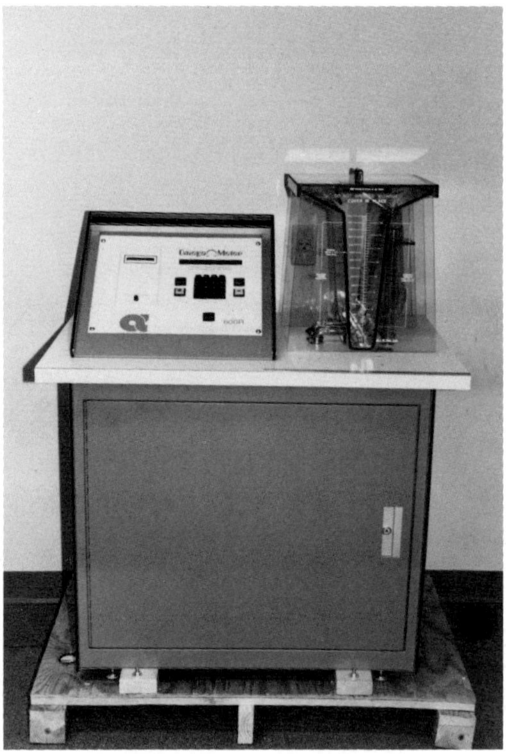

Figure 5-3: Equipment for testing electronic parts for ionic contamination by measuring the resistivity of an alcohol/water solution in which the part is submerged. (Courtesy of Kenco Industries, Inc.)

In hybrid microcircuits that are encased in hermetic, sealed packages, particles are detected by particle impact noise detection (PIND) testing. The PIND tester vibrates the circuits and detects the sound of a particle rattling around inside. When a particle is detected the package is opened and the circuit inspected for the contaminant, which is then removed.

Particles may be analyzed according to procedures described in the literature. Known volumes of solvent are filtered through a membrane, and the particles remaining on the filter are counted for each size range.[1] Typical results for cleaning solvents and solutions are given in Table 5-2. Identification of particles after they have been separated requires considerable skill and experience. A widely used procedure is to examine the particles under a microscope and match them against known standards. Descriptions and photographs of hundreds of particles may be found in "The Particle Atlas".[2]

Table 5-2: Particulate Contaminants in Cleaning Solvents

Solvent	Number of particles >1μ per 100 mls Reagent grade+	Class 100*
Acetone	>2000	<200
Ethyl alcohol	1000	<200
Methyl chloroform	1760	<200
Trichloroethylene	420	<200

+All solvents as received; particles measured at Rockwell International Corporation. These values may vary considerably for the same solvent depending on processing and storage conditions.
*Packaged in a Class 100 clean room; numbers indicate the maximum number of particles allowed by vendor's specifications.

Prevention of Contamination

To prevent contaminants from depositing on highly sensitive devices, ultraclean areas such as Class 10,000 (or better) clean rooms, Class 100 laminar-flow work stations, and laminar-flow ovens and coating stations should be employed during assembly and manufacture.[3] The Class designation defined in FED-STD-209 indicates the maximum number of particles 0.5 microns or larger distributed in 1 cubic foot of air. For example, there are 100 or fewer particles of at least 0.5 micron size in a Class 100 clean area. Also, before, during, and after assembly, parts should be stored in dry nitrogen-filled cabinets. In addition, ultrapure, filtered solvents and chemicals should be used during processing.

Fingerprints and saliva are two of the most pervasive and deleterious contaminants encountered in an electronics production facility. They can be prevented easily, however, by operators wearing gloves or finger cots and face masks and use of protective shields at microscope work stations.

Removal of Contaminants

Solvents and Solutions Used for Cleaning. Cleaning solvents and solutions may be placed in three categories, as shown in Table 5-3. The *hydrophobic* types are those which will effectively remove water-insoluble constituents such as greases, oils, and organic material. The *hydrophilic* solvents will easily remove water-soluble contaminants such as ionic salts. Since both types of contaminants are normally present on electronic assemblies, a safe practice is to use both types of solvents in a two-step process or to combine them in a *hydrophobic-hydrophilic* mixture.

Table 5-3: Cleaning Solutions and Solvent Types

Chemical type	*Examples*
Hydrophobic	
Organic solvents	Naphthas, xylene, toluene
Fluorocarbons	Freon TF, Freon TMC
Chlorinated hydrocarbons	1,1,1-Trichloroethane, perchloroethylene, trichloroethylene
Hydrophilic	
Organic solvents	Acetone, methyl ethyl ketone (MEK), methanol, ethanol, isopropanol
Ionic	Alkaline, acid, and detergent water solutions
Nonionic	Detergent-water solutions
Water	Tap, deionized, or distilled
Hydrophobic-hydrophilic	Alcohol naphtha (50:50 mixture), Du Pont TWD-602 (a Freon-water emulsion containing a surfactant), Freon TA (an azeotrope of Freon TF and acetone), Freon T-E35 (a blend of Freon TF and ethyl alcohol)

The hydrophobic-hydrophilic solvent combinations have become more important as one-step cleaning processes gain popularity. A few of these solvent mixtures are listed in Table 5-4. Most of these combinations are azeotropes, (mixtures having constant boiling temperatures) so that the solvent mixture does not change its composition when boiled, allowing them to be used in heated tanks or vapor degreasers. These solvents are particularly adapted for the removal of water-soluble and solvent-soluble solder fluxes.

A dramatic example of what can happen if only one type of solvent is used is shown in Figure 5-2. When only the hydrophobic solvent was used (Figure 5-2*a*), considerable blistering of the coating occurred after humidity exposure. When both hydrophobic and hydrophilic (aqueous solution) solvents were used, no blistering occurred (Figure 5-2*b*).

Solvent Purity. The solvents used to clean electronic assemblies prior to coating can themselves contain varying amounts of particulate contaminants and nonvolatile residues and may contribute to the deterioration of the assembly. The amount can be determined by evaporating a known volume of the solvent and weighing the residue, using a microanalytical balance. The nature of the residue may be determined by one or more microanalytical techniques. Residue weights are extremely small, usually in the parts-per-million range, but even these small amounts may be critical to the electrical performance of the assemblies. In general, adequate rinsing after cleaning is sufficient to remove such residues, but when ionic or nonionic detergent cleaners have been used, repeated rinsing with distilled water or fresh solvent, together with some means of agitation, is important. Several tests may be employed to establish the initial purity of cleaning solutions or solvents. The presence of ionic or ionizable impurities may be determined by evaporating to dryness a given volume of solvent, extracting the residues with isopropyl alcohol/water, and measuring the electrical resistivity of the extract as described previously. Other tests include measurement of the nonvolatile residues and particle analysis.

Table 5-4: Common Hydrophobic-Hydrophilic Solutions

Solution/Solvent	*Examples*
1,1,1 trichlorotrifluoro-ethane-water-detergent	Freon* T-WD 602 Genesolv** DW
1,1,1 trichlorotrifluoro-ethane-ethanol	Freon TE, TES, T-E 35 Genesolv DE, DES, DE-15, DE-35
1,1,1 trichlorotrifluoro-ethane-isopropanol	Freon T-P 35 Genesolv DI-15, DI-35
1,1,1 trichlorotrifluoro-ethane-methanol	Freon TMS
1,1,1 trichlorotrifluoro-ethane-methanol-methylene chloride	Genesolv DTA
1,1,1 trichloroethane-alcohols	Prelete***

*Freon is a trade name of Du Pont
** Genesolv is a trade name of Allied Chemical
*** Prelete is a trade name of Dow Chemical

Many federal and military specifications are available for the procurement of solvents and chemicals, but few provide sufficient control over very low levels of particulate contamination. Many firms have therefore found it necessary to supplement available specifications with their own tighter requirements. For space applications, some firms are specifying that deionized water and other solvents must contain no particles above 20 microns in size. Ultraclean solvents (filtered to 0.2 micron, < 1 ppm residue) are available through firms that supply chemicals to the semiconductor industry. Processes and assembly operations as well as the handling and storage of these solvents must be in an ultraclean environment (Class 100 or better) to maintain this low level of contamination.

The requirement that solvents contain no particles greater than 20 microns can also be met by filtering the solvents through a series of three consecutive membrane filters of 10-, 2-, and 0.5-micron pore sizes, respectively.[4] It is a safe practice in all microelectronic fabrication processes to use solvents and cleaning solutions that have been filtered through membrane filters immediately before use. Such filters are available in various pore-diameter sizes, starting from 0.1 micron.

Cleaning Methods. There are essentially five mechanisms of removing contaminants from a surface:

1. Solution. For example, sodium chloride is soluble in water.
2. Chemical reaction, which transforms the contaminants into soluble

products which can then be washed away. Examples include the removal of metal oxides and scale by acid or alkaline treatments.

3. Mechanical dislodging and removal of contaminant particles by a liquid or gas stream. Examples include metal and fiber particles removed by jet spray or by ultrasonic energy.
4. Plasma cleaning. This is an effective method of removing thin organic residues (< 1 mil) on parts.
5. Vacuum baking—removes volatile contaminants including water.

In practice, cleaning methods which employ combinations of these mechanisms are the most efficient.

The number of general cleaning methods is limited, but numerous modifications and combinations of the basic methods have been developed. The basic cleaning methods may be classified as follows:

1. Solvent spray
2. Solvent dip
3. Brush
4. Vapor degreasing
5. Ultrasonic immersion
6. Plasma cleaning
7. Vacuum baking

Vapor degreasing. Vapor degreasing is a widely used and efficient method for cleaning electronic parts. The part is suspended in a specially designed chamber (commercially available) which allows the vapors of a heated solvent to condense on the surface and flush it clean. The part is repeatedly washed with freshly distilled solvent, as opposed to hand scrubbing or other manual cleaning methods, where the solvent is stagnant and soon becomes contaminated. There are several variations of the vapor-degreasing method. The basic method utilizes vapor alone and is efficient when only small amounts of oil and grease contaminants are present. A variation is to combine vapor degreasing with a liquid-spray technique (Figure 5-4). The part to be cleaned is suspended in the solvent vapors until condensation ceases and is then sprayed with clean solvent for about 30 sec and returned to the vapor. Detailed instructions for vapor degreasing are found in the literature.[5]

The two most widely used solvents for removing greases and other hydrophobic contaminants by vapor-degreasing techniques are 1,1,1-trichloroethane and trichlorotrifluoroethane (Freon TF). They may be purchased from several manufacturers under different trade names. In many states, trichloroethane is considered an environmental and a health hazard whereas Freon TF is considered safe. Both are relatively stable but trichloroethane contains stabilizers to prevent hydrolytic breakdown, whereas Freon TF, because of its less reactive nature, requires little or no stabilization. The recently discovered hole in the upper atmospheric ozone layer is attributed to chlorofluorocarbons (CFC's) making these chemicals environmentally undesirable. Manufacturers of CFC's are rapidly developing substitutes for

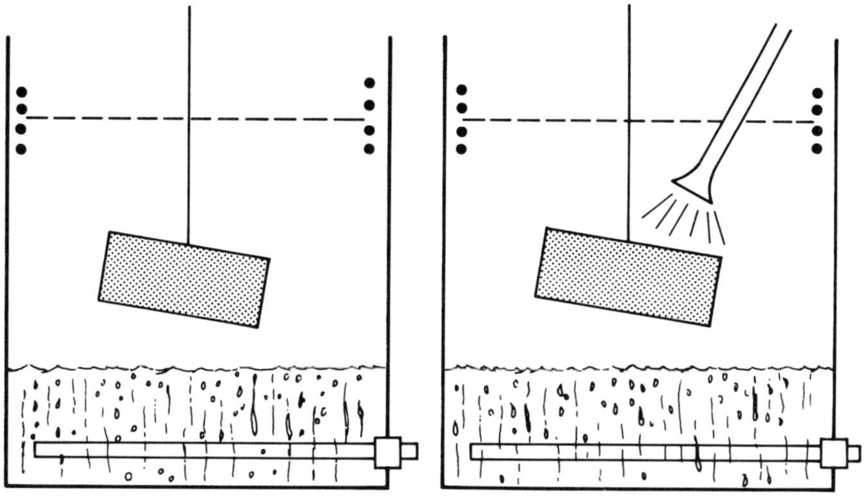

Figure 5-4: Diagrams of two types of commonly used vapor degreasers.

these compounds. As with other cleaning methods, the compatibility of these solvents with the substrates should be established before specifying them for production use. For example, it is reported that with some circuit-board laminates cleaning with trichloroethane causes leaching of resin, delamination, and discoloration.

Ultrasonic cleaning. One of the most effective cleaning methods entails immersion of the part in an ultrasonically agitated solvent bath. Because of the high energy imparted to the solvent, contaminants are rapidly and easily removed even from difficult-to-reach areas. Sound vibrations at frequencies higher than 20,000 hz (usually 20 to 40 khz) cause *cavitation,* or the rapid formation and sudden collapse of thousands of microscopic bubbles, in the liquid solvent or cleaning solution. The transfer of such energy to the part to be cleaned results in a very strong scrubbing action. Ceramic (lead zirconate or barium titanate) or magnetostrictive transducers are employed to generate the ultrasonic energy. The generator, usually an electronic oscillator, draws the required power from a 60-cycle line and converts it to useful output at the resonant frequency of the particular transducer. The transducer, which is excited by this output, expands and contracts at the same frequency, generating the desired ultrasonic vibration. Ultrasonic cleaning is a very rapid process, requiring only a few minutes per part, and lends itself well to automated procedures.

Almost any solvent or solution may be used for the liquid bath. A large number and variety of off-the-shelf solvents and cleaning solutions are available, and many others have been formulated for removal of specific contaminants. The choice of solvent is limited only by its corrosive effect on the tank, its toxicity or flammability, and its vibration-damping characteristics. Some liquids damp ultrasonic vibrations more than others. Concentrated detergent solutions and acetone have high damping characteristics and are not as effective as Freons, for example. For the most

efficient operation, the solvents (especially organic solvents) should be maintained at about 40°C. Cooling may be necessary, because ultrasonic agitation will heat the solvent after several hours of continuous operation.[6] Water detergent or alkaline solutions are more effective at temperatures of 40 to 60°C. In these cases the heat generated by the bath itself will be sufficient to maintain these temperatures.

Although ultrasonic cleaning is one of the most effective cleaning methods, it has several limitations which should be recognized. For example, because of its high efficiency, the bath becomes contaminated, and the solvent must be filtered or purified frequently and replaced periodically. Second, the ultrasonic vibration forms standing waves in the cleaning solvent. This phenomenon causes areas of concentrated energy called "hot" spots and areas of little energy called "cold" spots. The consequences are erratic cleaning and possible part damage. To avoid this, some manufacturers of ultrasonic equipment have used either multiple transducer elements to lessen this phenomenon or an electronic sweep that produces random harmonics thereby eliminating hot and cold spots.[7] Finally, a number of precautions should be taken in cleaning assembled circuits. It has been noted that fragile components such as glass diodes and other glass-sealed semiconductor devices may fail mechanically or electrically owing to cracking of the glass or lead separation. Also, wires and wire bonds in hybrid microcircuits are susceptible to damage. Even when such failures do not occur immediately, there is always the question of the effect that ultrasonic stresses may have on the long-term reliability and life expectancy of the device. Accordingly, ultrasonic cleaning has found its greatest use in the cleaning of bare substrates circuits prior to assembly with devices, and packages.

Ultrasonic vapor degreasing. A popular cleaning sequence consists of vapor degreasing, immersion in a heated ultrasonic tank, and spray rinsing. The part is first suspended in the solvent vapor, then immersed in the heated ultrasonic tank for about 3 minutes, followed by a final spray rinse. This process is easily automated by employing a robotic arm that is controlled by a microprocessor (Figure 5-5). To prevent contamination of parts due to particles the ultrasonic vapor degreaser continuously filters the solvent and may have water separators to remove any water that is dragged into the tank. Water separators are particularly important when using azeotropes of hydrophobic-hydrophilic solvents which tend to absorb water from the air.

In-line cleaning. As more production lines become automated, in-line processes become more important. For printed circuit boards there are a variety of in-line cleaners that use either water plus detergent or solvents as the cleaning solution. These cleaners can be placed "in-line" with a wave-soldering or vapor-phase soldering system to remove the flux and other contaminants immediately after soldering. Generally, in the aqueous cleaners the parts are placed on a conveyor belt that carries them through a pre-wash spray, a heated spray wash with detergent, two heated spray rinses, and finally hot air knives and infrared heaters for drying (Figure 5-6). The solvent-based cleaners use the same solvents that are used for vapor degreasing. On a conveyorized belt the parts proceed through a spray rinse then a vapor rinse and heated drying area. For difficult-to-clean parts solvent cleaners may be used in a heated immersion tank with optional ultrasonics prior to the spray rinse.

Plasma cleaning. Plasma-cleaning techniques are used to remove thin coat-

Figure 5-5: Ultrasonic vapor degreaser with a robotic conveyor system. (Courtesy of Delta Sonics, Inc.)

Figure 5-6: Automatic aqueous cleaning and drying system for printed circuit board assemblies. (Courtesy of Gram Corporation)

ings of organic impurities. Plasma is often the final cleaning step for metal packages to enhance the adhesion of the adhesives used to bond parts into them. It is also used prior to wirebonding devices on hybrid microcircuit substrates to assure clean bonding surfaces.

The most common plasmas used are oxygen, argon, and tetrafluoromethane (CF_4). An oxygen plasma works by forming ozone which in turn forms high energy oxygen atoms.[8] These atoms react with organic residues to form oxidation products such as carbon dioxide and water. Tetrafluoromethane works on the same principle; the high energy plasma consists of CF_3 and fluorine free radicals that react with an organic residue causing it to decompose into gaseous species.[8] An argon plasma is more gentle, breaking up an organic material by impingement; and it is inert, making it safer for delicate devices and plastic substrates. An argon plasma also takes about twice as long as a reactive plasma to remove a residue.

Vacuum baking. Volatile contaminants can be removed by placing the part into a vacuum bake oven that is heated under reduced pressure. This is more efficient than heating in an oven because the combination of heat and reduced pressure results in a higher effective temperature. Vacuum baking is commonly used to remove water from parts such as hybrid microcircuits prior to sealing in an inert atmosphere.

Verification of Surface Cleanliness. Confirmation that a surface is free of contamination is very important, particularly if the surface is to be subsequently coated. There are test methods to verify the cleanliness of a surface. Previously discussed were tests to determine ionic contamination which can also be used to confirm complete removal of residues such as plating salts and fluxes. The following test methods can be used to determine whether any organic films remain on a surface after cleaning.

Hydrophobic Surface Films by the Water-Break Test (ASTM F 22). This method can detect molecular layers of organic contaminants, especially on a smooth surface. The test is based on the ability of organic (hydrophobic) materials to prevent wetting of the surface by water. The part to be tested is withdrawn vertically from a container overflowing with water. The water must remain in a thin continuous film for 1 minute after withdrawal from the water to be considered clean.

Hydrophobic Surface Films by the Atomizer Test (ASTM F 21). This method can detect fractional molecular layers of organic contaminants, especially on a smooth surface. The test is also based on the ability of hydrophobic materials to interfere with surface wetting. A dry surface is sprayed with a fine mist. If the surface is free of organic films the water droplets will spread immediately to form a continuous water film. On a contaminated surface the droplets will remain as such.

Adhesive-Bonded Surface Durability of Aluminum (Wedge Test, ASTM D 3762). This method determines the environmental durability of adherent surface preparations by simulating the forces and effects on an adhesive bond joint. The test is intended for aluminum-to-aluminum applications but it may be used for other metal or plastic surfaces. After the two surfaces have been bonded, a wedge is inserted into the bondline. The test specimen is exposed to an aqueous environment, usually at an elevated temperature for a predetermined period of time (often 1 hour). An evaluation is done of the resulting crack growth with time and failure modes.

APPLICATION METHODS

Because of the wide variety and complexity of electrical and electronic parts which require coatings, numerous application methods are now in use. The coating of minute semiconductor junctions or the application of an ultrathin, continuous dielectric layer on the foil wrap of a capacitor, at one extreme, to the impregnation of huge motor rotors and stators or the coating of thousands of feet of wire, at the other, has necessitated the development of ingenious and varied deposition and polymerization techniques.

There are several basic methods of application. Some are suitable for rapid, mechanized, high-volume situations, whereas others are highly specialized and suitable to the coating of only one part at a time. A discussion of the coating processes most commonly used in the electronics industry is given in the following sections. Ten of these processes, their advantages, disadvantages, and typical applications are summarized in Table 5-5.

Spray Coating

The most widely used method of applying coatings to electrical parts is spraying. Spraying, includes not only the commonly used compressed-air vaporization technique, but also airless pressure spray, hot spray, hot-vapor impelled spray, electrostatic spray, dry-powder resin spray, electrostatic powder spray, and even the ubiquitous aerosol spray can.

Pressurized Liquid. Most spray methods are based on the principle of atomizing the paint into a fine mist, which is then directed to the surface to be coated. Atomization is accomplished by the use of compressed air, pressurized volatile solvents, or a high-velocity jetstream. With the proper technique, a thin, continuous wet layer of lacquer or enamel may be deposited without drips, sags, "curtains," "orange peel," or other unsightly surface imperfections. A skilled spray operator can apply a uniform coating and can usually touch up defective areas during the process or later.

In applying coating to substrates which have discrete components, such as circuit boards, the total desired thickness should be built up by spraying from four different directions in order to cover the top and all sides of the components and force the coating beneath the components. Avoiding pinholes, skips, or voids should be a major consideration in the application of environmental protective coatings. Turntables, either hand or power rotated, are useful in orienting the work for directional spraying. Conveyors may be utilized to move parts past a spray painter or automatic sprayheads, as in fully mechanized painting lines. Parts platforms or holders can be mounted on rotating spindles to make all surfaces accessible to the paint spray.

All paint spraying should, of course, be done in a ventilated booth which exhausts the overspray and solvent vapors to the outside, or preferably to a fume and solvent trap system. Not only are the fumes and paint aerosol noxious, but they may present a fire hazard, or may be toxic, especially after prolonged inhalation.

Airless Spray. A main disadvantage of air-pressure spraying is the tendency for the blown cloud of atomized paint to be swirled out of recesses, holes, and corners by the accompanying air stream, leaving these areas lightly coated or bare. Airless spray equipment has been developed which does not depend on atomiza-

Table 5-5: Application Methods for Coatings

Method	Advantages	Limitations	Typical Uses
Spray	Fast, adaptable to varied shapes and sizes; low equipment cost.	Difficult to obtain complete coating of complex parts and to obtain uniform thickness and reproducible coverage.	Motor frames and housings, electronic enclosures, circuit boards, electronic modules.
Dip	Thorough coverage even on complex parts such as tubes and high-density electronic modules.	Viscosity and pot life of dip must be monitored; speed of withdrawal must be regulated for consistent coating thickness.	Small and medium-sized parts, castings, moisture and fungus proofing of modules, temporary protection of finished machined parts.
Brush	Brushing action provides good "wetting" of surface, resulting in good adhesion. Lowest equipment cost.	Poor thickness control; not for precise applications; high labor cost.	Coating of individual components, spot repairs, or maintenance.
Roller-coat	High-speed continuous process; provides excellent thickness control.	Large runs of flat sheets or coil stock required to justify equipment cost and setup time; high equipment cost.	Metal decorating of sheet to be used to fabricate cans and boxes, application of photoresist to circuit boards.
Impregnate	Complete coverage of intricate and closely spaced parts; seals fine leaks or pores.	Requires vacuum or pressure cycling or both; special equipment usually required.	Coils, transformers, field and armature windings, metal castings; sealing of porous structures.
Fluidized bed	Thick coatings can be applied in one dip; uniform coating thickness on exposed surfaces; use of dry materials saves cost of solvent.	Requires preheating of part to above fusion temperature of coating; this temperature may be too high for some parts.	Motor stators; heavy-duty electrical insulation on castings; metal substrates for circuit boards, heat sinks.

(continued)

Table 5-5: (continued)

Method	Advantages	Limitations	Typical Uses
Screen-stencil	Deposits coating in selected areas through a mask; good pattern deposition and controlled thickness.	Requires flat or smoothly curved surface; preparation of screens is time-consuming.	Circuit boards, artwork, labels; maskant for etching solution; spot insulation between circuitry layers or under heat sinks or components.
Spin	Reproducible coatings of uniform thickness; good control of thickness.	Requires flat, smooth surfaces.	Photoresists for semiconductor devices and thin-film circuits; polymeric dielectrics for multilayer thin-film circuits.
Electrocoat	Good control of thickness and uniformity; parts wet from cleaning need not be dried before coating.	Limited number of coating types; compounds must be specially formulated ionic polymers; coatings often porous, sometimes nonadherent.	Primers for frames and bodies; complex castings such as open work; motor end belts.
Vacuum deposit	Ultrathin pinhole-free films possible; selective deposition can be made through masks.	Thermal instability of most plastics; decomposition occurs on products; vacuum control needed.	Experimental at present; potential use is in microelectronics for capacitor dielectrics.
Electrostatic spray	Highly efficient coverage and use of paint on complex parts; successfully automated.	High equipment cost; requires specially formulated coatings.	Heat dissipators; electronic enclosures; open-work grills and complex parts.

tion by high-velocity air, but produces a spray by forcing the material under high pressure through tiny-orifice nozzles. The coating material, which is circulated continuously in the storage system, may be heated to reduce its viscosity and to volatilize solvents more rapidly. The airless spray method is highly efficient, especially for coating parts having many recesses, cavities, or projections.

Hot-melt Flame Spraying. In hot-melt flame spraying, a cloud of powdered resin suspended in air is introduced into a special spraygun and is melted by being blown through the flame-shrouded nozzle of the gun. The melted plastic solidifies immediately upon contact with the object being coated. This technique is limited to plastics which are stable above their melting temperatures. Coatings applied by this technique are usually thermoplastic types such as polyethylenes, polypropylenes, and polytetrafluoroethylenes. Techniques have also been developed for spraying some unstable plastics that have sufficiently long induction periods before decomposition occurs.[9]

Other Spray Equipment. Epoxy and polyurethane coatings provide a high degree of chemical and solvent resistance, but have heretofore been difficult to apply in production because they are two-component systems and have relatively short pot lives. Spray equipment is available which facilitates the application of these two-component coatings. This equipment proportions and mixes the two components continuously and feeds the mixture to the sprayhead or extrusion nozzle for application.

The *aerosol spray can* is a very convenient way to package and dispense coatings, and almost every conventional paint, varnish, enamel, or one-component coating has been successfully packaged in an aerosol can. Automatic packaging equipment capable of packaging over 100 cans per minute has greatly reduced the price per can. Aerosol spraying is most economical for touchup coatings and for the priming or coating of a few parts. For high-volume production applications, automated spray or dipping equipment is the most economical method. A main limitation of aerosol packaging of coatings for electronics is that two-component systems, such as many electrical-grade epoxies and polyurethanes, cannot be packaged. At present the aerosol method is useful only for such one-component coatings as acrylics, vinyls, some silicones, epoxy acrylics, oil-modified polyurethanes, and some moisture-cured polyurethanes.

Electrostatic Spray Coating

Several modifications of the spray method have been developed and are in use. One of these, the electrostatic spray process, developed by the Ransburg Electro-Coating Company, is used extensively to coat objects having intricate shapes or a high proportion of open areas. The technique consists in depositing negatively charged plastic powder onto a metal part. The powder particles are rendered electrostatic as they are propelled through a diffuser at the front of the spray gun and are then rapidly attracted to the part (Figure 5-7). Subsequent curing or fusion of the applied coating may be performed in as little as 20 to 60 sec at 205°C or at lower temperatures for a longer period of time. Well over 95 percent of thermosetting powder coatings are applied by electrostatic spray techniques.

The coating is applied with an electrostatic spray gun. The power supply operates on 110 volts, 60 cycles, with a single-phase voltage having a power consumption of about 71 watts. The electric current is 210 microamps from the

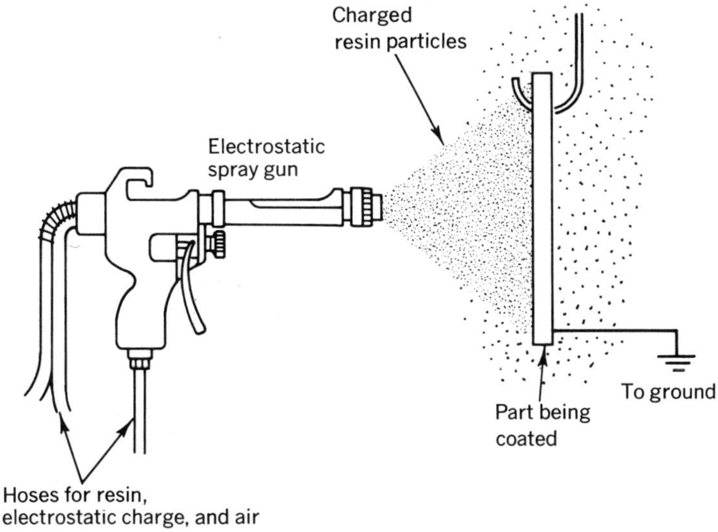

Figure 5-7: Electrostatic spray coating.

exposed voltage contact when it is grounded. The power output voltage does not exceed 90,000 volts direct current when the gun is actuated. Air used to propel the powder is passed through a drying bed to remove moisture prior to use. The powder is emitted from the nozzle by a rotating diffuser driven by a small air motor located in the handle. Adequate electrical protection is incorporated into the system to protect the operator from electric shock. Coatings ranging in thickness from 1 mil to more than 50 mils may be applied by this technique, depending on the plastic powder selected. In general, the same proprietary powder coatings used for fluidized-bed or powder-spray applications (described below) are also used for electrostatic spraying. Some types of coatings and their properties are listed in Table 5-6.

The advantages of the electrostatic spray process over other spray processes include the high deposition efficiency, with very little material lost as overspray, and the high film buildup that can be obtained in one application, reducing both time and cost. Paint-material savings of 25 to 50 percent over conventional spray methods are not uncommon. The process is also adaptable to automated production and hence less susceptible to human errors. This technique lends itself well to the coating of a wide variety of metal articles. In the electronics industry, for example, it is valuable for the electrical-insulation and corrosion-protection of heat dissipators (Figure 5-8), metal brackets, black-box enclosures, circuit-board housings, and motor windings.

Dip Coating

As the name implies, dip coating is the immersion of a part into a liquid composition (or dry powder, as in fluidized-bed coating), withdrawal, and drying or curing. It is a highly desirable method for achieving a continuous automated manufacturing operation, and it has the advantage over spray coating of assuring coverage of all areas and crevices. Dip coating is most suitable for one-component plastic sys-

Table 5-6: Powder Coating Types and Typical Properties[17]

Typical Properties	Std-cure Epoxy	Epoxy Polyester Hybrid	Acrylic	Polyester	Polyester TGIC*-cured	Fast-cure Epoxy
Application thickness	20-200 μ	20-200 μ	20-70 μ	20-70 μ	20-200 μ	20-200 μ
Bake cycle (metal temperature), °C, min	230, 3 165, 20	230, 3 165, 25	205, 10 180, 25	205, 7 175, 17	205, 7 155, 20	120, 30 ----
Outdoor Weatherability	Poor	Poor	Good	Good	Very good	Poor
Pencil Hardness	HB-5H	HB-2H	H-4H	HB-3H	HB-2H	HB-5H
Impact Resistance, in/lbs	60-160	60-160	20-100	60-160	60-160	60-160
Adhesion (w/o primer)	Excellent	Excellent	Excellent	Excellent	Excellent	Excellent
Chemical resistance	Excellent	Excellent	Excellent	Excellent	Excellent	Excellent
Solvent resistance	Excellent	Good	Good	Good	Good	Excellent

*Triglycerol isocyanurate.

Figure 5-8: Heat-sink rail electrostatically insulated with epoxy coating.

tems that have sufficiently long pot lives. Thus its use in the electronics industry is limited, because most high-reliability coatings are two-component types, and their pot lives after mixing are limited in some cases to fractions of an hour. Even in systems with pot lives of several hours,* polymerization and/or evaporation of volatile solvents results in continual viscosity changes, which cause variations in

*Some two-component systems are reported which have pot lives of up to 3 months.

thickness and adhesion and thus affect the performance of the cured coating. Such systems are therefore difficult to use on a production scale. Techniques used to maintain a stable two-component dip bath include replenishing solvents lost by evaporation to maintain constant viscosity and refrigerating the dip bath or maintaining it at a low temperature to extend its pot life.

The final thickness of a dip-coated part (which may range from a fraction of a mil to several hundred mils) depends on the viscosity of the bath, the rates of immersion and withdrawal, the temperature of the bath, and the number of dips. In the case of solvent-containing dip coatings, thicknesses of a fraction of a mil to about 2 mils may be obtained after one dip cycle. The thickness can, of course, be increased by multiple dipping, but when this is not economical, a nonsolvent-containing (100-percent-solids) coating should be used. The use of 100-percent-solids coatings usually results in buildups of 8 to 20 mils in a single dip.

Upon removal from the dip tank, the coated part must be drained, allowed to dry, and cured at an elevated temperature. Adequate drainage must be provided to avoid trapping liquid coating in pockets or recesses. Usually, resin flow continues after withdrawal of the part and can result in a thick buildup of coating, referred to as "tears," along the edge of the part. Tears may be mechanically "touched-off" or removed electrostatically. Drainage may be improved by designing the part to permit an easy flowoff.

Adequate air drying to remove solvents is recommended, since entrapped volatiles cause the coating to blister during a subsequent oven cure. A quick way of determining the adequacy of drying is a finger-touch test for tackiness. The degree of dryness and amount of residual solvent left in the coating may be determined more accurately by analytical tests. Such tests include the determination of weight loss versus time and the use of infrared spectrometric analyses.

With some systems, *hot dipping* may be used to ensure greater wettability and to minimize pinholes. This variation of dip coating consists in preheating the part to an elevated temperature (about 75°C) and dipping it in a room-temperature bath. This technique, of course, presupposes that the part can tolerate the preheating temperature. Because of the repeated exposure of the bath to heated parts, care must be taken to maintain a relatively constant bath viscosity. Preheated parts may be dipped, for example, in vinyl plastisol, which becomes gelled by the heat, leaving a solid layer on the surface. The cooler outer layer picked up on withdrawal remains in liquid form and drains off. An alternate technique entails dipping parts at ambient temperature into a molten resin. The coating solidifes when the part is removed from the bath and cooled. Resins used for hot-dip coating are normally thermoplastic types.

With high-viscosity coatings, such as the 100-percent-solids types, numerous air bubbles may be entrapped when the two components are mixed. Vacuum degassing is commonly used to eliminate this problem. Evacuation of the mixture at 600 to 700 torr (about water-aspirator pressure) for several minutes is recommended.

In summary, dip coating involves many variables, among which are the viscosity of the bath, temperature of the bath, rate of withdrawal, temperature of the part, and temperature and humidity of the environment. In spite of these variables and the many controls which are necessary, dip coating offers an economic solution for large-scale production and automatic coating of parts.

Fluidized-bed Coating. A unique variation of dip coating that is finding increasing use is the *fluidized-bed process,* sometimes called the aerated-bed

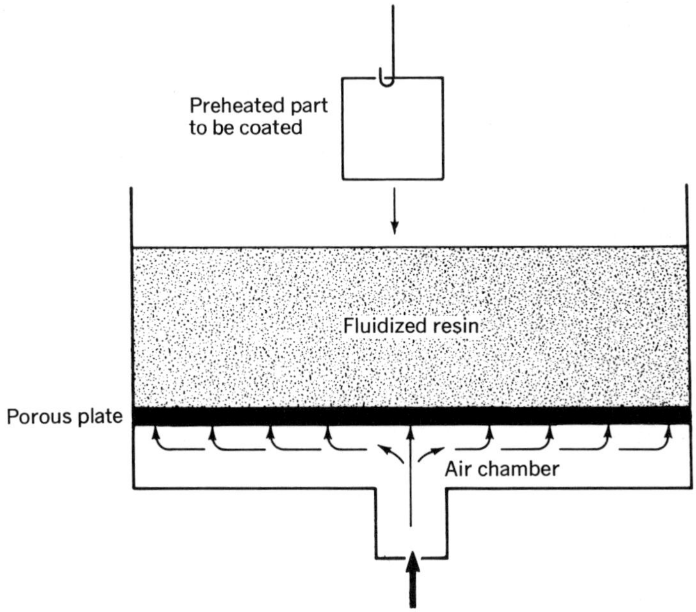

Figure 5-9: Fluidized-bed coating.

process. The basic process and modifications of it are the subject of numerous patents, both in the United States and in other countries.[10] The fluidized-bed process was invented in Germany by E. Gemmer, and a patent was issued in 1955 to Knapsack Griesheim.[11] In the same year, it was introduced in the United States by the Polymer Corporation. In this method, dry powdered resins or partially reacted powdered coatings are placed in a container and set in motion by controlled-velocity air or inert gas introduced from the bottom of the container through a screen and porous membrane (Figure 5-9). The resultant low-density powder is in constant motion and behaves as a low-viscosity low-surface-tension fluid. The part to be coated is preconditioned by one of several methods to promote adhesion of the particles to the surface and is then dipped into the fluidized-solid bed. The principal pretreatment methods are:

1. Preheating the part (usually metal) to a temperature slightly above that required to melt and fuse the plastic powder.
2. Applying a thin adhesive primer (a proprietary process with the advantage over preheating that areas of a part may be selectively coated).
3. Applying an electrostatic charge to the part to attract oppositely charged coating particles.

In all three cases, after the part is removed from the fluidized bed, it is normally

heated in an oven. For epoxy and other thermosetting coatings, this postbaking step further fuses and cures the coating, improving its physical and electrical properties. For thermoplastic coatings, the postbaking operation is important to smooth the surface and to obtain a less porous coating.

Process parameters. The number of process parameters affecting fluidized-bed coatings is very large; at least 20 have been reported.[12] However, once these parameters have been fixed and controlled, the process can be automated and gives highly reproducible results. Comprehensive studies of the many variables have been made by Pettigrew[13] and by Gaynor.[14]

The thickness of the deposited coating depends on many factors, including immersion time, heat retention of the part, mass of the part, and characteristics of the powder. Probably the most important parameter is the ability of the part to retain heat, which in turn is a function of the specific heat, thermal conductivity, and thermal stability of both the part and the coating material. The thickness, initial coating properties, and postbaking conditions depend on these thermal characteristics. Thicker coatings can be achieved with metal substrates having high specific heats and low thermal conductivities and with organic coatings having low specific heats. A matching of specific heats on a volume basis (specific heat × specific gravity) for the substrate and the coating powder is therefore desirable. The similarity of these values for aluminum or magnesium with some plastic powders is shown in Table 5-7.

Table 5-7: Specific-Heat Relationships Between Metals and Plastics Used in Powder Coatings[12]

Material	*Specific heat*	×	*specific gravity*	=	*specific heat (volume basis)*
Metals					
Aluminum	0.235	×	2.70	=	0.63
Steel, low carbon (0.06 C, 0.4 Mn)	0.132	×	7.87	=	1.04
Magnesium	0.265	×	1.74	=	0.46
Copper	0.096	×	8.96	=	0.86
Plastics					
Vinyl	0.30–0.50	×	1.35	=	0.41–0.68
Epoxy, unfilled	0.25	×	1.20	=	0.30
Cellulose acetate butyrate	0.30–0.40	×	1.18	=	0.35–0.47
Nylon	0.58	×	1.04	=	0.60

The time and temperature required in the postbaking step also vary with the mass of the part, its thermal conductivity, and the type of coating used. For example, postcures of several minutes to ½ hr at 205°C are required for epoxies, whereas thermoplastics such as polyethylene require a postbake of several minutes at 320°C. The necessity for postbaking depends on the size of the part and its ability to retain heat. Parts of sufficient mass to retain residual heat at 205°C or higher for several minutes generally require no postbaking, whereas small parts may require 5 to 30 min of postbaking.

The particle size of the resin powder is not a critical parameter, but a fairly narrow size range (100 to 150 microns) is reported to give optimum results. If the particle-size spectrum is wide, the greater air pressure required to fluidize the coarser particles causes the fine particles to rise above the bed as a dust cloud.[15]

Coating materials. Both thermoplastic and thermosetting coatings can be applied by fluidized-bed techniques. Some thermoplastic materials, notably vinyls, can be completely fused by the heat retained in the preheated part; thermosetting coatings usually require postcuring after the dipping step. Epoxies, chlorinated polyethers, polyethylene, and silicones are examples of coatings commonly applied by fluidized-bed methods. Although Teflon TFE and FEP powders can be applied by fluidized-bed techniques, very few practical applications have been found, probably because of the very high preheat temperature required and poor adhesion of the coatings to substrates. Poor adhesion may also be a problem with other coatings, and in such cases primers should always be used. Examples include cellulose acetate butyrate, nylons, and vinyls. Coating types such as epoxies, polyethylenes, and chlorinated polyethers normally do not require the use of primers.

Some powder coating types and critical deposition parameters are given in Table 5-8. A more complete list has been given by Landrock.[16] Some properties of these formulations are given in Table 5-6.

Table 5-8: Summary of Plastics for Fluidized-Bed Coating[16]

	. Fluidizing Conditions .			. . . Fluidized-Bed Powder . . .		
Resin	Preheat Temp. (°C)	Cure or Fusion Temp. (°C)	Cure or Fusion Time (min)	Maximum Operating Temp. (°C)	Adhesion	Weather Resistance
Epoxy	120–230	120–230	1–60	95–205	Excellent	Good
Vinyl	230–290	205–315	1–3	105	Poor	Good
Cellulose acetate butyrate	260–315	205–290	1–3	105	Poor	Good
Nylon	290–425	345–370	1	150	Poor	Fair
Polyethylene	260–315	205–315	1–5	105	Fair	Good
Polypropylene	260–370	205–315	1–3	125	Poor	Good
Hercules Penton	260–345	230–315	1–10	175	Poor	Good
Du Pont Teflon	425–540	425–480	1–3	260	Poor	Good

Advantages and limitations. The main advantage of the fluidized-bed process is that it permits the buildup of rather thick coating films (5 to 25 mils) in only one or two dips, a result not readily achieved with other coating methods. Rather complex parts with sharp edges and corners are easily and uniformly coated. The extra-thick coatings thus produced solve many chemical, solvent, erosion, and corrosion problems. Other advantages over the conventional methods of dipping and spraying include:

1. Lower cost, owing to the avoidance of waste material.
2. Greater ease of handling, because the powder resins are one-component systems.

3. Time savings, since no weighing, mixing, or deaerating are necessary as for the two-component coatings.
4. Reduction of fire hazard, because no solvents are employed.

The disadvantages should also be recognized:

1. Parts that can be coated are limited to metals or other materials having a high heat retention.
2. Heat-sensitive electronic components and assemblies cannot be coated, since parts must be capable of withstanding preheat temperatures of 205 to 345°C.
3. Thin pinhole-free films less than 5 mils thick are difficult to achieve.
4. Some coating powders are deteriorated by the high processing temperatures used.
5. The process is economical only for large production runs.

Because of the high speed and automation possible with the fluidized-bed process, it is used for many diverse electronic and electrical applications, such as the coating of electronic enclosures, housings, covers, brackets, wire harness guides, heat sinks and dissipators and the insulation of transformers and motor rotors and stators. Fluidized-bed coatings are also used for the moisture protection and packaging of resistors, capacitors, varistors, and thermistors.

Sprayed-powder Process. The sprayed-powder process entails the spraying of powdered coatings onto preheated metal parts by means of flock-coating sprayguns. As in the fluidized-bed process, powder particles contacting a hot surface will fuse and form an integral coating. The same powders sold for fluidized-bed coating are normally also used for spray-powder coating.

Electrostatic Fluidized-bed Process. As the name implies, this process combines elements of both the fluidized-bed and the electrostatic-spray processes. The normal fluidized bed equipment is used, but in addition, the plastic particles are given a negative charge by applying a high-voltage (about 90,000 volts) direct current. The part to be coated is then electrically grounded and suspended above the bed, so that the charged particles can be attracted toward the part. This process has several advantages over the fluidized-bed process:

1. The part to be coated does not need to be immersed in the bath.
2. Preheating of the part is not necessary for particle adhesion.
3. Large fluidized beds to contain the part are not required for coating large objects.
4. The process can be used to coat thin, low-mass parts such as wires, which cannot be coated by the fluidized-bed process because of their poor heat retention.

The process has been found useful for coating either large, complex parts or thin-gauge wire.

Vacuum Impregnation. *Vacuum impregnation,* still another variation of dip coating, employs a vacuum or vacuum-pressure cycling during the dipping opera-

tion. This technique is particularly useful for the thorough coating (impregnation) of intricate parts containing numerous small cavities, such as coils, windings, porous metal, and ceramic surfaces. Associated hardware such as porous-metal castings may be sealed to maintain vacuum, hold pressures, or contain certain gaseous environments. Thorough impregnation is important in many electrical-insulation applications because any small voids may act as moisture traps or corona regions, resulting in gradual insulation breakdown. This is especially true for high-voltage-transformer insulation.

Porous specimens can be impregnated simply by soaking them in the resin or coating; however, vacuum-evacuation techniques aid in reducing the amount of entrapped air and moisture. The air is removed and then replaced with the impregnating material. In order to do this more effectively, it is desirable to evacuate the specimen before immersing it in the resin. After immersion, the vacuum is again applied and then interrupted periodically to force the liquid into the pores. A sequence of vacuum followed by pressure is even more efficient for impregnating very closely wound coils. The amount of vacuum and time of immersion required are determined by the temperature, viscosity, volatility, and pot life of the resin.

Screen Printing

Coatings formulated with thixotropic and wetting agents can be applied by *squeegee* through a silk or stainless-steel screen. To apply coatings by screening, the surfaces must be fairly flat. The thickness of the coating may be varied from a fraction to a mil to more than 6 mils, depending on the screen mesh size and on the number of coats applied. After operating parameters have been established, very reproducible thicknesses are obtained. To eliminate pinholes and other imperfections, it is advisable, as with other coating methods, to apply and superimpose two or three layers of coating, with each layer partially or fully cured prior to application of the next one. By using screens with a photoresist pattern, insulation coatings can be applied to substrates, such as ceramics or epoxy laminates, in selected areas. This is often necessary in isolating a thin- or thick-film conductor or resistor from an active device which is to be positioned above it. Screenable epoxies, polyurethanes, and other coating types are commercially available. Typical screenable epoxy formulations may be cured at 80 to 150°C and provide 1-mil-thick coatings when squeegeed through a stainless-steel screen.

The screening process for applying conductors, resistors, and capacitors in selected areas for fabricating thick-film circuits is a highly developed art. Automated equipment (Figure 5-10) for mass production is available. Screens of silk, nylon, or stainless steel are formed into masks to yield the desired pattern. Photoemulsion compositions containing gelatin chromate or photoresist are used in preparing the screens. Line thicknesses of 10 mils with 10-mil spacings are easily attainable, but advances are being made in improving definition to 2-mil lines and 2-mil spacings.[18]

Spin Coating

Spin coating is the standard process for applying reproducible, uniformly thin (microns) coatings to substrates. The two main applications for spin-on coatings are photoresists used in the processing of integrated circuits and interlayer dielectric polymers such as polyimides used in the fabrication of multilayer thin-film

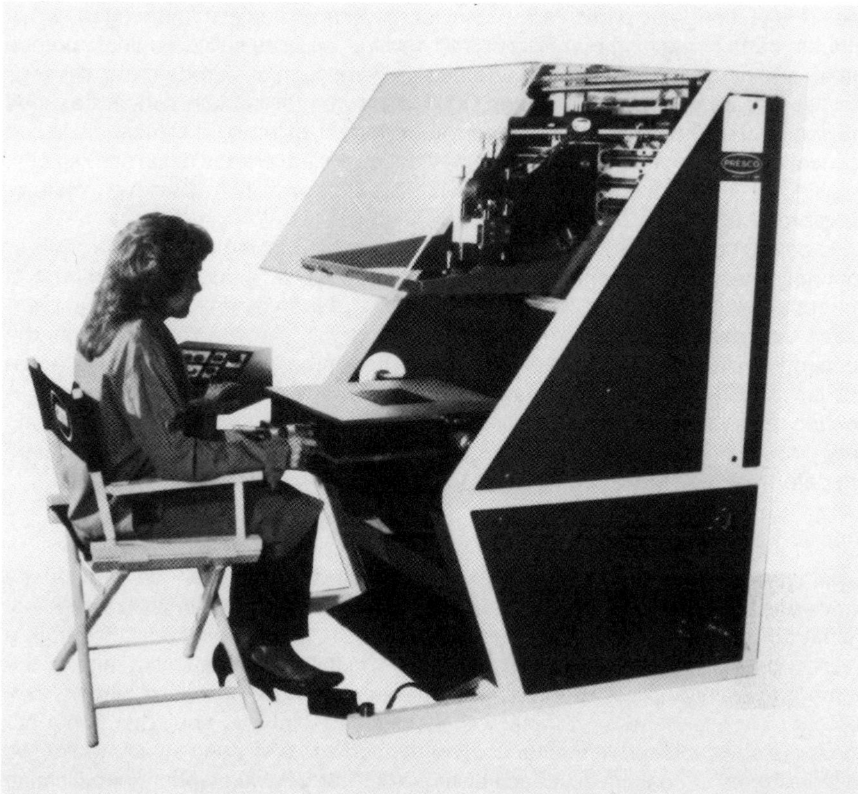

Figure 5-10: Automatic screen printer for coating application. (Courtesy of Affiliated Manufacturers, Inc.)

circuits. In semiconductor device manufacturing facilities, spin coating is part of the automated in-line production process.

The first step in spin-coating consists of placing the substrate on a motor-driven rotary vacuum chuck mounted in a bowl. The coating is then dispensed onto the center of the substrate and the spinner turned on causing the coating to be evenly distributed by centrifugal force. In general, excess solution is used to assure good substrate coverage. The resulting coating thickness is a function of the solution viscosity and the spinner speed and acceleration; a thicker coating can be obtained by increasing the viscosity of the solution or reducing the spin speed or both. Typically, thickness is a hyperbolic function of the spin speed; thinner coatings are obtained at higher speeds. Regulation of these parameters allows control of coating thickness.

The substrates most often used for spun-on coatings are circular silicon wafers, although square or rectangular substrates can also be used. In the latter cases coatings are thicker at the edges and corners and the uniformity is harder to obtain. It is also important that the substrates be very smooth (often polishing is necessary) and flat to get reproducible coating thicknesses.

Flow and Curtain Coating

Flow coating and *curtain coating,* which are related processes, are high-volume continuous operations. In either case, the coating is applied from above the part and allowed to flow over the part on an open-mesh moving conveyor belt, which passes through a bake oven. Flow coating is used for objects having considerable open space or for small parts which can be handled easily on a moving conveyor. The coating is showered onto the conveyor through a series of spray nozzles; the excess is caught in a pan and is recycled through a filter back to the nozzles.

In curtain coating, the paint flows in a thin, continuous layer from a slit that spans the width of the coating line. This technique is adaptable for coating flat sheets or continuous strips. Neither flow coating nor curtain coating is used extensively in the electronics industry except for volume production of small, low-cost insulated metal parts.

Roller Coating

Roller coating, a highly mechanized and closely controlled method for coating flat stock, provides a means of applying a fixed amount of liquid coating to one or both surfaces of any sheet or strip requiring an even coating. Thus it is used for applying photoresist coatings to copper-laminate stock on which etched circuitry is to be developed and for applying thin films on metal foil for use in capacitors.

In the roller-coating process, the liquid coating material is transferred by means of a series of metering and printing rolls from a reservoir pan to the flat surface. The printing roll may be cut out to lay down a pattern, or the metering roll may be lithographed to print a functional or decorative design. After the coating operation, the material passes through a zoned drying and curing oven, and the sheets are stacked or the strip is rerolled. Roller coating is also a universally used method for decorating the metal stock for such end products as cans, bottle caps, toys, television trays, automobile instrument panels, and lighting fixtures.

Brushing

Brushing is not the usual method of applying coatings to electronic assemblies. However, when small portions of a part or rather inaccessible areas require coating, a skillfully handled brush is often the answer. Brushing is also used extensively for touchup and repair work, such as repairing skips or damaged areas of coating, covering spots where components have been removed and replaced, and covering pinholes left from electrical test probes. Applying the coating with a small camel's-hair brush is ideal for these small touchup jobs. The thickness of brush coatings is difficult to control, and outlines cannot be strictly observed unless masking is used, but the job is done quickly with a minimum investment.

Automated Coating Processes

Automation is important in large-scale production processes. Both spray and dip coating have been automated and are widely used as in-line processes, particularly for printed circuit assemblies. Spin coating is an in-line process in semiconductor device fabrication.

Spray coating. Conveyorized spray coating is the most expeditious method for coating assembled printed circuit boards. Uniform and reproducible coatings of acrylics, epoxies, urethanes, and silicones can be applied. Two-part materials are

metered and mixed with a static mixer. Also, moisture sensitive coatings are stored in nitrogen-purged pressure pots. With automated equipment, coatings may be sprayed and cured in a continuous in-line operation (Figure 5-11). The parts to be coated are loaded on a conveyor belt which carries them through the spray chamber in which the spray nozzles are angled for three-dimensional coating and then through the drying tunnel where the coating is cured. Both convection and infrared radiation are used to cure the coating.

Figure 5-11: Conformal spray coater with an in-line infrared convection curing oven for coating printed circuit board assemblies. (Courtesy of Integrated Technologies, Inc.)

Dip coating. Conveyorized dip coating systems are designed to coat printed circuit boards with epoxies, silicones, urethanes or acrylics (Figure 5-12). The viscosity of the coating in the dip tank is monitored and both the solvent and the coating are automatically replenished. The part is automatically transported through the processing sequence on a continually moving conveyor belt. After loading, the part is dipped, proceeds to the drip chamber and then the drying tunnel where the coating is cured. Both convection and infrared radiation can be used to cure the coating.

Figure 5-12: Automatic dip coating system with an in-line infrared convection curing oven for applying conformal coatings to printed circuit board assemblies. (Courtesy of Integrated Technologies, Inc.)

CURING AND POLYMERIZATION OF SOLID FILMS

After application of a coating it must be cured to attain its final mechanical and electrical properties. Curing involves driving off solvents and other volatile constituents and polymerizing the coating. Each type of coating has a unique cure mechanism as described in Chapters 1 through 3. The properties of a coating can be varied somewhat by the curing process which affects the degree of polymerization. Characteristics such as hardness, electrical resistivity, and outgassing can be modified by varying the cure parameters. There are two methods of curing typically used, heat and radiation.

Heat Curing

Organic coatings are generally cured with heat. The temperatures and time required for the different types of coatings are described in Chapters 1 through 3. By changing the cure temperature and time the properties of a coating can be varied.

Step curing. A single temperature is not always the optimum cure schedule for a coating. Step curing involves curing a coating at different temperatures (usually from low to high) and for different times; sometimes a vacuum is used. Step cures are particularly useful in limiting the outgassing of a coating. One device manufacturer found that by doing a five-step cure, each step at a higher temperature, the amount of moisture in an encapsulant was reduced by a factor of four.[19]

Equipment. The choice of equipment used to heat cure coatings depends largely on the desired properties of the cured product and the level of facility automation.

Convection oven. For batch processes and single-temperature curing an oven is all that's necessary. Convection ovens maintain an even temperature by circulating heated air or nitrogen through the chamber.

Belt-conveyorized furnace. For assembly line processes, where parts are produced in series, a conveyorized furnace is used. An oven loses its temperature stability each time the door is opened, a problem not encountered with a conveyorized system. Conveyorized furnaces have heated zones that can be maintained at several temperatures so that step curing is easily performed. By changing the belt speed, the dwell time in each heated zone is adjusted. A conveyorized belt furnace can also be placed in-line with a spray or dip coating system as shown in Figures 5-11 and 5-12.

Infrared furnace. A variation on the conveyorized belt furnace is the infrared (IR) furnace. Infrared energy rather than radiant heat is used to cure the coating. The energy is absorbed by the carbon-carbon bonds of the molecular backbone as well as carbon-hydrogen and other bonds. The bonds begin to vibrate at a faster rate causing localized heating of the coating. If the wavelength of the IR energy is narrow, only the organic coating will be heated making the process cooler overall than radiant heating.

Photocuring

Photocurable coatings have been used for over a century as photoresists primarily in photoengraving and printing processes.[20] However, these materials have only recently gained favor in the electronics industry. Properties such as rapid cure, reduced solvent emissions, and the ability to coat heat-sensitive devices and circuits make photocurable products attractive as protective coatings for many electronic applications. In addition, photocurable coatings often have enhanced physical and mechanical properties,[21] and, because of their short cure times (seconds), lend themselves to completely automated lines.[22]

Ultraviolet light (UV), rather than heat, is used to drive the polymerization process. Optimal curing occurs in the range of 250 to 400 nanometers with a radiant output of at least 150 milliwatts per square centimeter.[23]

Resins that have been modified for photocuring include acrylics, urethanes, epoxies, polyimides, and silicones. These materials are used for conformal coatings, adhesives, sealants, encapsulants, solder masks, polymer thick films, and marking and printing inks.

Photoinitiation

All photocurable coatings require a photoinitiator to start the polymerization process. The photons from the ultraviolet source react with the photoinitiator to

generate active catalysts that join and crosslink monomers and oligomers into high molecular weight polymeric structures that are insoluble and chemically resistant. There are two types of photo-induced polymerization: (1) free-radical and (2) cationic.

Free-radical polymerization requires high intensity UV, provides rapid cure and permits application of a wide range of thicknesses. Although photocurable coatings based on free-radical polymerization processes are the most widely used systems, coatings using cationic photoinitiators are beginning to gain attention. Cationic polymerization needs only low intensity UV and curing continues after the radiation source has been removed.[23] Also, the cationic processes will cure areas that are not directly exposed to the light source, called "dark" or "shadow" cure, whereas the free-radical process requires complete exposure to the ultraviolet energy for effective curing. The choice of free-radical vs cationic cure process depends largely on the application. For example, shadow curing would be beneficial for curing a die-attach adhesive but undesirable where precise patterns of tight tolerances are to be formed.

Free-radical photopolymerization. Free radicals used for photoinitiation are generated by one of two methods: (1) intermolecular hydrogen-abstraction and (2) intramolecular photocleavage.[24] Examples of each type are presented in Figure 5-13.

Both types of photoinitiators require absorption of light energy, followed by excitation to produce active free radicals. The free radicals react with monomers and oligomers to initiate polymerization. Competing with polymerization is the decay and recombination of the free radicals and reactions of both the free radicals and the growing polymer chain with air oxygen. A general reaction scheme is shown in Figure 5-14. Oxygen inhibition can be avoided or controlled by addition of oxygen scavengers in the coating formulation or by effecting the cure in an inert atmosphere.

Cationic photopolymerization. In cationic photopolymerization compounds that convert to Lewis acids on exposure to ultraviolet light are used as photoinitiators. Among these compounds are aryldiazonium salts, diaryliodonium salts, and triarylsulfonium salts, specific examples of which are shown in Figure 5-15. The Lewis acids generated are long-lived and continue to catalyze polymerization after the radiation source has been removed. This ensures curing in shadowed areas and under components.[25] A further benefit of this process is that an inert atmosphere is not necessary since the reactions are not inhibited by oxygen.

Properties

Photocurable coatings are generally one-component, 100-percent solids, thus eliminating solvent emissions. Their shelf life ranges from six months to several years when stored properly. Cationic photoinitiated products generally have longer shelf lives than the free-radical types. They can be applied by a variety of conventional methods including dipping, spraying, spinning, or screen-printing. The ability to cure in a few seconds to a few minutes renders photopolymerizable coatings amenable to modern assembly lines. However, after the initial radiation induced cure, a postcure of 5 to 30 minutes at elevated temperatures (50 to 150 C) is generally recommended so that optimum adhesion and hardness are

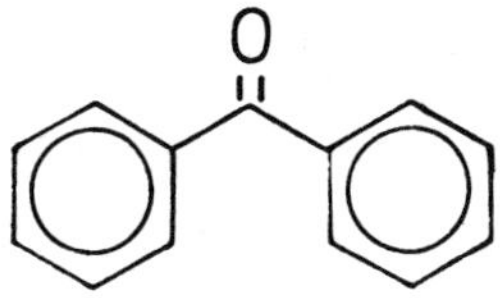

Benzophenone

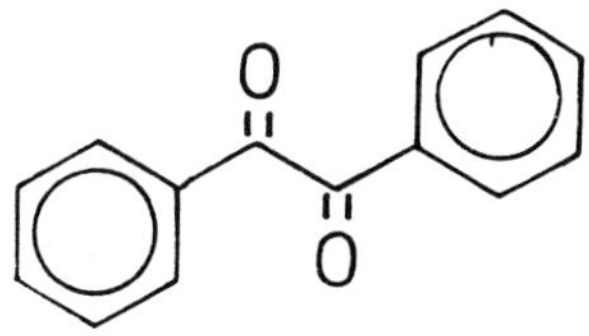

Benzil

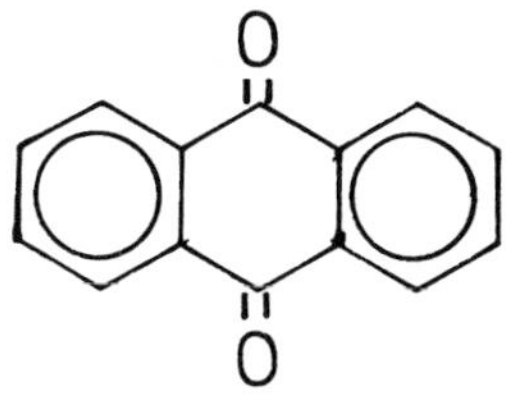

9,10 Anthraquinone

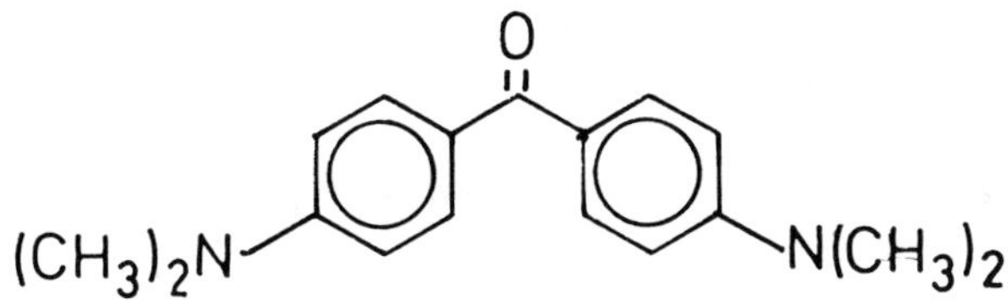

4,4'-bis(N,N dimethylamino)benzophenone
(Michler's ketone)

(1) Compounds that undergo intermolecular H-abstraction.

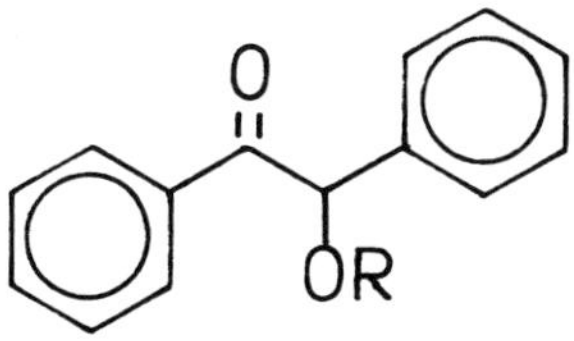

Benzoin ether

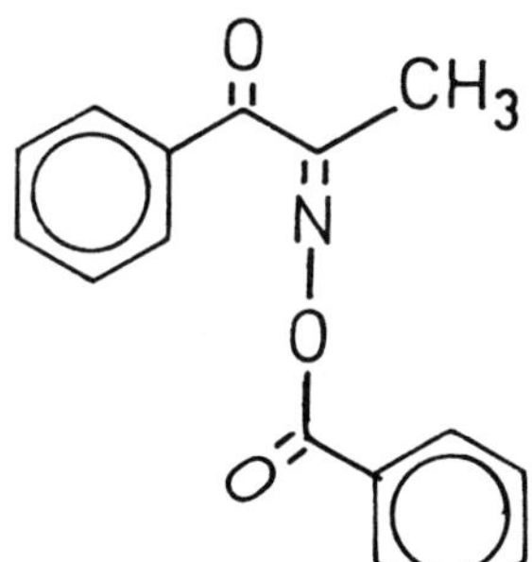

Phenyl-1,2-propanedione-2-o-benzoyl oxime

(2) Compounds that undergo intramolecular photocleavage.

Figure 5-13: Typical free-radical photoinitiators and their structures.[24]

```
(1) initiator + light -------------> free radical
(2) free radical + monomer --------> polymer initiation
(3) free radical + free radical ---> recombination
(4) polymer + monomer -------------> polymer propagation
(5) polymer + polymer -------------> polymer termination
(6) polymer + oxygen --------------> oxygen inhibition
(7) free radical + oxygen ---------> oxygen scavenging
```

steps (1), (2), (4), and (5) generate polymer chains; steps (3), (6), and (7) inhibit chain growth.

Figure 5-14: General reaction scheme for free radical photopolymerization.

Aryldiazonium Salts

Diaryliodonium Salts

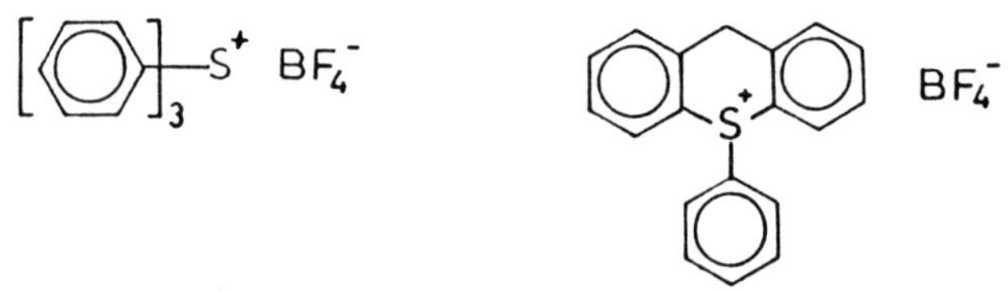

Triarylsulfonium Salts

Figure 5-15: Typical cationic photoinitators and their structures.[24]

achieved.[26] The electrical and mechanical properties of each photocurable coating depends on the resin used. Overall, photocurable coatings tend to be highly crosslinked and therefore very durable and chemical resistant. Some examples of UV-curable materials and their characteristics are summarized in Table 5-9.

Equipment. To properly cure photocurable coatings the quality of the UV source is crucial. Generally, medium pressure mercury vapor lamps are used which operate at 200 W/in of arc length and at wavelengths between 250 and 400 nanometers.[23] Many of the companies that sell UV-curable coatings also sell curing equipment. The UV-curing systems range from conveyorized-belt equipment for conformal coating and ink curing of printed circuit boards and components to portable, hand-held wands for spot curing (Figure 5-16).[27]

THIN-FILM DEPOSITION

Thin polymeric films are defined, for purposes of this discussion, as solid organic polymer coatings which are less than 3 microns thick (30,000 Å, or about ⅛ mil). Commercial applications of such films have been confined largely to capacitor dielectrics; however, they are presently being applied as microcircuit insulation, as corrosion-protective coatings, and as barrier coatings for semiconductor devices. One of the key problems in the fabrication of microelectronic circuits is the difficulty in depositing continuous ultrathin pinhole-free dielectric coatings in layers or in specific patterns. To date, the inorganic materials have dominated the semiconductor-device and microcircuits industry, but as new high-temperature-stable organic materials are developed and as techniques are refined for depositing thin films, more uses will be made of organic thin films. Among the advantages of ultrathin organic coatings are their high dielectric-breakdown strengths, low dielectric constants, their ability to form pinhole-free layers, and in some cases the high purity of the coating, which avoids tunneling and leakage of current. Among their disadvantages are the thermal and oxidative instability inherent in so many of the presently available polymers.

Thin defect-free polymer films generally cannot be deposited by the customary methods of spraying or dipping one- or two-component resin systems. Thus only a limited number of coating types and techniques lend themselves to thin-film formation. Among the most important techniques are the following:

1. Spin Coating
2. Vacuum or vapor deposition
3. Ultraviolet-light polymerization
4. Electron-beam polymerization
5. Glow-discharge/plasma polymerization
6. Gamma irradiation

Among coatings, the thermoplastic types are most amenable to thin-film deposition, especially those which polymerize by a free-radical-addition mechanism. Hence styrene, divinylbenzene, and other vinyl and acrylic monomers are very suitable and widely used.

Table 5-9: Typical Ultraviolet-Curable Coatings and Encapsulants

Trade name	Manufacturer	Chemical Type	Characteristics
Impruv 363	Loctite	Modified acrylate	A low viscosity compound designed for shallow potting and thin-film encapsulation.
Impruv 365	Loctite	Modified acrylate	A medium viscosity sealant and potting compound; dual cure (UV and anaerobic) permits deep potting.
Shadowcure 361	Loctite	Modified acrylic	A conformal coating formulated to provide moisture, solvent and environmental protection to printed circuit boards; qualified to MIL-I-46058C.
Uniset UV-156	Emerson & Cuming	Modified acrylic	A high flexibility compound for shallow potting.
Uniset UV-900	Emerson & Cuming	Modified acrylic	A sealant and potting compound with excellent environmental resistance.
Uniset UV-920	Emerson & Cuming	Modified acrylic	A low viscosity conformal coating for printed circuit boards; qualified to MIL-I-46058C.
605A	UVEXS	Methacrylated/acrylated urethane	A thixotropic encapsulant with excellent electrical properties.
Uniset UV-230	Emerson & Cuming	Modified epoxy	A heat resistant capacitor coating with excellent dielectric properties.
Cyracure	Union Carbide	Cycloaliphatic epoxide	A complete conformal coating system that allows user to choose viscosity and flexibility.
UV 10	Master Bond	Modified epoxy	A low viscosity compound for coating, potting and encapsulating; good physical strength and chemical resistance.
UV 15	Master Bond	Modified epoxy	An adhesive and conformal coating with superior insulation and chemical resistance.

(continued)

Table 5-9: (continued)

Trade name	Manufacturer	Chemical Type	Characteristics
UV 14-3	Master Bond	Modified epoxy	A solvent soluble adhesive and coating for temporary bonds and coatings.
RSV S-8	Wacker	Modified silicone	A flexible resilient, conformal coating.
RSV S-10	Wacker	Modified silicone	A protective coating for both rigid and flexible printed circuit boards
Probimide 348	Ciba-Geigy	Polyimide	A high-solids solvent-soluble coating used as a general circuit passivant and alpha-particle barrier
Probimide 337	Ciba-Geigy	Polyimide	A solvent-soluble coating used as an interlayer dielectric, general passivant and ion-implant mask
Probimide 400	Ciba-Geigy	Polyimide	A fully imidized solvent-soluble system for use as an intermetal dielectric, general passivant, planarization, and ion implant mask
Pyralin PD P1-2701-D	Du Pont	Polyimide	A solvent-soluble one-part coating for use as a general passivant, interlayer dielectric and in thick-film packaging applications
Pyralin PD P1-2702-D	Du Pont	Polyimide	A solvent-soluble one-part coating for use as a general passivant and interlayer dielectric for thick film packaging applications

Spin Coating

Spin coating is used extensively for the application of photoresist as part of the fabrication process for semiconductor devices and thin-film circuits. However, it is also gaining wide use as the preferred application method for polymeric interlayer dielectrics such as polyimides in the fabrication of multilayer thin-film circuits. Typical film thickness range from a few tenths of a micron to a mil depending on the viscosity of the coating solution and the spinner speed and acceleration. Spin coating was previously described under "Application Methods".

Vapor Deposition

The vapor or vacuum deposition of coatings, though a highly desirable method, can be employed only for a few polymer types. Unlike metals, most plastics cannot

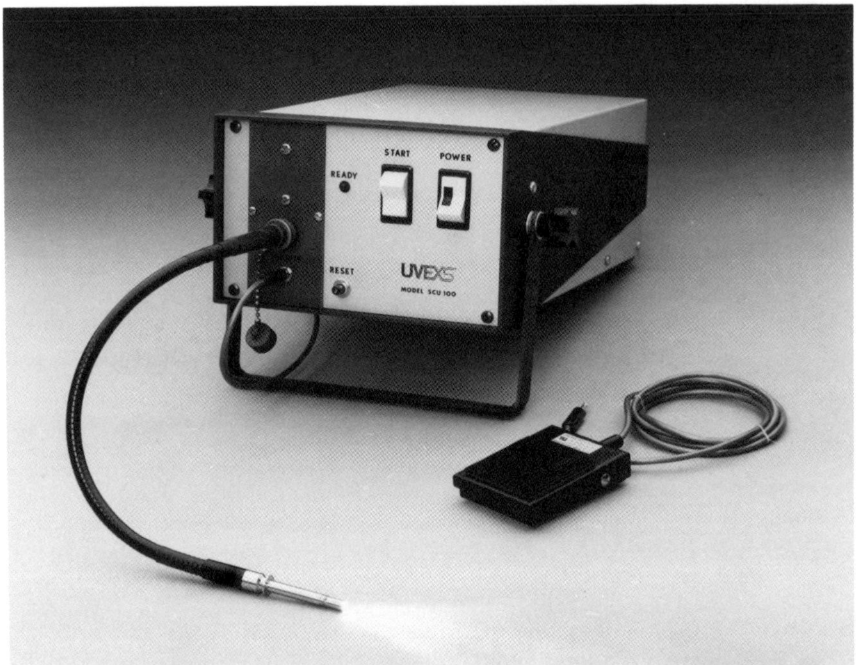

Figure 5-16: Portable ultraviolet curing system with flexible wand for spot curing of ultraviolet-radiation curable adhesives, coatings, sealants, encapsulants, and marking inks. (Courtesy of UVEXS, Inc.)

be heated to their vaporization temperatures without gross decomposition and changes in their physical and electrical properties. This is particularly true of the two-component highly cross-linked thermosetting plastics (epoxies, polyurethanes, phenolics). With these systems, charring or carbonization occurs at elevated temperatures before evaporation can take place.[28] The vaporization method is most successful with low molecular weight polymers; however, the low molecular weight species will recombine with little or no interaction, and only soft films with poor physical properties are obtained.

Polyethylene can be deposited as film 900 to 2500 Å thick by vaporizing solid polyethylene at 290°C and 10^{-4} torr.[29] It was found that high deposition rates resulted in poor films which had small pieces of solid plastic embedded in them. This situation could be corrected by inserting a baffle above the vapor source and directly below the part to be coated. The dielectric constant of these polyethylene films was 2.02, slightly lower than that for the bulk plastic, which was 2.25. There have been some reports of Teflon vapor deposition. Teflon was depolymerized at 538°C in a vacuum of 10^{-5} to 10^{-6} torr and reported to repolymerize as an adherent film on a cold substrate.[30] However, the greatest success in vapor deposition has been achieved with the polyxylylenes. Solid dimers of *p*-xylene are pyrolyzed under moderate vacuum, forming gaseous diradicals which in turn

combine on a substrate into long-chain linear polymers. The equipment for this process consists basically of a pyrolysis tube and a vacuum chamber (Figure 5-17). Details of the chemistry and the process are described in Chapter 2.

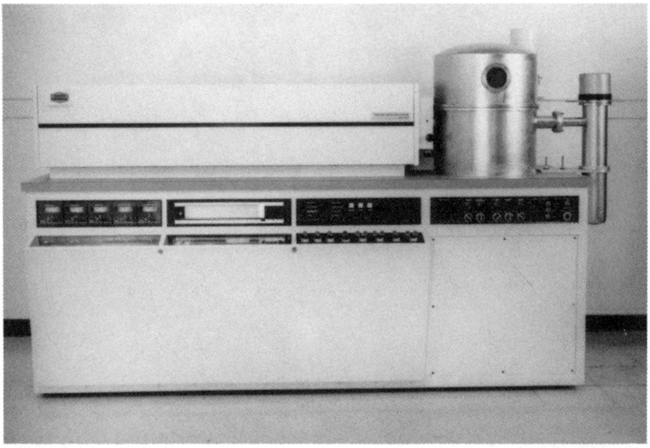

Figure 5-17: Equipment for the vapor deposition of parylene coatings. (Courtesy of Union Carbide Corporation)

Ultraviolet-Light Polymerization

Gaseous monomers adsorbed onto a surface may be polymerized by exposure to ultraviolet light in either the presence or absence of air. The most effective ultraviolet energy range has been found to be between 2,000 and 3,500 Å and may be generated from a 700-watt medium-pressure Hanovia mercury arc lamp. The deposition chamber is normally evacuated to about 10^{-6} torr to remove water vapor, and the gaseous monomer is introduced and maintained at about 4 torr during irradiation.

Monomers most suitable for ultraviolet polymerization are those containing double bonds (unsaturation). Such bonds open up by a free-radical mechanism, allowing a head-to-tail coupling, with the formation of thin solid films. Styrene, butadiene, methylmethacrylate, acrolein, and divinylbenzene are but a few examples of monomers that have been successfully polymerized in this manner.

The ultraviolet polymerization of methylmethacrylate was reported as early as 1938. The deposition of a solid film on the walls of a vessel containing the gaseous monomer on exposure to ultraviolet light was reported by Melville.[31] More recently other work has been done on the polymerization of methylmethacrylate, acrolein, divinylbenzene,[32] butadiene,[33] tetrafluoroethylene, and hexachlorobutadiene.[34] Pinhole-free and highly adherent films a few millionths of an inch thick have been prepared by investigators at General Electric.[34, 35] By employing a mask in front of the substrate during ultraviolet irradiation, very fine, well-defined insulation layers, which are useful in microcircuits, can be deposited (Figure 5-18). White[36] reported

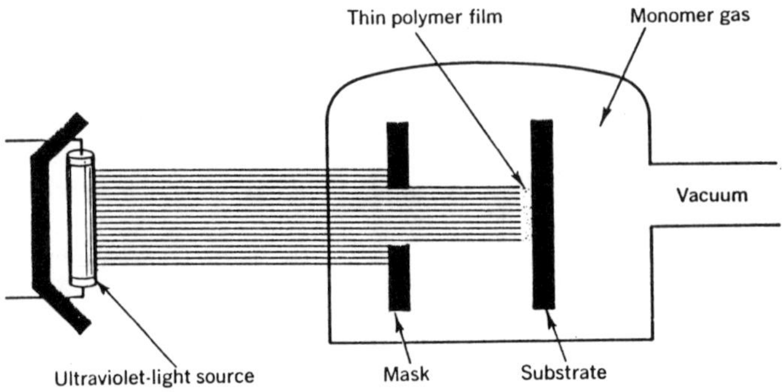

Figure 5-18: Photodeposition process. A substrate is placed inside a vacuum chamber, the air is pumped out, and a monomer gas is introduced into the system. When the system is irradiated with ultraviolet light, a polymer film is deposited upon the surface of the substrate. The thickness of the film depends on the time of the ultraviolet irradiation. If a mask is placed in front of the ultraviolet light source, a pattern can be formed on the substrate. (General Electric Research and Development Center)

the ultraviolet deposition of films 50 to 100 Å thick having a low incidence of pinholes. Films derived from chemisorbed gas molecules possessed better physical properties than those derived from physically adsorbed molecules.

In ultraviolet polymerization gaseous molecules are first adsorbed onto the surface, either by chemisorption or physical adsorption, and are then energized by ultraviolet light. Free-radical formation is the main mechanism, with subsequent formation of long-chain polymers through coupling of the free-radical species. When an ultraviolet source having a broad wavelength spectrum (2,000 to 4,000 Å) is used, some undesirable precipitation of polymer may occur in the gaseous phase. Some investigators have avoided this by using a filter to screen out the more energetic lower-wavelength photons, thus controlling and restricting the deposition to the exposed surface only.

Ultraviolet light has a distinct advantage over both electron-beam and glow-discharge methods. The ultraviolet photon energy (4 to 5 ev) is in the range necessary for polymerization by the free-radical mechanism, whereas the energies of the electron beam or glow discharge are several hundred times greater, making it more difficult to control the exact composition of the coating. Scission and ionic fragments generated by such high-energy bombardment result in a heterogeneous composition.[37]

Another advantage of ultraviolet polymerization is that equipment requirements are quite simple and inexpensive. The equipment consists of a mercury arc or other source of ultraviolet light, a quartz tube or a bell jar with a quartz window to allow entrance of the ultraviolet light, and a means for evacuating, purging, and introducing the gaseous monomer into the chamber. The high-vacuum systems necessary for electron and glow-discharge polymerization are therefore not required.

Electron-bombardment Polymerization

The polymerization of gases by the electron beam was discovered accidentally. During the use of electron tubes and electron-beam equipment, it was noted that thin films formed on the inside of the tube. These films were identified as resulting from trace amounts of contaminant gases entering the chamber, such as those of silicone oil used in the vacuum pump. For many years, these films were considered contaminants and nuisance.[38] In 1954, however, it was suggested that electron-beam polymerization could be useful as a new way of depositing ultrathin films;[39] and in 1958 the significance of this technique in fabricating microelectronic circuits and devices was reported.[40]

In effecting polymerization, the electron beam forms free radicals, ion radicals, and other activated species from liquids or from gases adsorbed on a surface. These activated species are short-lived and quickly combine into long chain solid polymers. In this respect, electron-beam polymerization is similar to ultraviolet polymerization. Thin films can be deposited selectively by exposing only certain areas of a substrate. By employing a mask or by programming and directing the beam, very fine lines and areas may be defined.[41]

Gases or liquids which may be polymerized by electronic bombardment are normally those containing double bonds (ethylenic groups). Thus monomeric tetrafluoroethylene has been polymerized with high-speed electrons. Under the same energy conditions, copolymers of tetrafluoroethylene with hexafluoropropene, polyvinylidene fluoride, and other ethylenic polymers have been prepared.[42] The continuous and rapid curing of styrene, polyesters, diallylphthalate, and vinyl carbazole monomers to 1-mil coatings by means of an electron source between 50 and 250 kev was reported by Davison.[43] Styrenated polyesters, polyvinyl chloride plastisols, epoxy acrylics, and acrylic monomers have also been polymerized in films from 0.5 to 125 mils by an electron source of 300 kv, 25 ma and a total output of 7.5 kw.[44] Other ethylenic monomers, including acrylic, methacrylic, acrylonitrile, styrene, and organopolysiloxanes, have been polymerized by electrons at 1×10^7 roentgens/sec and have been applied as coatings to electronic devices. Germanium diodes coated with the liquid monomers and cured with high-energy electrons for periods up to 3 min at a total dose of 10^6 showed little change in reverse-current characteristics.

The advantages of applying and curing coatings by the electron-beam method are the rapidity of curing and the fact that heat is not required, which obviates the possible deterioration of heat-sensitive devices.[45] However, the process has found limited application because of the high initial cost of the equipment required to produce high-energy electrons, subsequent high operating costs, and shielding problems. Depending on the energy of the electrons, overheating and damage to the substrate on which the coating is applied may also occur. Still another problem—encountered not only with electron-beam curing, but with the curing of most ethylenic monomers by a free-radical mechanism—is the inhibition of cure that can occur in the presence of atmospheric oxygen.[44] Films may often remain only partially cured and tacky. Techniques to prevent this condition by removing oxygen during the cure include evacuation, the use of an inert-gas environment, or the use of oxygen scavengers.

Glow-discharge/ Plasma Polymerization

The *glow-discharge method* for depositing and polymerizing thin films, also known as the *electric-* or *silent-discharge* method, entails striking an electrical discharge between two electrodes in the presence of a monomer gas. A potential of several hundred to several thousand volts between electrodes is employed. Under dc conditions the substrate to be coated is either made the anode or attached in some manner to the anode. In ac discharge, the substrate is placed between the anode and the cathode. A third variation of glow discharge makes use of radio-frequency-excited molecules and does not require the use of electrodes. This method is referred to as *electrodeless discharge* or *plasma polymerization.* The desire to use organic materials as dielectrics and passivants on electronic devices has made plasma polymerization an area of intense study.[46-48]

In plasma polymerization a monomer is excited by radiofrequency or microwave energy to form active species, primarily free radicals and ions. These activated species combine to form a thin polymer film on a surface. The physical and electrical properties of the polymer formed depend on the energy frequency,[49] the monomer gas pressure,[50] reactor design,[51] the monomer itself, and the surface on which the coating is to be deposited. Some surfaces lend themselves well to the formation of strong chemical bonds, whereas others are rather passive, and only weak adhesion results. The energy used varies from low radiofrequencies (13 MHz) to high microwave frequencies (2.5 GHz).[52] Typical pressures used are about 100 millitorr, although higher pressures are being studied.[50]

A key advantage of the glow-discharge or RF plasma methods is the complete coating coverage provided to the part. The large spread in the angle of incidence with which the activated gas molecules strike the surface allows the polymers to form within and behind all crevices and to cover all surface imperfections, resulting in an integral coating.[53] Other advantages include low substrate temperature (the substrate may be kept even at room temperature), the fact that a very high vacuum system is not necessary, as in other methods, and the wide choice of monomer starting materials which can be used.

To polymerize, the monomers must have active groups such as double bonds, oxygen, fluorine, amines, or combinations of these. Mixtures of two or more monomers may be used which result in new classes of polymers having unique properties.[54]

A large number of monomers have been polymerized by glow discharge, yielding films from several hundred to 10,000 Å thick. Bradley and Hammes[55] reported the formation of thin films from more than 40 different monomers and have provided considerable data on electrical resistivity at various temperatures and activation energies for electrical conduction. However, no data were given on the exact characterization of the polymer films. Thirty different monomers were investigated by means of radio-frequency plasma by Smolinsky and Heiss[56] as potential thin-film capacitor dielectrics. Of these, the best films were those derived from dimethylpolysiloxane, triethylsilane, diethylvinylsilane, vinyltrimethylsilane, heptene-2, cyclohexene, styrene, valeronitrile, and 2,5-dimethyl-2,4 hexadiene. Films were defined in terms of electrical and physical properties; but again, as with other investigations, no structural or compositional characterization of the polymer films was performed.

The glow-discharge technique has also been used to produce thin polymeric

films from halogenated olefins, including fluorinated derivatives.[57-60] These films, found suitable as dielectric and electronic-device coatings, were formed by the application of a potential of several hundred volts between electrodes.[61, 62] Finally, a series of hydrocarbon and fluorocarbon films was prepared by the electrodeless-discharge method, with isobutylene, ethylene, acetylene, diacetylene, toluene, trifluoroethylene, and tetrafluoroethylene used as starting materials.[63] Films deposited from ethylene, acetylene, and diacetylene possessed structures analogous to that of highly cross-linked polyethylene. Films derived from toluene showed structures (by infrared spectroscopy) similar to that of polystyrene, and films derived from the fluorinated monomers were assumed to be Teflon-like in structure.

The properties of plasma-polymerized thin films have been difficult to reproduce because of the large number of variables influencing their formation. Often free radicals remain in the thin film causing instability and degradation.[46] Though not widely used today, plasma polymerization is under extensive study and practical applications for electronics are sure to follow.

Plasmas can also be used to induce surface changes in polymers. Exposure of a polymer to a plasma produces changes in surface properties such as wettability, molecular weight of the surface layer, and chemical composition of the surface.[64] Plasmas can produce hydrophobic or hydrophilic properties depending on the species that is "grafted" onto the surface.

Fluorocarbon plasmas (CHF_3, CF_4, C_2F_6, etc) are often used and produce a Teflon-like surface on the organic polymer[65] rendering it hydrophobic. Other functional groups such as hydroxyl or amino groups can be added to a polymer surface by using allyl alcohol or allylamine plasmas, respectively,[66] rendering the surface hydrophilic.

Gamma-radiation Curing

As in the case of other forms of radiation energy, gamma radiation can be used to polymerize or copolymerize certain gaseous, liquid, or solid monomers. Gamma-radiation-induced reactions are simliar in nature and mechanism to those induced by ultraviolet radiation. Consequently, the same classes of polymer resins used for ultraviolet curing are amenable to gamma curing. These include the general class of ethylenic compounds such as the vinyls, acrylics, allylics, and polyesters. Epoxy resins, however, are difficult to polymerize by gamma radiation. High dosages of 4×10^8 rads and higher are required, and even then only a small percentage increase in molecular weight occurs.[67] However, indirect techniques may be used for radiation curing of epoxies, or the resin may be modified to incorporate vinyl or ethylenic groups. Gamma ray exposure can then induce polymerization through the vinyl groups, and the resulting polymer will contain an epoxy-type backbone structure. Ethylenic polymers such as polybutadiene may be so treated, and the resulting resins can then be polymerized by gamma irradiation.[68]

Another important use for gamma radiation is in grafting gaseous or liquid monomers onto the surfaces of solid plastics or coatings. The surfaces may be altered to achieve improved or unique properties. The wetting characteristics, antistatic properties, adhesion properties, or surface resistivity of coatings and plastics may thus be changed or engineered to meet desired requirements. As an example, acrylic acid molecules have been grafted onto Teflon film to give it antistatic properties.[69] Other important applications in which the moisture resist-

ance, chemical resistance, and radiation stability of polymers have been improved by gamma-radiation grafting are described by DasGupta.[70]

Commercial gamma-radiation equipment utilizes any one of three gamma sources: cobalt 60, cesium 137, and iridium 192. Dose rates may range from 5×10^4 rads/hr to 6.4×10^4 rads/hr.

Although extensive experimental work has been performed on the use of gamma radiation for polymerizing monomers, very little is reported on the practical applictions of this technique for electronics. However, gamma-radiation curing, as well as other radiation-curing methods, appears to hold much promise for future electronic developments. Among some potential uses are the curing of insulating coatings or capacitor dielectric films when heat curing is not feasible, the fabrication of detailed microinsulation patterns, and the development of materials with precise bulk or surface electrical properties.

GENERAL PROPERTIES OF THIN-FILM POLYMER COATINGS

Many properties of thin polymer films are independent of their chemical structure or composition. The most important of these is the very high dielectric-breakdown strength which all defect-free thin films possess; dielectric-breakdown voltages are markedly higher than those for identical polymers in bulk form. The increase of dielectric strength with a decrease in film thickness is a well-known phenomenon, previously described in Chap. 4 and in other works.[71]

Many coating processes yield films that are pinhole free and continuous. The lack of pinholes and continuity of the film are largely dependent on the coating method, surface treatment, cleanliness of the surface, and the ambient in which the coating is applied. Dust and contaminants normally contained in air or derived from processing chemicals may be 3 microns or larger, and it is obvious that such particles can impair the continuity and integrity of very thin films. Extreme care should therefore be taken to coat the parts in ultraclean areas such as class 100 clean rooms and laminar flow clean benches to use purified or filtered solvents and chemicals.

Still another property common to most thin films is their high degree of purity. With adequate controls, the polymerization reaction may be directed so that no side reactions or by-products are formed. For example, ultraviolet polymerization involves a coupling or addition reaction in which no by-products are possible. Vapor-deposition techniques also result in highly pure films. High-energy techniques such as the glow-discharge and electron-beam methods require greater control to prevent the occurrence of alternate reactions.

Thin films can conduct nonohmic currents by a number of mechanisms. Currents may be due to ion movement or to electron tunneling, impurities, space charges, or field emission.[72] According to the *tunneling effect,* the current density J decreases exponentially with the thickness of the film, x:

$$J = R\left(\frac{V}{x}\right)^2 \frac{1}{\varphi} \exp - \frac{k\varphi^{3/2}x}{V} \tag{1}$$

where φ = potential energy barrier
V = voltage across the dielectric film
k = dielectric constant of film
R = gas constant
J = current density
x = film thickness

However, J is negligibly small at thicknesses greater than 100 Å. As most organic films are deposited in thicknesses greater than 100 Å, currents due to the tunneling effect are not considered to be significant. A further discussion of the tunneling mechanism for conduction has been given by Holm.[73]

Current flow by electron emission, referred to as the *Richardson-Schottky effect,* is attributed to induction by a large electric field. The applicable equation is

$$J = AT^2 \exp \frac{(q^3V/kx)^{1/2} - \varphi}{RT} \tag{2}$$

where A = Richardson constant
T = temperature
q = electronic charge
and the remaining symbols are as defined for Eq. (1).

and the remaining symbols are as defined for Equation (1).

Finally, conduction due to impurity centers, called the *Frenkel-Poole effect,*[74] consists in the increase in electrical conductivity of insulators prior to breakdown in a high electric field according to

$$\sigma = \sigma_0 \exp \frac{(e^3E/\epsilon)^{1/2}}{kT} \tag{3}$$

where E = applied electric field = > 10^5 volts/cm
σ = conductivity with electric field
σ_0 = initial conductivity without field
ϵ = electronic component of dielectric constant

REFERENCES

1. Austin, P. R., and S. W. Timmerman: "Design and Operation of Clean Rooms," chap. 9, Business News Publishing Co., Detroit, 1965.
2. McCrone, W. C., editor: "The Particle Atlas," Vol. I–IV, Ann Arbor Science Publishers, Ann Arbor, Mich., 1974.
3. FED. STD. No. 209, "Clean Room and Work Station Requirements."
4. Heuring, H. F.: Cleaning Electronic Components and Subassemblies, *Electron, Packaging Prod.,* June, 1967.

5. Werner, M.: Solvent Vapor Degreasing and Recovery, "Metal Finishing Guidebook," Metals and Plastics Publications, Westwood, N.J., 1967.
6. Jeffrey, L. R.: Ultrasonic Cleaning, "Metal Finishing Guidebook," Metals and Plastics Publications, Westwood, N.J., 1967.
7. *Delta Sonics, Inc.,* '9 + 1 Reasons Why . . .', *Technical Bulletin.*
8. Bunshah, R. F.: "Deposition Technologies for Films and Coatings—Developments and Applications," Noyes Publications, Park Ridge, N.J., 1982, p. 48.
9. Von Fischer, W., and E. G. Bobalek: "Organic Protective Coatings," Reinhold, New York, 1953.
10. Davis, W. J.: Methods and Apparatus for Applying Protective Coatings, U.S. Patent 3004861, October, 1961: Dettling, C. J.: Fluidized Bed Coating Method, U.S. Patent 2974060, March, 1961; Dettling, C. J., and R. E. Hartline: Process and Apparatus for Producing Continuous Coatings, U.S. Patent 2987413, June, 1961; Gemmer, E.: Fluidized Bed Coating Process, U.S. Patent 2974059, March, 1961; Gemmer, E.: Fluidized Bed Coating Process for Coating with Thermosetting Materials, U.S. Patent 3090696, May, 1963.
11. Gemmer, E.: Process and Apparatus for the Preparation of Protective Coatings from Pulverulent Synthetic Thermoplastic Materials, German Patent 933019, September, 1955.
12. Landrock, A. H.: The Coating of Aluminum with Plastics by the Fluidized Bed and Electrostatic Powder Techniques, *Plastec Note* 18, Plastics Technical Evaluation Center, Picatinny Arsenal, N.J., February, 1968.
13. Pettigrew, C. K.: Fluidized Bed Coatings, *Mod. Plastics,* part 1, vol. 43, no. 12 (August, 1966), and part 2, vol. 44, no. 1 (September, 1966).
14. Gaynor, J., A. H. Robinson, M. Allen, and E. E. Stone: Variables in Fluidized Bed Coatings, *Mod. Plastics,* vol. 43, no. 5 (January, 1966).
15. Greensitt, E. A.: Epoxy Powder Coatings, *Paint Technol. (London),* vol. 28, no. 9 (September, 1964).
16. Landrock, A. H.: Fluidized Bed Coatings with Plastics: Technology and Potential for Military Applications, *Plastec Rept.* 13, Plastics Technical Evaluation Center, Picatinny Arsenal, January, 1964.
17. Ramig, A.: Powder Coatings Technology: Status and Future Opportunities, *Proceedings of ACS Division of Polymeric Materials: Science and Engineering,* Vol 55, 1986.
18. Short, O. A.: Conductor Compositions for Fine Line Printing, *Electron, Packaging Prod.,* February, 1968.
19. Lowry, R. K.: Moisture Process Control and Mass Spectrometry Correlations, "Proceedings of RADC/NBS International Workshop on Moisture Measurement and Control for Microelectronics (IV)," 1986.
20. DeForest, W. F.: "Photoresist Materials and Processes," McGraw-Hill, New York, 1975.
21. Hoyle, C. E.: Photocurable Coatings, *Proceedings of the ACS Division of Polymeric Materials: Science and Engineering,* vol. 55, 1986.
22. Puder, A. T.: The Role of Ultraviolet Curing in PC Board Assembly Automation, *Electrionics,* June 1986.
23. Schick, S.: Ultraviolet Curing: A New Path to Productivity, *Proceedings of the International Society of Hybrid Microelectronics,* 1986.
24. Pappas, S. P.: "UV Curing: Science and Technology," Vol 1, Technology Marketing Corporation, Connecticut, 1978.
25. Koleske, J. V., and Austin, T. M.: A New UV-Curable Conformal Coating and an Accelerated Moisture Resistance Test Method, Presented at the IPC 29th Annual Meeting, 1986.
26. Morgan, C. R., et al: UV-Curable Conformal Coatings, *Polymers in Electronics,* ACS, Washington, D.C., 1986.
27. Schick, S.: Portable Ultraviolet System Provides High Intensity Cures, *Adhesives Age,* April 1986.

28. White, P.: Thin Film Dielectrics, *Insulation,* September, 1963.
29. White, M.: Vacuum Evaporation of Polythene, *Vacuum,* **15:** 449 (1965).
30. *Proc. Seventh Conf. Soc. Vacuum Coaters,* pp. 78–81, 1964.
31. Melville, H. W.: *Proc. Roy. Soc. (London),* **A163:** 511 (1937); Melville, H. W.: *Proc. Roy. Soc. (London),* **A167:** 99 (1938).
32. Gregor, L. V., and H. L. McGee: Electrical Insulation Conference, *Natl. Acad. Sci./Natl. Res. Council Publ.* 1680, 1963.
33. White, P.: *Electrochem. Technol.,* **4:** 468 (1966).
34. "Thin Film Polymers," G. E. Public Information Unit, Schenectady, N.Y., July, 1967.
35. Wright, A. N.: Surface Photopolymerization of Vinyl and Diene Monomers, *Nature,* **215** (5104): 953–955 (1967).
36. White, P.: *Proc. Chem. Soc.,* 337 (1961).
37. Caswell, H. L.: Fabrication of Thin Film Insulation, Electrical Insulation Conference, *Natl. Acad. Sci./Natl. Res. Council Publ.* 1238, 1965.
38. Poole, K. M.: *Proc. Phys. Soc. (London),* vol. B66542 (1953).
39. Ennos, A. E.: *Brit. J. Appl. Phys.,* **5:** 27 (1954).
40. Buck, D. A., and K. R. Shoulders: *Proc. Eastern Joint Computer Conf., Philadephia,* 1958.
41. White, P.: Preparation and Properties of Dielectric Layers Formed by Surface Irradiation Techniques, *Insulation,* May, 1967.
42. Muehlberg, P. E., et al.: Process for Making Tetrafluoroethylene Polymers, U.S. Patent 3170858, February, 1965.
43. Davison, W. H. T.: Irradiation of Materials, Australian Patent 248844, January, 1964.
44. Radiation Dynamics: Radiation Curing of Substrates, *Prod. Finishing,* vol. 20, no. 5 (May, 1967).
45. Schimtz, J. V., and E. J. Lawton: Process for Applying Protective Coatings by Means of High Energy Electrons, U.S. Patent 2900277, August, 1959.
46. Numata, K., et al: Characteristics of Ionized Cluster Deposition of Polyethylene Films, *Proceedings of the ACS Division of Polymeric Materials: Science and Engineering,* Vol 56, 1987.
47. D'Agostino, R.: Plasma Polymerization of Fluorocarbons, *ibid.*
48. Ratner, R. D. and Y. Haque: Properties of the Plasma and Plasma-Deposited Films of Perfluoropropane and Ethylene Oxide, *ibid.*
49. Claude, R. and M. Moisan: Comparison of Microwave and Lower Frequency Discharges for Plasma Polymerization, *ibid.*
50. Amouroux, J., et al: Comparison Between Atmospheric Pressure and Low Pressure Plasma Polymerization, *ibid.*
51. Yeh, Y.-S., et al: Influence of Reactor Design Factor on the Deposition of Plasma Polymerization, *ibid.*
52. Wertheimer, M. H. and J. Moisan: *J. Vac. Sci. Technol.,* A3(6), 2643 (1985).
53. Caswell, H. L.: Fabrication of Thin Film Insulation, Electrical Insulation Conference, *Natl. Acad. Sci./Natl. Res. Council Publ.* 1238, 1965.
54. Sadhir, R. K., et al: Deposition and Characterization of Plasma Polymerized Films of Hexafluorobenzene and Ammonia, *Proceedings of the ACS Division of Polymeric Materials: Science and Engineering,* Vol 56, 1987.
55. Bradley, A., and J. Hammes: Electrical Properties of Thin Organic Films, *J. Electrochem. Soc.,* January, 1963.
56. Smolinsky, G., and J. H. Heiss: Thin Organic Films Formed by a Glow Discharge Technique, *Proc. ACS,* vol. 28, no. 1 (April, 1968).
57. Goodman, J.: *J. Polymer Sci.,* **44:** 551 (1960).
58. Weisz, P. B., R. D. Pittman, and R. D. Goodwin: Gaseous Polymerization by Electrical Discharge, U.S. Patent 2676145, April, 1954.
59. Bamford, C. H., et. al.: Uses of Electric Discharge in Polymer Science, *Soc. Chem. Ind. Monograph* 17, 1963.

60. Otozai, K.: *Bull. Chem. Soc. (Japan),* **27:** 476 (1954).
61. Coleman, J. H.: Polymerizing Method and Apparatus, U.S. Patent 3068510, December, 1959.
62. Coleman, J. H.: Polymerizing Method and Apparatus for Curing the Same, U.S. Patent 3069283, February, 1959.
63. Redmond, J. P., and A. F. Pitas: Investigation of Polymeric Coatings by Electrodeless Discharge, *Final Rept. NASA Contract* NASW-1625, May, 1968.
64. Hudis, M.: "Techniques and Applications of Plasma Chemistry," Wiley, New York, 1974, p. 113.
65. Klausner, M., et al: Surface Modification of Polymers with Fluorine Containing Plasma: Deposition vs. Replacement Reactions, *Proceedings of the ACS Division of Polymeric Materials: Science and Engineering,* Vol 56, 1987.
66. Gombotz, W. R. and A. S. Hoffman: Functionalization of Polymeric Films by Plasma Polymerization of Allyl Alcohol and Allylamine, *ibid.*
67. Khimchenko, Yu. I., et al.: "Effect of Gamma Irradiation on the Structure of Epoxy Resin and Organometallic Polymers Based on Epoxy Resin," Academy of Sciences U.S.S.R., Institute of General and Inorganic Chemistry, translation by J. Kourilo, Aerospace Technology Division, Library of Congress, June, 1964.
68. Wheelock, C. E.: Solid Resins from Irradiated Epoxidized Liquid Conjugated Diene Polymers, U.S. Patent 2959531, November, 1960.
69. Nakagawa, K., and K. Manaka: Antistatic Treatment of Teflon Films by Radiation Induced Graft Polymerization, Genshiryoku Kogyo **11**(1): 47–50 (1965) (Japan); *Chem. Abstracts,* 95733r, 1967.
70. DasGupta, S.: The Use of Cobalt-60 for the Production of Industrially Important Materials, *Canadian Nuclear Assoc. Conf., Montreal,* May 27–29, 1963.
71. Whitehead, S.: "Dielectric Breakdown of Solids," Clarendon, New York, 1951: O'Dwyer, J. J.: "Theory of Dielectric Breakdown of Solids," Clarendon, New York, 1964.
72. Gregor, L. V.: New Materials in Electronics: Polymer Dielectric Films, *IBM Rept.* MP22-0116, June, 1967.
73. Holm, R.: *J. Appl. Phys.,* **22:** 569 (1951).
74. Frenkel, J.: *J. Phys. Rev.,* **54:** 647 (1938).

6

Circuit Coatings

CIRCUIT-BOARD COATINGS

Types of Coatings

The coatings most commonly used for circuit-board insulation and protection are the acrylics, polyurethanes, epoxies, silicones, and polyxylylenes. The choice among them depends largely on the operating and storage requirements of the circuit boards. For high-temperature applications, for example, only silicones should be used, whereas for reworkability polyurethanes are outstanding, and epoxies are preferable for adhesion and resistance to moisture. The most widely used circuit-board coatings are covered by military specification Mil-I-46058, which defines four types as follows:

Type AR	Acrylic
Type ER	Epoxy
Type UR	Polyurethane
Type SR	Silicone
Type XY	Polyxylylene

Types ER and UR are general-purpose coatings. Type SR is used for applications requiring resistance to high temperatures, and Type XY is used when a uniform, pinhole-free coating is necessary.

Many excellent circuit-board formulations are on the market, some of which have been qualified to one of these classifications. Several of these and their main characteristics are listed in Table 6-1. Some companies, however, have found it necessary to develop their own coating formulations and to generate specifications which go beyond the requirements of the military specifications.

Table 6-1: Typical Coatings Designed for Circuit-Board Protection

Trade name	Chemical type	Characteristics
Conap Conathane CE-1170	Acrylic	One component; cures in 1 hour at 60 C; listed on QPL of MIL-I-46058C.
Conap Conathane CE-1171	Acrylic	One component; cures in 1 hour at 60 C; higher viscosity than CE-1170; listed on QPL of MIL-I-46058C.
Humiseal 1B31	Acrylic	One component; easily repairable and flexible; fluoresces under UV; qualified to MIL-I-46058C.
Humiseal 1B73	Acrylic	One component; fast air drying with high resistance to tackiness at elevated temperature; fluoresces under UV light; available in aerosol spray cans; qualified to MIL-I-46058C.
Miller-Stephenson MS-475	Acrylic	Good abrasion resistance; provides insulation and moisture resistance; solderable for ease or repair; removed by solvent systems; qualified to MIL-I-46058.
Tech Spray 1710 Fine-L-Kote AR	Acrylic	High surface gloss transparent coating; cures in 24 hours at room temperature; available in aerosol or applied by dip, brush or spray; meets MIL-I-46058C.
Hysol PC20 STD	Acrylic	A one-component solvent-based clear coating for continuous operation up to 110 C; easily removed with hot soldering iron, M version conforms to MIL-I-46058C.
Furane Epocast	Epoxy	A 100-percent-solids; low temperature cure; meets MIL-I-46058C requirements.
Humiseal 2A53	Epoxy	Two component, modified epoxy system; good resistance to solvents and chemicals; contains fluorescent tracer; qualified to MIL-I-46058C.
Hysol PC12-007	Epoxy	A 100-percent-solids epoxy coating suitable for operation up to 125 C; widely used for commercial printed circuits; M version conforms to MIL-I-46058C, Type ER.
Hysol PC17 STD	Epoxy	Solvent based; may be air dried or heat cured; may be used as a varnish or as a primer to improve the adhesion of subsequent encapsulants; maximum continuous operating temperature is 125 C; M version conforms to MIL-I-46058C, type ER.

(continued)

Table 6-1: (continued)

Trade name	Chemical type	Characteristics
Hysol PC16 STD	Epoxy	A 100-percent-solids coating similar to PC12-007, except modified to increase film thickness, shelf stability, and appearance; may be applied with vapor-spray; M version conforms to MIL-I-46058C, Type ER.
Emerson & Cuming Uniset 2081-31B	Epoxy	Colorless, one-component, solvent-based coating; good resistance to acid, base, solvents, salt spray and detergents; meets MIL-I-46058C requirements, type ER.
Insulcast 270	Epoxy	Two-component, 100 % solids, general purpose coating suitable for continuous operation up to 125 C; conforms to MIL-I-46058C.
Insulcast 275	Epoxy	Two-component, thixotropic dip compound; purpose coating suitable for continuous operation up to 175 C.
Conap Conathane CE-1164	Polyurethane (moisture-cure)	One component; cures in 4 hours at 60 C; listed on Qualified Products List (QPL) of MIL-I-46058C.
Conap Conathane CE-1166	Polyurethane (moisture-cure)	One component; cures in 5 hours at 60 C; listed on QPL of MIL-I-46058C.
Conap Conathane CE-1155	Polyurethane	Two component; cures in 3 hour at 60 C; very low viscosity; listed on QPL of MIL-I-46058C.
Conap Conathane CE-1165	Polyurethane	Two component; cures in 5 hour at 60 C; very low viscosity; listed on QPL of MIL-I-46058C.
Humiseal 1A20	Polyurethane (moisture-cure)	One component; resistant to solvents and chemicals; contains fluorescent tracer; qualified to MIL-I-46058C.
Humiseal 2A64	Polyurethane	Two component solvent based; high solids, low viscosity for easy application; contains fluorescent tracer, meets MIL-I-46058 requirements.
Furane Uralane 8267	Polyurethane	One component; room temperature or low temperature curing for high volume PC board assemblies.
Furane Uralane 5750-A/B	Polyurethane	Two component; soft, repairable; low outgassing; meets MIL-I-46058C requirements.
Miller-Stephenson MS-470	Polyurethane	A durable, solvent resistant coating; solderable for ease of repair; qualified to MIL-I-46058.

(continued)

Table 6-1: (continued)

Trade name	Chemical type	Characteristics
Hysol PC15 STD	Polyurethane	A two-component coating that is flexible at room temperature and at low temperatures; easily repaired by soldering through; contains fluorescent dye for inspection.
Hysol PC18 STD	Polyurethane	A one-component solvent-based coating; may be air dried or cured at low temperature; may be repaired by soldering through; M version conforms to MIL-I-46058C, type UR.
PRC PR-1538	Polyurethane	A 100-percent-solids two-component coating which meets MIL-I-46058C, type UR; suitable over a temperature range of -55 to 150 C.
PRC PR-1568	Polyurethane	One-component, moisture-curing thin-film coating imparting good Q properties electrical system.
Emerson & Cuming Eccocoat LN 1461-96	Polyurethane	Two-component, solvent-based coating with good moisture and abrasion resistance; suitable at continuous operating temperatures up to 130 C.
Tech Spray 1711 Fine-L-Kote UR	Polyurethane	Clear flexible coating with good dielectric and chemical resistance properties; cures in 12 hours at room temperature; available in aerosol or applied by dip, brush or spray; meets MIL-I-46058C.
Union Carbide Parylene	Polyxylylene	One-component vapor-deposited coating (see Chap. 2 for properties).
Dow Corning 1-2577	Silicone	A 50-percent solution in xylene; air dries in 2 hours; removable with xylene.
Dow Corning 997	Silicone varnish	A 50-percent solution in xylene; cures in 4 hours at 100 C.
G.E. SS-4175	Silicone	A 30-percent-solids in xylene; cures quickly at 125 to 150 C.
Emerson & Cuming Amicon SC-920	Silicone	Highly repairable, medium viscosity coating that offers good resistance to moisture and thermal shock; optimized for high speed automated production.
Insulcast R.T.V.S. 16	Silicone	Room temperature vulcanizing, specially stabilized for high temperature operation to 260 C continuous, 315 C short term.
Tech Spray 1709 Fine-L-Kote SR	Silicone	Flexible coating; cures in 72 hours at room temperature; available in aerosol or applied by dip, brush or spray; meets MIL-I-46058C.

(continued)

Table 6-1: (continued)

Trade name	Chemical type	Characteristics
Tech Spray 1712 Fine-L-Kote HT	Silicone	High temperature resistant coating up to 350 C; cures in 12 hours at room temperature; available in aerosol or applied by dip, brush or spray; meets MIL-I-46058C.
Miller-Stephenson MS-460	Silicone	Solderable coating for ease of repair; good dielectric properties and moisture resistance; curing may be heat accelerated; qualified to MIL-I-46058.

Reliability

Printed-circuit boards are usually coated as the last step, after the components have been assembled and joined. Application of a conformal coating greatly improves the reliability of the entire assembly. This is particularly important for circuit boards used in military and space electronics, where the expense of a completely assembled board may be in the $5,000-to-$10,000 range and the storage or operational life expectancies may be as high as 10 years. For commercial applications, such as television and radio, organic coating of the circuit boards also provides longer service life. It has often been demonstrated often that unprotected electronic assemblies exposed to severe environments fail because of moisture penetration, degradation of insulation properties, electrical shorting, or corrosion. An example of the effects of moisture is depicted in Figure 6-1. On an uncoated circuit board there was excessive corrosion on components, solder, and conductor lines after only 2 days of exposure at 95 percent relative humidity.[1] A similar board coated with 1.5 mils of polyurethane showed no signs of corrosion and passed all required electrical tests after 6 months in 95-percent-relative humidity (Figure 6-2). Figure 6-3 shows the corrosive effects on an uncoated circuit board after 50 hr of exposure to salt spray and the protection afforded by a 1-mil-thick polyurethane coating under the same conditions. Thus for the required protection against both humidity and salt spray a coating or some type of encapsulant protection is mandatory.

To ensure maximum reliability of circuit boards and other electronic assemblies, defense contracts specify that hardware shall meet the stringent environmental tests called out in specifications such as MIL-E-5272, MIL-E-16400, and MIL-STD-202. In addition to humidity and temperature tests (Figure 6-4), these specifications define salt-spray, abrasion, impact, fungus, and other tests. Table 6-2 lists the tests defined in MIL-STD-202. Many of these are applicable to coatings alone or in conjunction with electronic components.

The use of conformal coatings for printed circuits allows the design engineer greater freedom in achieving narrower conductor lines and closer spacings. Without coatings, impurities, moisture, and other contaminants can bridge the conductors, causing decreases in the insulation resistance or arcing between conductors.

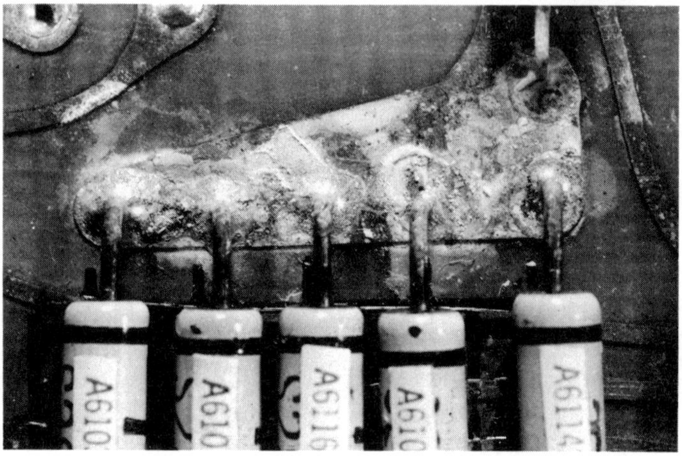

Figure 6-1: Uncoated circuit-board section showing extensive blistering and corrosion after only 2 days of humidity exposure.

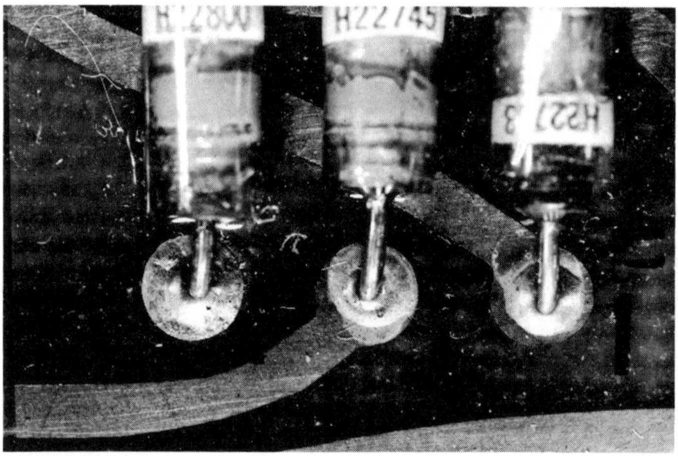

Figure 6-2: Polyurethane-coated circuit-board section after 6 months of humidity cycling.

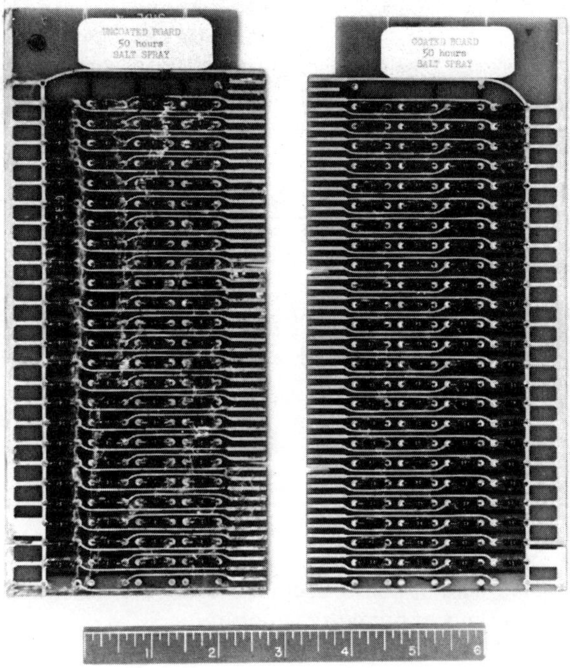

Figure 6-3: Effects of 50 hr of salt spray on uncoated circuit board (left) and coated circuit board (right). (Rockwell International Corporation)

The dielectric-breakdown voltage across a clean, dry surface is very high, but it is extremely difficult to maintain a surface in this condition. The increased electrical insulation afforded by an epoxy coating on a G-10 laminate board with 1/16-in. conductor spacings was shown by McGuiness.[2] Flashover voltage for the coated board was 5,000 volts, compared with 1,600 volts for an uncoated board, in tests at 27°C in 50 percent relative humidity. Allowable spacings between conductors for various applied voltages are given in Table 6-3. The reduction in spacings made possible by the use of conformal coatings is also apparent from this table. These data are specified in MIL-STD-275.

Coatings also serve other functions which contribute to the overall reliability of the assembly. Among these is the rigidity imparted to thin leads, solder joints, and components, which prevents their breaking or lifting during normal handling or vibration testing. Table 6-4 provides a sampling of this reduction of failures, where the failure mode was the breakage of resistor soldered leads, followed by the complete separation of the resistors from the board. These data show that although coatings are helpful, they should not be relied on completely to prevent vibration damage. Auxiliary filleting with thickened epoxies or polyurethanes is necessary, especially for large components.

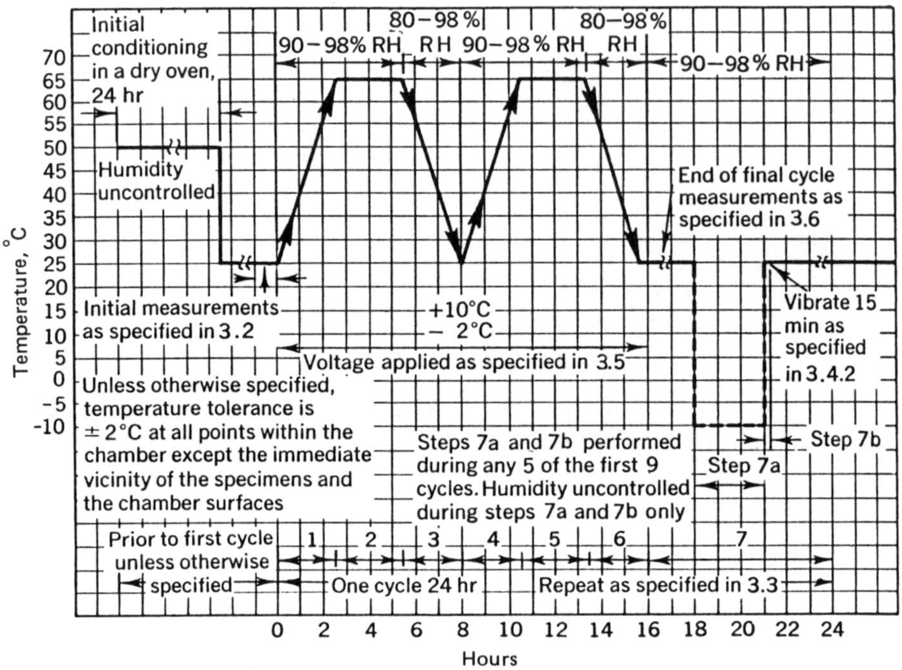

Figure 6-4: Graphical representation of moisture-resistance test, MIL-STD-202, Method 106.

Failure Due to Moisture

Several modes of failure can result from the absorption and permeation of moisture in circuit boards. A basic one is the marked decrease in the electrical insulating properties of circuit-board laminates when they are exposed to warm humid environments. Decreases of as much as nine decades in insulation resistance values have been reported for uncoated laminates. With a suitable protective coating, however, the decrease is only one to three decades under the same conditions (Table 6-5). Coatings afford similar protection to epoxy-glass laminates under the more severe stresses of combined temperature, humidity, and voltage (Table 6-6).

The sharp and rapid decrease in the insulation resistance of some coatings has been found to occur primarily during the first humidity cycle, with some recovery in subsequent cycles.[3] It has been postulated that coatings absorb water during the first humidity cycle, and that acceleration by an applied voltage causes hydrolysis of some of the coating constituents. This then results in a breakdown of the coating in the region of the "hot" conductors and causes the insulation resistance to drop off rapidly. Not all coatings undergo this initial sharp drop in resistance because of differences in their moisture-absorption and permeability characteristics (Figure 6-5).

Hydrolysis is not the only mechanism for electrolytic reactions. Coatings may contain ions or ionizable impurities that can migrate under an imposed potential. Hence a coating between two conductors may function as an electrolyte, causing a

Table 6-2: MIL-STD-202 Test Methods

Test	Method
Environment (100 Class)	
Salt spray (corrosion)	101D
Humidity (steady state)	103B
Immersion	104A
Barometric pressure	105C
Moisture resistance	106E
Thermal shock	107G
Life (at elevated ambient temperature)	108A
Explosion	109B
Sand and dust	110A
Flammability (external flame)	111A
Seal	112D
Physical characteristics (200 Class)	
Vibration	201A
Shock (specimens weighing not more than 4 lb)	202B
Shock (specified pulse)	213B
Random drop	203B
Vibration, high frequency	204D
Vibration, random	214A
Shock, medium impact	205E
Life (rotational)	206
Shock, high impact	207A
Solderability	208F
Resistance to solvents	215E
Electrical characteristics (300 Class)	
Dielectric withstanding voltage	301
Insulation resistance	302
DC resistance	303
Resistance-temperature characteristic	304
Capacitance	305
Q factor	306
Contact resistance	307
Current-noise test for fixed resistors	308
Voltage coefficient of resistance	309
Contact-chatter monitoring	310
Life, low level switching	311
Intermediate current switching	312

Table 6-3: Conductor Spacing (from MIL-STD-202)

Voltage between conductors DC or AC peak (volts)	Minimum spacing	
	Surface layers	Internal layers
0-100	0.005 inch (0.13 mm)	0.004 inch (0.10 mm)
101-300	0.015 inch (0.38 mm)	0.008 inch (0.20 mm)
301-500	0.030 inch (0.76 mm)	0.010 inch (0.253 mm)
Greater than 500 1/	0.00012 inch (.00305 mm) per volt	0.0001 inch (.0025 mm) per volt

1/ For reference only, voltages greater than 500 should be evaluated for the specific design application.

Table 6-4: Coating Variations versus Component Failures Under Vibrational Stress[1]

Condition	Input, g					Total failures
	5	10	15	20	30	
Uncoated	0	1	5	4	2	12
Polyurethane coated, no component filleting	0	1	0	0	4	5
Epoxy coated	0	0	0	9	1	10
Polyurethane coated, epoxy-polyamide filleting	0	0	0	0	0	0

Table 6-5: Humidity Effects on Insulation Resistance of Circuit Boards[4]

Sample	Insulation resistance, ohms		
	Initial	After cycle 5	After cycle 10
A			
Uncoated	1.5×10^{12}	3×10^{7}	5×10^{4}
Coated	1×10^{12}	3×10^{9}	2×10^{9}
B			
Uncoated	3×10^{12}	5×10^{7}	5×10^{3}
Coated	3.1×10^{12}	5×10^{11}	4×10^{11}
C			
Uncoated	3×10^{12}	5.5×10^{7}	1.5×10^{6}
Coated	3×10^{12}	3×10^{11}	4.1×10^{10}

local cell to form in which one of the conductors becomes oxidized and corroded.

In addition to minimizing the breakdown of electrical properties in laminates, coatings can also minimize or prevent discoloration caused by moisture and processing conditions. Some laminates, including epoxy glass, will whiten, or "measle," when they undergo many processing steps and are then subjected to a warm humid environment. Although this effect may not be structurally or electrically serious, it has resulted in the rejection of many circuit boards because of poor appearance. Coated, etched, copper-clad laminates which had not undergone processing were found to pass humidity cycling with no whitening. The use of a coating with good humidity-resistance properties was found to minimize or eliminate this defect.

The curing schedule for a coating has a significant effect on its moisture-barrier properties. In Table 6-7, boards A2, B2, and C2 offered the best resistance to humidity, and it was concluded that an extended cure at elevated temperature enhanced the performance of the coating. This improvement is probably due to elimination of residual solvents and further cross-linking of the polymer. The boards in group D showed no blistering or whitening. These results indicate that laminate

Table 6-6: Effects of Humidity, Temperature, and Voltage Stressing on Insulation Resistance of Coatings*[3]

Coating	Chemical type	Initial reading	Insulation resistance after exposure, ohms 1 day	3 days	8 days	10 days
Control	Uncoated epoxy substrate	10^{15}	10^4–10^5	10^4–10^5	10^4–10^6	10^4–10^6
3M EC1981	Fluorocarbon block copolymer	$>10^{14}$	2×10^9	5×10^8	5×10^8	1×10^9
3M L-4595	Fluorocarbon block copolymer	$>10^{14}$	1×10^{10}	5×10^7	1×10^8	1×10^8
3M L-4466	Fluorocarbon block copolymer	$>10^{14}$	2×10^{10}	5×10^8	1×10^9	2×10^9
Dow Corning DC-X9-6326	Silicone	$>10^{14}$	6×10^{10}	2×10^9	4×10^9	5×10^9
Humiseal 1B31	Acrylic	$>10^{14}$	2×10^{10}	8×10^7	8×10^8	8×10^8

*Coating applied to "Y" conductor pattern on GF epoxy laminate and tested according to MIL-I-46058C

whitening and blistering of the coating are intensified during the operations of mounting, fluxing, and soldering the components.

Coating Thickness and Coverage

Theoretically, the thicker the coating, the better its humidity-barrier properties, because the amount of moisture permeating a coating is inversely related to thickness. This holds true fairly well for the 100-percent-solids coatings. However, it is not always true in the case of solvent-based coatings because the probability that solvent volatiles will be entrapped in the cured coating is greater with thicker coatings. In a thinner coating these volatiles are more easily released. A thick coating with entrapped solvent molecules produces a more porous structure in which water can be more readily absorbed and transmitted, causing blistering, corrosion, and large decreases in insulation resistance values. However, solvent-based coatings may be formulated that contain various additives permitting rapid release of solvents on curing. Cure cycles can be optimized to achieve the same results. Thus, with optimum application techniques, both 100-percent-solids and solvent-based coatings will afford better protection against humidity if their thickness is increased.

Most circuit-board manufacturers use coatings 0.5 to 3 mils thick. Thicker coatings cause components such as glass diodes or glass-sealed resistors to crack. Cracking is attributed to stresses from shrinkage of the coating when the solvent evaporates, shrinkage from polymerization during curing, or large differences in the coefficients of expansion between the glass and the plastic coating. Cracking may occur soon after curing, or later during testing or rework operations,

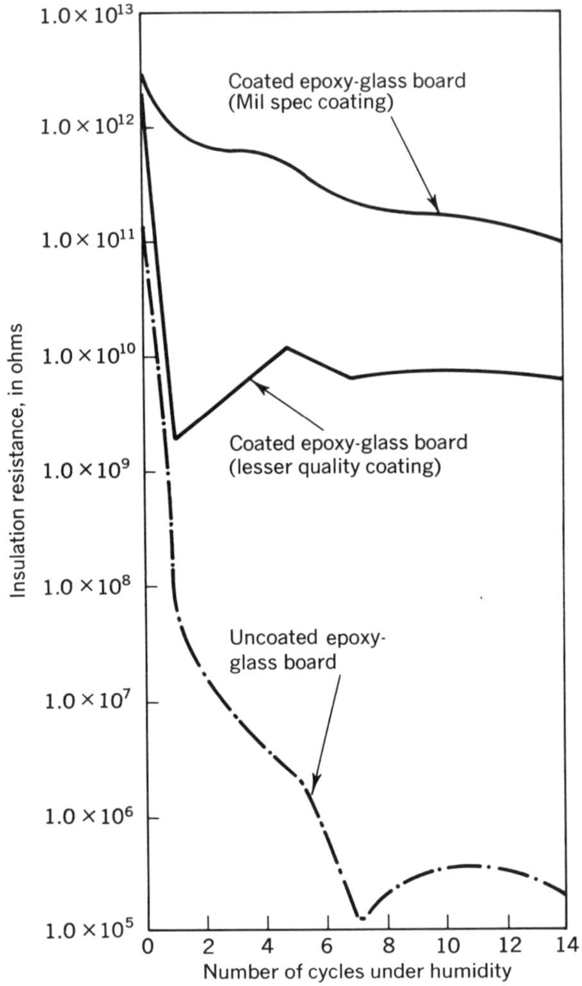

Figure 6-5: Insulation resistance of coated and uncoated circuit boards as a function of humidity cycling in accordance with MIL-I-46058 (3).

when additional stresses are imposed. For stress-sensitive components it is therefore important to avoid very thick coatings and, if filleting is used, to avoid bridging between components.

Most circuit-board manufacturers use protective coatings containing small amounts of fluorescent pigments to allow visual or ultraviolet-light inspection and to assure that all areas have been coated thoroughly. Two examples of fluorescent indicators used in polyurethane formulations are rhodamine B, which is pink to red in both the visual and ultraviolet regions of the spectrum, and 2,6-distyrylpyridine, which is colorless in the visual region but intensely blue in the ultraviolet region. Fluorescent or visual indicating pigments may be incorporated in the formulation in

Table 6-7: Effect of Coatings on Laminate Whitening*

Board No.	Dry Schedule After Detergent Wash	Cure Schedule	Appearance of Board After Humidity Test
High density of components			
A1	1 hr at 50°C	1 hr at 50°C	Heavy whitening of board material
A2	1 hr at 50°C	24 hr at 75°C	Tiny blisters noted, small amount of whitening of board material
A3	24 hr at 75°C	24 hr at 50°C	Tiny blisters noted, small amount of whitening of board material
A4	24 hr at 75°C	1 hr at 75°C	Heavy whitening of board material
Medium density of components			
B1	1 hr at 50°C	1 hr at 50°C	Heavy blistering
B2	1 hr at 50°C	24 hr at 75°C	Good appearance
B3	24 hr at 75°C	24 hr at 50°C	Heavy blistering
B4	24 hr at 75°C	1 hr at 75°C	Heavy blistering
Low density of components			
C2	1 hr at 50°C	24 hr at 75°C	Good appearance
C4	24 hr at 75°C	1 hr at 75°C	A small amount of whitening of board material on solder side
C5	1 hr at 50°C	24 hr at 50°C	Heavy whitening of board material
No components			
D1	1 hr at 50°C	1 hr at 50°C	Excellent appearance, no whitening, no blistering
D2	1 hr at 50°C	24 hr at 75°C	Excellent appearance, no whitening, no blistering
D3	24 hr at 75°C	24 hr at 50°C	Excellent appearance, no whitening, no blistering
D4	24 hr at 75°C	1 hr at 75°C	Excellent appearance, no whitening, no blistering
D5	1 hr at 50°C	4 hr at room temp.	Excellent appearance, no whitening, no blistering

*Courtesy of Autonetics, Division of Rockwell International Corporation.

small concentrations of 0.5 to 2 percent to give the desired results. Bare spots, pinholes, and other discontinuities can easily be detected, which facilitates quality control. Fluorescent indicators are also used to determine whether particles of coatings have migrated and contaminated other portions of a system. Particle migration is especially critical to the functioning of parts in the immediate vicinity of the circuit board, such as gyros, accelerometers, bearings, and rotating memory disks.

Reworkability of Coated Assemblies

Many commercial electronic modules are designed as throwaway items, and hence no coating or, at most, a relatively inexpensive varnish or polyester is employed. For more complex and expensive equipment special coatings are needed for long-term reliability. Because of the probability that one or more defective components or solder joints will have to be repaired, it is imperative that the module be reworkable. Hence, in addition to meeting the numerous engineering and manufacturing requirements, the coating must be easily removable so that defective components may be replaced. The removal technique must be one which does little or no damage to adjacent components, surfaces, and markings. Combining these desirable features in one material is, of course, a very difficult problem. Where ambient or service temperatures are less than 135°C, polyurethanes are popular because, being thermoplastic, they can be melted in localized areas with a hot soldering iron. Epoxy polyamides and epoxy amines can be softened and removed with a hot soldering iron, but this is much more difficult. If heat is applied for too long a period of time, decomposition and carbonization of the epoxy polymer result (Table 6-8). Such dark-colored residues not only give a poor appearance, but also may affect subsequent solderability. Epoxies cured with amines or with other catalysts behave similarly; many of them will char immediately, without going through a softening stage. Silicone-elastomeric coatings are soft enough to be removed easily with an X-acto knife. Because they are thermosetting, they will not soften or melt with heat, but will eventually decompose. Silicones and other thermosetting plastics may swell on prolonged contact with chlorinated or fluorinated solvents such as methylene chloride or trichloroethylene, but they do not truly dissolve.

In general, there are four problems encountered in removal with solvents:

1. Solvents do not dissolve the plastic. The plastic can absorb large amounts of solvent and swell; and the softened material may then be removed by mechanical means.
2. The solvent cannot be localized; usually the entire assembly must be immersed, which then presents the risk of damaging other areas.
3. Swelling of the plastic can generate high stresses which may then induce other types of failure.
4. The process is slow, requiring immersion for 1 or 2 days.

There have been some attempts to localize the solvent by preparing a thixotropic form of it with an additive such as Cab-O-Sil. Thixotropic compositions containing over 90 percent solvent will not flow under normal conditions and assume the

Table 6-8: Removability Characteristics of Circuit-Board Coatings

Coating	Composition Part A	Composition Part B	Pbw	Cure Schedule (hr/°C)	Removability with Hot Soldering Iron (315°C tip, 20-sec dwell)
Epoxy polyamide	Epon 828 Ketone	Versamid 115, Shell Curing Agent Z	50:50	3/65	Softens and becomes cheesy
Epoxy amine	Epon 1001	Diethylene triamine	60:15	3/65	Softens and becomes cheesy
Epoxy polyamide	Epon 1001	Versamid 115, Shell Curing Agent Z	100:30	3/65	Softens and becomes cheesy
Epoxy amine	Epon 828 Ketone	Versamid 115, Shell Curing Agent Z	–	2/80 + 2/150	Softens and becomes cheesy
Polyurethane	Mobay Mondur CB-60	Castor oil	–	3/65	Liquefies

Note: All coatings were formulated with suitable solvents for spray or dip applications.

consistency of solids. This method is also extremely slow, however, and presents the additional problem of solvent evaporation and the need for its frequent replenishment.

For thin organic coatings (< 1 mil) plasma etching (described in Chapter 5) is a useful removal technique. Plasma etching is the only method capable of removing solvent-resistant coatings such as polyxylylene derivatives. Because reactive plasmas cannot be localized the entire assembly must be exposed; however, plasma etching is a dry process so no residues are left behind. Thicker coatings can also be removed but the time required becomes prohibitive—several hours for coatings greater than 1 mil thick. Another disadvantage is that reactive plasmas are not selective and therefore care must be taken to prevent etching and damage to other plastic parts such as the printed circuit laminate, plastic-encapsulated devices, and markings. Plasma etching is ideal for removal of plastic coatings on ceramic circuit boards such as those used for hybrid microcircuits.

Multilayer-circuit Insulation

Another interesting use for organic coatings is in the fabrication of multilayer circuit boards. Epoxy coatings may be used in lieu of prepreg as the permanent insulation between circuit-board conductor layers, or as the insulation between conductors within the same layer, for leveling the surface. Coatings formulated with suitable thixotropic and wetting agents may be squeegeed through a screen, so that the interconnection hole areas are left open. A number of layers are superimposed, and each one is either partially or completely cured before the next layer is applied, reducing the chance of pinholes and other discontinuities. Insulation thicknesses of a fraction of a mil to more than 6 mils may be attained, depending on the screen mesh size and the number of applications. The epoxy insulation may then be sensitized and plated, and the next layer of conductor lines etched by the usual procedures. A four-layer multilayer circuit board prepared in this way is shown in Figure 6-6.

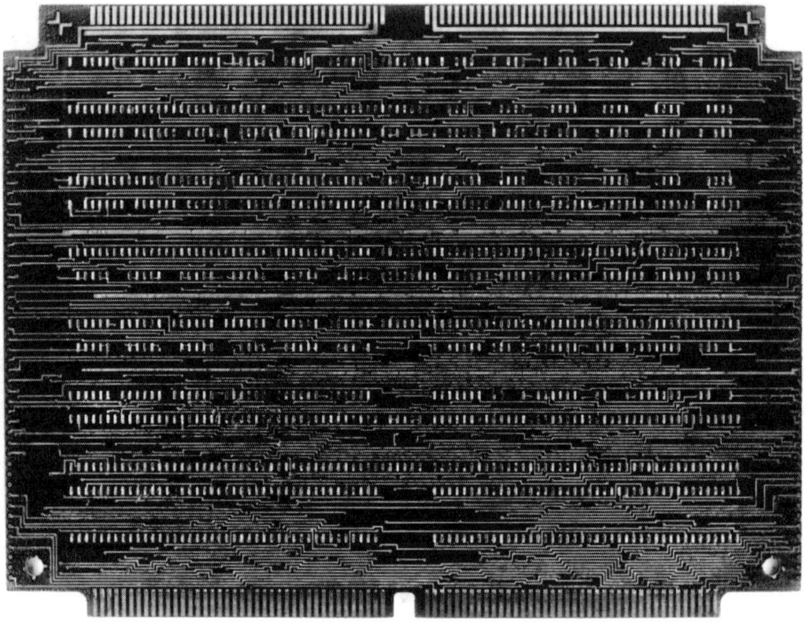

Figure 6-6: Four-layer multilayer circuit board prepared by epoxy screen-printing.

RESIST COATINGS

Photoresists

Organic materials are often used as temporary coatings in the fabrication and processing of printed circuits. Photosensitive coatings, referred to as *photoresists,* are used as maskants for etching and plating operations and constitute the most important category of temporary coatings. Depending on their chemical structure, they may serve as either negative or positive resists and may either polymerize (harden) or decompose into soluble constituents upon exposure to high energy radiation.

There are four lithographic processes used to activate resists:[5] (1) short wavelength (UV) photolithography (1–5 micron resolution), (2) electron beam lithography (<1 micron resolution), (3) x-ray lithography (0.02 micron resolution), (4) ion beam lithography (<1 micron resolution). The resolution of features is ultimately limited by the wavelength of the exposing radiation. This factor becomes critical in the fabrication of semiconductor devices where submicron geometries are required. However, for printed-circuit board technology with geometries on the order of mils, UV photolithography is the most commonly used process.

A resist that is hardened by ultraviolent light is highly resistant to etching or plating solutions, so that metal may be removed or deposited selectively. As an example, in fabricating printed-circuit boards from copper-clad plastic laminates, the photoresist is first applied to the entire copper-clad surface and allowed to dry.

The coated surface is then exposed to light through a photocopy master, a transparent pattern of the conductor circuit printed on a Mylar film. After exposure, the master is removed, and the board is immersed in a solvent that removes only the unexposed, unpolymerized photoresist coating, leaving bare copper in these areas which is then removed by immersion in an etchant solution. The exposed photoresist covering the rest of the copper surface is then removed with another solvent, leaving the desired pattern of copper circuitry on a plastic substrate. The steps for etching a simple circuit pattern by means of a negative-acting resist are given in Figure 6-7.

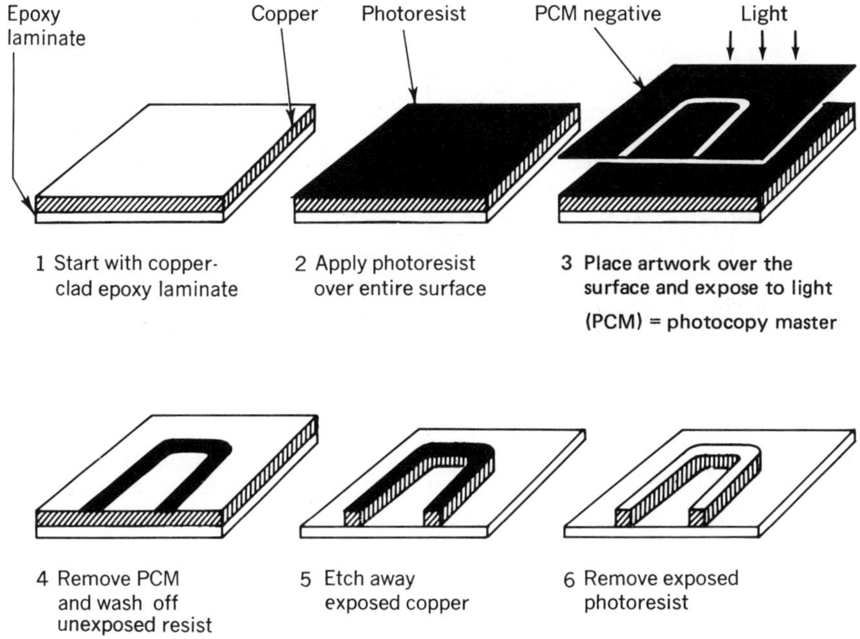

Figure 6-7: Photoetching steps for printed circuits.

Besides their use in printed-circuit fabrication, photoresists are essential in the manufacture of thin-film circuits, semiconductor devices, integrated circuits, and other microelectronic assemblies. The application dictates the type of photoresist to be used. For example, photoresists designed for large printed-circuit boards will not normally produce the very detailed micropatterns required for transistors, integrated circuits, and other miniaturized devices and circuits. For these applications specially developed and highly refined photoresists are available. Most photoresists either are covered by patents or their compositions are proprietary.

Negative Resists. Negative photoresists consist of a resin that is capable of polymerizing, thereby becoming insoluble to solvents, through reaction with a light sensitive compound.[6] It is not necessary for the resin itself to be photosensitive although many are. In addition, the resists contain solvents and additives to improve processing and stability.

In general, however, the negative-acting types contain compounds or polymers with ethylenic groups capable of cross-linking by a free-radical mechanism when they are sensitized by ultraviolet or high-intensity light. Examples of light-sensitive compounds are the cinnamic esters of polyvinyl alcohol or other hydroxy-containing compounds,[7] cinnamic derivatives of active hydrogen-containing compounds such as acetophenone (commonly referred to as chalcones),[8] polymeric quaternary ammonium salts,[9] the azide polymers,[10] and diazonium salt compounds.

Table 6-9: Photoresists and Their Properties

Trade name	*Manufacturer*	*Characteristics*
KPR	Kodak	A negative working resist having low viscosity (11.6-13.0 centistokes), applied by spraying, whirling or dipping. Used on printed circuits and for photoetching of copper.
KPR-3	Kodak	A negative working resist that is specifically designed to be used for single dip-coating operations and as an electroplating resist. Viscosity is 32-36 centistokes.
KPR-4	Kodak	A negative working resist designed for roller-coating application. Used on printed circuit boards and copper laminates. Viscosity of 435-515 centistokes.
KTFR	Kodak	A negative working high-resolution resist for photochemical etching of aluminum, copper, tool steel, and stainless steel. Viscosity is 540-618 centistokes.
Dynalith EPR-5000	Dynachem	A positive working high resolution resist designed to give 1 to 2 micron lines. Applied by spin-coating at a viscosity of 18 centipoises.
Dynalith OFPR 800 PG	Dynachem	A positive working resist in a non-ECA* based solvent system.
Dynalith OFPR-800	Dynachem	A positive working resist for general processing.

(continued)

Table 6-9: (continued)

Trade name	Manufacturer	Characteristics
Dynalith OMR series	Dynachem	Negative working resists for contact or projection printing.
Microposit 1300 series	Shipley	A positive working resist system for use in the fabrication of semiconductor devices, photo masks, thin film circuits, electrooptical displays, and other microelectronic applications. Viscosity ranges from 2 to 100 centistokes depending on the resist.
Microposit 1400 series	Shipley	A positive working resist system for use in the fabrication of semiconductors and photomasks.
Microposit 111 series	Shipley	A positive working resist system for use in the fabrication of semiconductors, thin films, and printed circuits.
Microposit 2400 series	Shipley	A positive working resist system for use in the fabrication of semiconductors.
Microposit 119S	Shipley	A positive working resist for roller coating applications.
Riston I series	Du Pont	A general purpose negative working dry film resist system, available in a number of thicknesses (1-3 mils). Solvent processable.
Riston II series	Du Pont	A general purpose negative working dry film resist system, available in a number of thicknesses (1-3 mils). Semi-aqueous processable.
Riston III series	Du Pont	A specialty purpose negative working dry film resist system for plating, etching and tenting processes. Available in a number of thicknesses (1-3 mils). Aqueous processable.
Vacrel	Du Pont	A negative working dry film resist system for solder masking and as a dielectric coating for printed circuit boards.

*Ethoxyethyl Acetate.

Positive Resists. Positive photoresists are also composed of resins, sensitizers, solvents, and additives.[6] However their chemistry is much different, involving the use of alkaline aqueous stripping and developing solutions rather than organic solvents. Upon exposure to light these resists change from solvent-soluble to water-soluble materials.

The most common sensitizers are derivatives of orthoquinone diazides. Unlike negative resist resins, positive resist resins contain saturated linear molecules which do not crosslink. Novolac resins are the most commonly used resin types. Often a copolymer is added to improve coating flexibility and adhesion; examples are polyvinyl ethers, styrenes, and melamine resins. Some commercially available photoresists and their characteristics are given in Table 6-9.

Generally, liquid photoresists are applied by spraying or spinning. Spraying under very controlled conditions in a laminar-flow clean-air enclosure has been found to be a reliable method for microelectronic applications. Other methods currently in use include dipping, and roller coating. In all methods, coatings may be applied in thicknesses from a fraction of a micron (for high-density microelectronic circuits) to over 0.1 mil for most printed-circuit-board details.

Dry Film Resists

Automated application techniques of dry film photoresists offer greater uniformity, better thickness control and greater reproducibility than spin or roller coating of liquid resists. Dry film resists are photosensitive polymer films sandwiched between sheets of polyester and polyolefin films for protection. The protective film is removed and the resist is applied to the substrate which can be flexible or rigid. Using automated equipment such as the cut-sheet laminator shown in Figure 6-8, the resist can be applied to one or both sides of a substrate and then roll or vacuum laminated. The dry film resist is processed like liquid resists: first the pattern is exposed, then developed and finally the unexposed resist is removed with a stripper. Du Pont, maker of the first solid film resist, offers four types of dry film resists: solvent processable, semiaqueous processable, aqueous processable, and a resist that can be both a solder mask and a dielectric coating (Table 6-9).

Screened Resists

Another method of producing printed circuits is with *screenable resists,* also referred to as *screenable inks.* Like the photoresists, they act as etching or plating maskants. The resist coating is applied to a metal-clad substrate board through a silk or metal screen, on which a circuit pattern has been defined by means of the well-known gelatin compositions.[11] The coating may be squeegeed through the screen either manually or by automatic screen printers, which are commercially available. After the resist has been applied, it is air-dried or cured at an elevated temperature, and the board is exposed to etching or plating solutions, depending on whether metal is to be removed or added. On completion of these processing steps, the resist is removed with solvents such as trichloroethylene or lacquer thinners. The screened-resist method for printing circuits is more direct than the photoresist method but is limited in its ability to produce fine-line definitions and spacings. The screened-resist method can safely yield line widths of 10 to 15 mils, whereas the photoresist method produces line widths of 5 mils or less, and line densities as high as 200 per millimeter have been achieved.[12]

Figure 6-8: Heated roll cut-sheet laminator for applying dry film photoresist to printed circuit boards. (Courtesy of DuPont)

To be suitable as plating resists, coatings must be stable to chemical plating baths normally employed in circuit-board manufacture, such as those for gold, copper, tin, nickel, or solder. To be effective as etching resists, coatings must be able to withstand the commonly employed etching solutions, such as ferric chloride or ammonium persulfate. Screened-resist coatings consist primarily of vinyl polymers, but acrylic, cellulosic, and asphaltic resins have also been employed.

Other Resists and Maskants

In assembling and processing electronic assemblies, resists or maskants may be needed to "stop off" certain areas from receiving conformal coating or solder. For example, in the conformal coating of printed-circuit boards, the edge conductor strip should be uncoated to allow electrical contact with an edge connector. For two-layer and multilayer circuit boards, the interconnecting through-holes must be kept free of contaminants. Contaminants in such areas can affect the reliability of subsequent steps, such as adhesion of plating and electrical continuity of interconnections.

In soldering, of course, solder splatter is undesirable, especially where bridging or shorting of closely spaced conductors might occur. To prevent solder wetting of such areas, solder maskants, also referred to as *solder stopoffs,* are employed. They are usually epoxy formulations because of the need to withstand exposure to solder temperatures of 250°C and higher. Maskants other than epoxies may blister, lift, outgas, or decompose.

Solder maskants are normally applied by screening, although they have also been applied by spraying through a stencil and by brushing. The main limitation of epoxy solder maskants is that once they have been applied and cured, repair is virtually impossible. A fully cured epoxy coating is strongly adherent and cannot be removed without damage to the substrate and circuitry.[13] The maskants useful for stopping off either conformal coatings or solder may thus take a number of forms:

1. Screenable coatings
2. Strippable dip or spray coatings
3. Adhesive-backed tapes
4. Molded rubber pressed into close contact by means of a clamp fixture

COATINGS FOR THIN- AND THICK-FILM CIRCUITS

Coatings, either alone or in conjunction with encapsulants, are being employed more frequently as a method of packaging both thin- and thick-film circuits. Great economies can be achieved over the normal packaging methods involving hermetically sealed metal cans. However, because of the close electrical tolerances to which thin- or thick-film elements such as resistors must be held, and because of the sensitivity of these elements to changes in ambient, hermetically sealed packages containing an inert gas are still mandatory for many applications. When organic coatings are planned, they should first be checked for compatibility with the resistors, capacitors, conductors, and other circuit elements. This may be established by:

1. Assessing changes in electrical values after coating.
2. Determining electrical values after environmental testing such as thermal cycling and humidity exposure.
3. Determining the degree of adhesion of the coating to the various substrates and surfaces comprising the circuit.
4. Checking for corrosiveness to metallic surfaces by long-term or accelerated aging.

Many thin-film resistors deposited on ceramic chips must maintain close tolerances (± 0.1 percent). Because most coatings effect changes greater than this, it is common practice to protect the resistors with special-composition glass or silicon monoxide. Then the circuit chip may be packaged in plastic by coating, transfer molding, or other suitable method.

Conductive, resistive, dielectric, and insulating coatings, also referred to as *inks, pastes,* or simply *compositions,* are employed in the fabrication of thick-film circuits. The pastes, ranging in viscosity from 20,000 to 250,000 centipoises, are of a thixotropic consistency, so that they can be squeegeed through a screen mask onto a ceramic or plastic substrate. They consist of organic vehicles in which are suspended fine particles of metal, metal oxides, or combinations of these, depending on whether conductive or resistive properties are desired. The use of a high-dielectric-constant filler will result in capacitive formulations. Pastes designed for

firing after screening also contain glass frit or ceramic as part of their composition and are often referred to as *cermets* or *cermet inks.* During the firing of the films, the organic portion is removed by pyrolysis, and the glass frit becomes fused, wetting the surface and providing adhesion and sealing of the metal or metal-oxide particles. Optimum adhesion is obtained by careful control of the temperature and time of firing. There are numerous books which discuss thick-film inks.[13]

These coatings become ceramics, once fired, and are not considered organic films. However, new pastes have been formulated in which the base is a polymer rather than a ceramic. These compositions are called polymer thick films.

Polymer Thick Films

Polymer thick films (PTFs), like cermet thick films, are screenable pastes which can form conductor, resistor, and dielectric circuit functions. However, unlike cermet films, PTFs contain polymeric resins which remain an integral part of the final thick film composition after processing. The key advantage of PTFs is that they can be processed at relatively low temperatures (120–165°C), namely at temperatures required to cure the resin. Cermet thick films, on the other hand, require temperatures of 850–1000°C not only to completely burn off all the organic binders but also to effect melting of the glass frit and sintering of the ceramic particles. A further advantage of PTFs is their low cost for both the material and processing. They have received wide acceptance in commercial products notably for flexible membrane switches, touch keyboards, automotive parts and telecommunications. Recently, there has been a resurgence of interest and activity in PTFs because of their application to surface mount devices resulting in replacement of the conventional double-sided plated through-hole boards with single-sided boards having no through-holes. Because of their low cure temperatures, PTFs may be processed on both ceramic and plastic substrates. The key advantage of using plastic substrates is that thick film resistors can be batch screened, avoiding the previous costs in attaching and solder connecting discrete chip resistors. To complement the PTF conductors and resistors, dielectric pastes are also available and are useful in fabricating multilayer interconnect substrates.

PTF Conductors. Polymer conductors are formulated with three major ingredients: a polymeric material, a functional material, and a solvent.

The polymeric material can be cured (hardened) at low temperatures and provides integrity to the film and adhesion to the substrate. A variety of polymeric binders may be used. For flexible printed circuits and membrane switches thermoplastic polymers such as acrylics, polyesters, or vinyl copolymers are used. For rigid substrates, the thermosetting polymers including epoxy, polyimide, and phenolic may be used.

The functional material is generally a metal that combines high electrical conductivity, good solderability, and low cost. The best compromise is silver. Silver PTFs are solderable using conventional tin-lead and tin-lead-silver solders but the contact time with the molten solder is critical. Extended contact time causes leaching of the silver and loss of adhesion. Some characteristics of a silver-based PTF are given in Table 6-10.

Commercially available conductor pastes are limited to silver compositions though a few silver-palladium pastes are also available. The amount of metal filler may range from 60 to 80% but 66 to 77% is typical. Cure temperatures range from

Table 6-10: Characteristics of PTF Conductors (Silver Based)*

Sheet Resistance	0.038 ohms/sq/mil
Adhesion	1,500 psi tensile on FR-4 epoxy board
Solderability	>95% coverage with RMA flux and 62/36/2 tin/lead/silver solder
Silver Migration	Less than for silver cermet conductors but may require polymer protective coating
Line Definition	7--10 mils when screened through a 200–250 mesh screen

*EMCA Silver PTF cured 30 minutes at 165°C.

120°C for 1 to 2 hours to 350°C for 30 minutes and may be performed in either convection or infrared furnaces. Curing in air is generally satisfactory except for some high temperature curing compositions where circuit materials such as copper may degrade through oxidation. Curing a polymer at temperatures above 250°C for extended times may also result in decomposition of the polymer and adverse changes in electrical and physical properties. In such cases, curing may be effected in a nitrogen ambient. PTF conductor pastes are generally screen printed through 200–250 mesh screens and provide line widths and spacings of about 7–10 mils. Cured thicknesses are 25–30 microns.

The solvent should have a high boiling temperature. It is used to dissolve the resin and control the rheological properties of the paste so that it is easy to screen. The solvent evaporates during the curing cycle.

PTF Resistors. Polymer thick film resistors are formulated similar to the conductors except that, instead of metals, resistive fillers are used as the functional component. Resistive fillers consist of carbon or graphite compositions. Resistor values may be varied and controlled by controlling the particle size and concentration of the filler, the type of polymer binder used, and the cure schedule.

Resistor pastes having sheet resistivities ranging from 10 ohms per square to 1 gigaohm per square are commercially available, however resistor stabilities are much better for the lower sheet resistance pastes. Reported TCRs are −100 to −200 ppm/C for the 100 ohm to 100K ohm resistor pastes, −500 ppm/C for the 1 Megohm pastes, and −800 ppm/C for the 10 Megohm pastes when measured at 125°C[14] (Table 6-11). Because of their low temperature curing properties, PTF resistors may be used on a wide variety of substrate materials including low cost plastic boards such as epoxy, polyimide, polysulfone, and phenolic. In contrast, cermet resistors can only be applied to ceramic or other high temperature stable substrates because of the high temperatures (>850°C) required to fire them. Resistance values for polymer thick films are much more sensitive to the nature of the substrate than resistance values for cermet resistors. Early applications of PTF resistors were plagued by large changes in resistance values that occurred during temperature cycling, elevated temperature burn-in, or aging. It was soon discovered that the glass transition temperature, Tg, of the substrate had a major influence on resistor stability. At and above the Tg the molecular structure of the plastic substrate relaxes and experiences a sharp increase in expansion coefficient. A thermal mismatch then occurs between the resistor and the substrate which degrades the resistor both physically and electrically. It is therefore important to select a substrate that has an expansion coefficient matching that of the resistors

Table 6-11: Characteristics of PTF Resistors*

TCR (25°–125°C)	±200 ppm/°C for 10, 100, 1K, 10K ohms/sq		
	±300 ppm/°C for 100K ohms/sq		
	±500 ppm/°C for 1 Megohm/sq		
	±800 ppm/°C for 10 Megohm/sq		
ΔR after cure	-5 to -15%		
ΔR after burn-in (85°C/10 hr)	-5 to -10%		
ΔR after burn-in and 85°C storage	1K	10K	100K
24 hr	0	0	0
100 hr	0	0	-0.5%
250 hr	0	0	-0.6%
ΔR after 1,000 hr at 85°C	<0.5%		
Power Rating	5 watts/sq in		

*EMCA Bulletin 515, 4000 Series Polymer Resistor Ink.

over the temperature range in which the circuit will be operated and tested. A Tg that is higher than any temperature to which the circuit will be subjected is desirable. Other substrate properties that have been found important in achieving reproducible and stable resistance values include low porosity, high surface smoothness, complete cure, and lack of moisture, contaminants or outgassing products in the plastic.

PTF Dielectrics. Several polymer thick-film dielectrics are commercially available and may be used as insulation layers in fabricating multilayer circuits or as overcoatings for resistors and conductors. Vias of 15–20 mils in diameter may be formed during screen-printing. The processing steps and conditions for dielectric pastes are very similar to those for conductors and resistors, for example, screen-printing through a 200 mesh stainless steel screen and curing for 30 min. at 150–175°C or for 10 min. if infrared heating is used. Volume resistivity is greater than 10^{12} ohm-cm and dielectric constants range from 4–8 at 1 KHz, depending on the composition. Breakdown voltages of greater than 1000 volts/mil have been reported.[15]

Conformal Coatings for Thick-film Circuits

Most of the criteria described as applicable to the selection of conformal coatings for etched printed-circuit boards also apply to thick-film circuit protection. Hence moisture resistance, stability of electrical insulating properties, adhesion, solder-through properties, and transparency are desirable attributes. Many coatings used for epoxy circuit boards, however, cannot be used for ceramic circuits because of differences in adhesion to the substrate. Organic coatings generally have poorer adhesion to ceramic than to epoxy substrates. Primers may be needed to achieve both adhesion and other desirable properties.

Polymer coatings or encapsulants used over thick-film circuits should also be chosen carefully for purity and stability. Outgassing or contaminants of a reducing nature, such as hydrogen gas, evolving from the polymer can lower resistance values, especially of the palladium—palladium oxide—silver inks. This is because the oxides in such compositions are reduced to free metal, which increases

electrical conductivity. Hence plastic overcoatings may have a damaging effect instead of the intended beneficial effect. Platinum-composition resistors are reported to be unaffected by hydrogen and less sensitive to impurities.[16] This is also true for ruthenium-based resistors.

Screened resistors are much more sensitive than screened conductors to stresses and to chemical interaction with overcoatings. Variations in resistance values after coating and after exposure to thermal, vibration, and humidity conditions should be determined experimentally for each resistor ink before it is used.

Particle Immobilizing Coatings. Though organic coatings are not generally used in hermetically sealed hybrid circuits, some NASA and military programs have required their use to assure that any particles that remain on the circuit after cleaning or that slough from the circuit after sealing do not migrate and cause electrical shorts. The function of the coating is to freeze in place all loose particles and to protect the circuit from any particles that may subsequently detach during vibration, constant acceleration testing, or actual use. Parylene coatings have received the greatest attention as particle immobilizing coatings for hybrid circuits. The Parylenes, a tradename of Union Carbide for highly purified p-polyxylylene polymers, are discussed in detail in Chapter 2.

Polymeric Dielectrics for High-Density, High-Speed Circuits

The semiconductor industry has achieved extremely high levels of integration and miniaturization in the development of LSI-VLSI-VHSIC devices. To reap the full benefits of these high-speed devices it is necessary to improve the next level of integration—the multichip substrate. The ultimate goal is *wafer-scale-integration* which is an entire circuit containing memories and microprocessors built into a single silicon wafer. However, in the late 1980s, the processing has been too complex and the yields too low to build such structures at a reasonable price.

In the interim, multilayer interconnect substrates can be fabricated using an inorganic substrate base (ceramic or silicon) with photodelineated metal conductors (aluminum, gold, or copper) and polymeric dielectrics for interlayer insulation. This technology is often called *hybrid-wafer-scale-integration, silicon-on-silicon,* or *silicon-integrated-substrate* (Figures 6-9 and 6-10). These structures have the following advantages over standard thick- or thin-film circuits: finer lines, shorter device-to-device signal paths, lower electrical resistance, shorter delay times, and better thermal conductivity (when silicon is used as a base).[17,18]

Silicon-integrated-substrate technology is one of the fastest growing areas in microelectronics. Much effort is being directed toward the study of polymers as dielectric materials for high-density, high-speed multilayer circuits.

In the late 1970s polymers were developed for use as interlayer insulation for semiconductor devices (Chapter 9). In the early 1980s many of the same materials and techniques used for fabricating electronic devices were extended to multichip substrates. The driving force for using polymers was the development of high- and very-high-speed integrated circuits whose performance is greatly affected by the electrical properties of the materials used to fabricate the substrate.

Candidate polymers must have a low dielectric constant (≤ 4), a low dissipation factor (≤ 0.005), and high electrical resistance ($\geq 10^{14}$ ohm-cm) over a wide frequency range (generally 25 MHz to GHz). The polymer must be able to withstand the high temperatures ($\geq 400°C$) often associated with metal deposition and must

Figure 6-9: A four-fold size reduction (6.3 in^2 vs 24 in^2) plus a three-fold weight reduction is achieved by replacing a printed circuit board covered with surface-mount devices with a silicon-on-silicon multilayer, multichip module. (Courtesy of Rockwell International Corporation)

Figure 6–10: High-Speed High-Density Multichip Data Memory Module; 2 x 4" substrate fabricated from low dielectric constant polyimide and multilayer thin film aluminum. (Courtesy Hughes Microelectronic Circuits Division, Newport Beach.)

be resistant to chemicals used for wet processing such as etching. Finally, fine-line metal conductors (≤1mil) must adhere to the polymer and well-shaped, small vias (≤1mil) must be easily made in the material.

Polyimide coatings are the most widely used dielectrics.[19-23] They combine high thermal stability (approximately 500°C) and chemical resistance with low dielectric constants (≤3) and high insulation resistance (≥10^{16} ohm-cm). Polyimides may be modified to suit a variety of applications, ranging from semiconductor device coatings to structural composites. They have great processing versatility and can be applied to a substrate by spinning, screen printing, or as a dry film. Some polyimides are photosensitive (see Chapters 2 and 9) allowing conductor lines and vias of a few microns to be fabricated.

However, the use of polyimides as interlayer dielectrics for large multichip interconnect substrates is not without some drawbacks. Most polyimides absorb 2 to 3% water which degrades their dielectric properties. Figure 6-11 shows the increase in capacitance of a photocured polyimide on exposure to 85% relative humidity at 85 °C.[24] The dielectric properties of polyimides also vary with frequency and can become electrically lossy for circuits operating at high speeds (> 1GHz)[25] The coefficients of thermal expansion of most polyimides range from 20 to 40 ppm/°C and presents a high mismatch with the CTE of silicon (3 ppm/°C). Solvent removal and curing also results in shrinkage of the dielectric. Thus as thicker

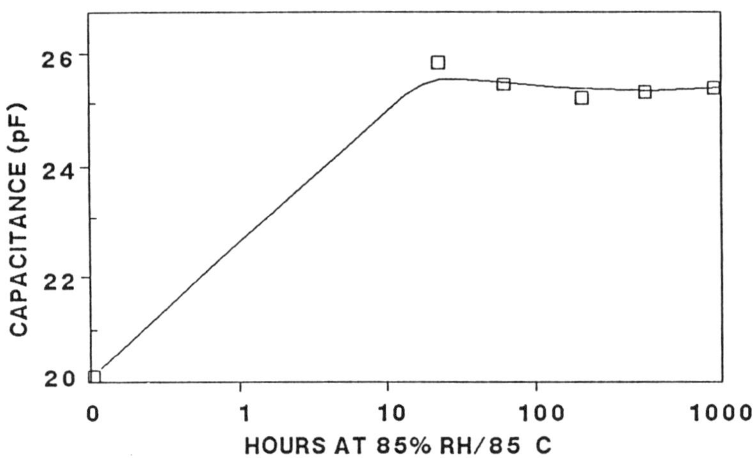

Figure 6-11: Degradation of the Dielectric Properties of Polyimide on Humidity/Temperature Exposure.[24]

polyimide layers are deposited on one side of the wafer to form the multilayer structure, bowing and stresses are imparted to the wafer. Fortunately, polyimide manufacturers have recently introduced highly densified polyimide formulations whose CTEs (3-5 ppm/°C) approximate that for silicon. These same formulations are also reported to have water absorptions of 0.5% and less.[26]

Besides polyimides, other polymers are being investigated. Parylene N (Chapter 2) has been suggested as a candidate because its dielectric constant (2.65) and dissipation factor (0.0002) are very low and remain nearly constant over a fre-

quency range of 60hz to 1Ghz. Vias can be formed in parylene using plasma or ion etching as with most dielectrics. Polybenzocyclobutane (PBCB)[27] and cycloaliphatic epoxies[28] are also being intensely evaluated as alternates to polyimide; both are reported to have attractive physical and electrical properties. Finally, there are several proprietary polymer formulations under study as insulating layers for multilayer interconnect substrates.[29,30]

REFERENCES

1. Etchason, P. T., and J. J. Licari: High Reliability Polyurethane Insulation Coating for Circuit Board Protection, *Insulation,* June, 1962.
2. McGuiness, E. W.: Improved Reliability for Printed Circuits with Protective Coatings, *Soc. Plastics Engrs. Tech. Paper* 7, January, 1961.
3. Weigand, B. L., et al: Manufacturing Technology for Conformal Coating, Technical Report AFWAL-TR-80-4139, September 1980.
4. Beccasio, A. J.: Conformal Coatings for Printed Circuit Assemblies, *Motorola Rept.* 2, DA 36-039-SC-89136, ASTIA Doc. AD 273080, November, 1961–January, 1962.
5. Thompson, L. F.; C. G. Wilson; M. J. Bowden, (eds): "Introduction to Microlithography," American Chemical Society, Washington, D.C., 1983.
6. DeForest, W. S.: "Photoresist," McGraw-Hill, New York, 1975.
7. Minsk, L. M.: Light Sensitive Unsaturated Esters of Polyvinyl Alcohol, U.S. Patent 2725372, November, 1955; Minsk, L. M.; Photosensitization of Polymeric Cinnamic Acid Esters, U.S. Patent 2670285, February, 1954.
8. Unruh, C., et al.: Vinyl Cinnamal Acetophenone Polymers, U.S. Patent 2716102, October, 1955.
9. Williams, J.: Photoprocess Using Light Sensitive Polymeric Quaternary Salts, U.S. Patent 2908667, October, 1959.
10. Merrill, S., et al.: Light Sensitive Compositions and Their Use in Photomechanical Products, U.S. Patent 2948610, October, 1960.
11. Eisler, P. (ed.): "The Technology of Printed Circuits," Academic Press, New York, 1959.
12. Coombs, C. F. (ed.): "Printed Circuits Handbook," McGraw-Hill, New York, 1967.
13. (a) Licari, J. J. and L. R. Enlow: "Hybrid Microcircuit Technology Handbook," Noyes Publications, New Jersey, 1988. (b) Rikoski, R. A.: "Hybrid Microelectronics Circuits; the Thick Film," John Wiley and Sons, New York, 1973. (c) Topfer, M. L.: "Thick-Film Microelectronics: Fabrication, Design and Applications," Van Nostrand Reinhold Company, New York, 1971.
14. Martin, F. W. and S. Shahbazi, Polymer Thick Film For Reliability Applications, *Hybrid Circuits Journal,* ISHM-UK, No.3, Autumn 1983.
15. *EMCA Bulletin* 531, 4552 Polymer Multilayer Dielectric
16. EPP Staff Report: Screenable Materials for Hybrid Circuits, *Electron. Packaging Prod.,* March, 1968.
17. Evans, A. A. and J. K. Hagge: Advanced Packaging Concepts-Microelectronics Multiple Chip Modules Utilizing Silicon Substrates, *Proceedings of First International SAMPE Electronics Conference,* June 1987.
18. Hagge, J. K.: Ultra Reliable Packaging for Silicon-On-Silicon WSI, *Proceedings of the IEEE/EIA Electronics Components Conference,* May 1988.
19. Takasago, H. et al: Advanced Copper/Polyimide Hybrid Technology, *Proceedings of the 36th Electronic Components Conference,* pp 481–487.
20. Homa, T. R. and A. J. Posocco: Reliability of Metallized Ceramic/Polyimide Substrates, *ibid,* pp 609–615.
21. Mittal, K. L.: "Polyimides: Synthesis, Characterization and Applications," vol. 1 and 2, Plenum Press, New York and London, 1984.

22. Shambrook, K. P. and Trask, P. A., High-Density Multichip Interconnect, *Proc. Electronic Components Conf.,* Houston TX, May 1989.
23. Licari, J. J., Fabrication and Packaging of High-Density Multichip Interconnect (HDMI) Substrates, *Proc. NAECON (National Aerospace and Electronics Conf.), Dayton OH, May 1989.*
24. *Chakravorty, K. K., Chien, C. P., Cech, J. M., and Tanielian, M. H., Hybrid Wafer Scale Integration Using Photosensitive Polyimide Dielectric and Electroplated Copper Conductor Lines, Proc. Third SAMPE Electronic Materials Conf.,* Los Angeles CA June 1989.
25. Wilson, A. M.: Use of Polyimides in VLSI Fabrication, *ibid,* pp 715–733.
26. DuPont Bulletin PI2611.
27. Kirchoff, R. A., et al: Benzocyclobutenes in Polymer Synthesis, *18th International SAMPE Technical Conference,* 1986.
28. Koleske, J. V. and T. M. Austin: A New UV-Curable Conformal Coating and an Accelerated Moisture Resistance Test Method, *Proceedings of the IPC 29th Annual Meeting,* April 1986.
29. Shiflett, C. C., et al: High-Density Multilayer Hybrid Circuits Made with Polymer Insulating Layers (Polyhic's), *Proceedings of the 1986 ISHM.*
30. Ohdaira, H. et al: A New Polymeric Multilayer Substrate, *Proceedings of the 1987 ISHM.*

7

Coatings for Space Electronics

Coatings used for space electronics, besides having to meet most ground and flight requirements, must also be able to meet an entirely new set of requirements:

1. Stability under high-vacuum and thermal-vacuum conditions
2. Low outgassing under thermal-vacuum conditions
3. Nontoxicity of outgassing products
4. Stability to the radiation of outer space
5. Stability to sterilization conditions such as thermal and ethylene-oxide exposure

Extensive work has been carried out in all these areas both by the National Aeronautics and Space Administration (NASA) and by many aerospace firms.

OUTGASSING

Plastic materials, because they are organic and usually of a heterogeneous nature, are more susceptible to outgassing than inorganic or metallic materials. They are inherently less stable at high temperatures and vacuum. The main mode of outgassing consists of the volatilization of additives such as plasticizers, flame retardants, diluents, and solvents normally used in compounding and formulation. Because these additives are of low molecular weight and are not molecularly linked with the main polymer chain, they are easily released under thermal-vacuum conditions. A second mode of outgassing is *sublimation,* or vaporization of a solid without going through the liquid state, followed by redeposition of the solid from the gaseous phase onto a cool surface. Still a third mode of outgassing is the actual decomposition by rupture of carbon bonds, resulting in the breakdown of the

polymer into smaller fractions having higher vapor pressure. Bonds that are susceptible to rupture may be represented by C—X where X may be another carbon (C), sulfur (S), nitrogen (N), oxygen (O), or hydrogen (H) atom.

Outgassing products from plastics are serious contaminants in spacecraft because of one or more of the following reasons:

1. They may be extremely flammable.
2. They may be noxious or toxic to astronauts.
3. They may interact with volatiles from other materials to yield even more toxic or flammable products.
4. They may be corrosive to metal surfaces or electronic devices.
5. They may deposit on electrical devices such as switches and relays, causing electrical opens.
6. They may deposit on slip rings, increasing noise levels and contact resistance.
7. They may deposit onto windows, optical lenses, or display units, decreasing their transparency and reflectance.

Although the deep-space vacuum is less than 10^{-12} torr, the usual screening test performed in engineering laboratories entails exposure at 10^{-6} to 10^{-8} torr, since these pressures are so much easier to achieve. An approximation of the space vacuum as a function of distance in space is given in Table 7-1.

Table 7-1: Gas Pressures as a Function of Distance in Space[1]

Altitude, miles	Pressure, torr	Molecules, atoms, or ions/cm^3
Sea level	760	2.5×10^{19}
20	10^2	4×10^{17}
125	10^{-6}	10^{10}
500	10^{-9}	10^6
4,000	10^{-13}	10^3
14,000	10^{-12}	10–100

A mathematical treatment of weight loss based on vapor-pressure data according to the Knudsen-Langmuir equation is useful for metals and inorganic materials[2] but is not generally applicable to plastic materials. In the case of plastics, it is necessary to resort to experimental data for each proprietary coating formulation. An attempt to collate and reduce data for comparative purposes has been made by Jaffe.[1] The temperatures required for an estimated weight loss of 10 percent/year are shown in Table 7-2, but designers are urged to use these data only as an estimated order of merit for the behavior of polymers in vacuum.

Table 7-2: Decomposition of Polymers in High Vacuum*

Polymer Type	Temperature for 10% Loss/Year in Vacuum (°C)
Polyamide, nylon	27–210
Polysulfide	38
Polymethacrylate	38–150
Polyester	38–240
Epoxy	38–240
Polyurethane	70–150
Polyvinyl chloride	90
Neoprene	93
Alkyd	93–150
Polymethylmethacrylate	105–200
Polyacrylonitrile	115
Styrene butadiene	130
Polystyrene	130–215
Phenolic	130–265
Polyvinyl alcohol	155
Polyvinyl acetate	160
Polypropylene	190–245
Silicone	205
Polystyrene, cross-linked	230–255
Polyvinyl fluoride	240
Polyethylene, low-density	240–280
Polybutadiene	255
Vinylidene fluoride hexafluoropropene	255
Polychlorotrifluoroethylene	255
Polyvinylidene fluoride	270
Poly-*p*-xylylene (parylene)	280
Polyethylene, high-density	295
Polytetrafluoroethylene (Teflon)	380
Methylphenyl silicone resin	>380

*For quality of data and references, see (1), from which these data were abstracted.

It is evident that generalizations cannot be made regarding epoxies, polyesters, and polyurethanes. Since there are literally thousands of formulations, experimental data are needed before any one is specified. Some polymer types, regardless of formulation, are inherently unstable in vacuum and should be avoided; among these are polyamides, acrylics, polysulfides, and neoprenes.

It is important to select coatings carefully to minimize outgassing. Even with very stable polymers such as epoxies or silicones, certain principles must be followed in selecting plastic that will release a minimum of volatiles. Solvent-based formulations should be avoided whenever possible, to eliminate the problem of poor solvent removal after application and curing of the coating. Several satisfactory solventless epoxies and polyurethanes have been developed and are approved by NASA.[3] Formulations containing plasticizers, additives with high vapor pressures, and other ingredients that are not chemically bound to the basic resin structure should also be avoided.

The amount of outgassing in polymer types that produce tightly cross-linked, rigid coatings is small and almost entirely limited to gases which are absorbed or adsorbed from the ambient, such as water or carbon dioxide.[4, 5] Coatings made flexible by the addition of plasticizers and elastomeric coatings, because of the

greater absorptive capacity and containment of incompletely polymerized volatile fractions, will normally outgas more than corresponding rigid systems.[6] Some inherently plasticized coatings are available. In these cases, the plasticizer is co-reacted with the resin or hardener and becomes an integral part of the structure, rather than a physical admixture. Such coatings will be flexible and still maintain low outgassing properties.

Several rules may be followed to reduce the percentage of volatiles in a coating by preconditioning either the liquid resin or the cured coating: In the case of solventless, (100-percent-solids) systems, volatile contaminants may be removed by evacuating the resin and hardeners separately at elevated temperature by means of a rotating evaporator. This relatively inexpensive equipment continuously exposes thin layers of the resin (thus achieving large surface area) to vacuum and greatly accelerates the removal of the volatiles. A diagram showing the principle of the rotoevaporator is given in Figure 7-1.

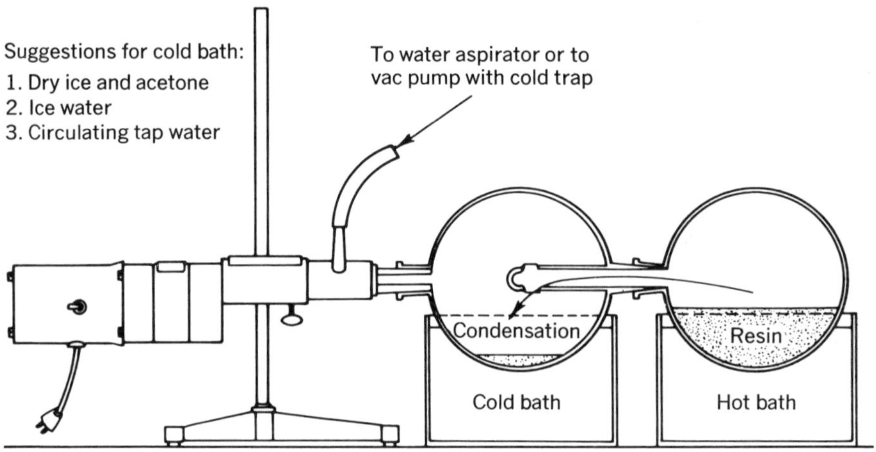

Figure 7-1: Diagram of Roto-Evaporator. (Rinco Instrument Co.)

For both solvent and solventless formulations, preconditioning of the coated assemblies before they are incorporated in an electronic system or subsystem will often render acceptable many materials that would otherwise be judged unacceptable. Preconditioning consists in postcuring the coating for a prolonged period or at elevated temperature in an open environment, or under vacuum. The effectiveness of this method has been demonstrated by Thorne, Whipple, and Boehm for silicones.[7] Postbaking a silicone insulating varnish at a high temperature and for a long period of time resulted in a 72 percent reduction in the amount of outgassing products (see Figures 7-2 and 7-3).

Requirements and Tests for Spacecraft Materials

The outgassing of volatiles from materials used in spacecraft, especially those used in the crew compartments of manned space vehicles, must be zero or extremely low. Outgassing products may consist of moisture, unreacted constitu-

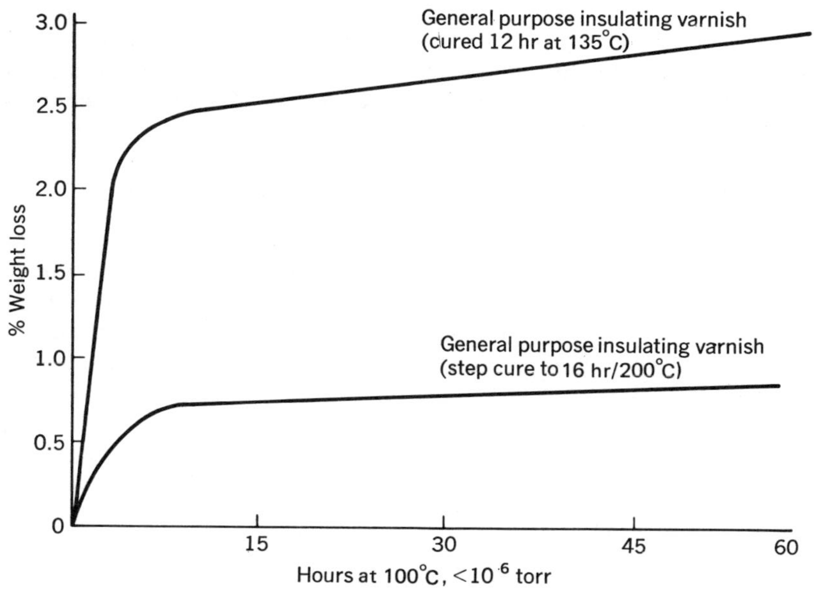

Figure 7-2: Vacuum performance of silicone varnish.[7]

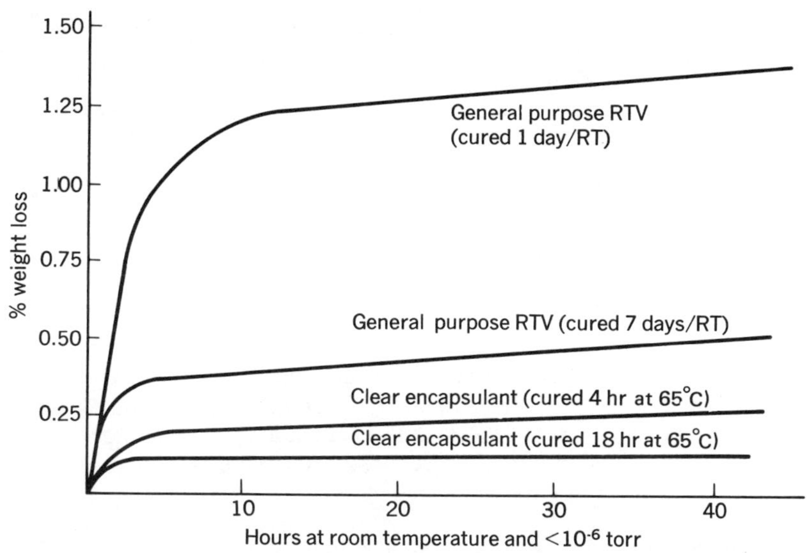

Figure 7-3: Vacuum performance of room-temperature-curing and low-temperature-curing silicones.[7]

ents, solvents, and low molecular weight polymers. These products may condense on critical surfaces such as mirrors, windows, lenses, and optical instruments or may be toxic to astronauts. Both quantitative and qualitative data for all plastics and organic materials used in spacecraft are therefore required. Ultra-sensitive and accurate analytical techniques such as gas chromatography and mass spectrometry are employed to fully characterize the outgassing constituents. Two test procedures are used to determine the outgassing characteristics of materials for use in NASA space flight programs. The maximum allowable outgassing levels for organic materials are defined in ASTM E-595, "Total Mass Loss and Collected Volatile Condensable Materials From Outgassing in a Vacuum Environment." This procedure is also detailed in Johnson Space Center Specification SP-R-0022A, "Vacuum Stability Requirements of Polymeric Material for Spacecraft Application," and is used by NASA's Johnson Space Center and Marshall Space Flight Center for qualification of nonmetallic materials. According to these specifications now used for testing almost all spacecraft materials, the cured material is subjected to 125°C at 5×10^{-5} Torr (or lower pressure) for 24 hours. The total mass loss (TML), less water vapor that is recovered, must not exceed 1% and the condensed volatiles (collected volatile condensable material, CVCM), must not exceed 0.1%.

The second test that must be run on all nonmetallic materials to qualify for use in manned spacecraft is the Toxic Offgassed Products Test and is described in NASA NHB 8060.1B, "Flammability, Odor, and Offgassing Requirements and Test Procedures for Materials." The material under test is placed in a chamber of known volume and subjected to 50° C for 72 hours. The atmosphere in the chamber is then sampled and analyzed qualitatively and quantitatively for offgas products. The test conditions, such as temperature, pressure and atmosphere are established for each program. The maximum allowable concentrations (MAC) are established by the NASA Toxicology Laboratory.

The degree of cure of a polymeric material is a key factor in whether it will pass the outgassing requirements. A material cured at a lower temperature than recommended or for a shorter period of time may fail, while cured at a higher temperature, longer period will permit it to pass. Hence, in the use of any material, the cure schedule is extremely important and must be specified. A list of polymeric materials and their cure schedules that pass the ASTM E-595 tests are given in E043-1784. Goddard Space Flight Center has compiled a list of over 2,600 polymers that have been tested.[8,9] The maximum allowable levels of outgassing for nonmetallic materials are also defined in NASA Handbook NHB 8060.1B.

In addition to the tests just described, extended thermal/vacuum testing is advisable when qualifying spacecraft materials. Most manufacturers and users of polymeric materials age their materials for extended periods at elevated temperature and high vacuum and measure any degradation of physical and electrical properties.[10]

Toxicity of Outgassing Products

In addition to the weight loss or percentage of volatiles, both the nature and toxicity of the outgassing products must be known. For both manned spacecraft and underwater vehicles, organic coatings must not release benzene, chlorinated solvents, or other toxic gases. The outgassing products of a coating may be high but consist essentially of water or other innocuous products. Other coatings may

release only trace amounts of gases, but these may be poisonous in concentrations as low as parts per million. Extensive toxicity data are available for gases and chemicals, but the amounts and nature of these constituents must be previously known.[11] Sophisticated analysis techniques such as gas chromatography/mass spectrometry are used to obtain this information. When there is no information, or when the synergistic effect of several gases is suspected, actual toxicity studies should be performed.

STERILIZATION

Sterilization Requirements

A key objective of the space program in the 1960s and early 1970s was to probe the surface of the moon and Mars for living organisms. The issue was how to assure that, in removing and analyzing samples, microorganisms in fact were not introduced from Earth-contaminated hardware. Thus considerable effort was spent on methods for sterilizing not only electronic subassemblies and assemblies but the entire spacecraft prior to leaving the Earth's atmosphere.

The sterilization of spacecraft for planetary missions, particularly the Mars mission (Viking), was considered important in obtaining an unequivocal understanding of the origin of life. The first planetary landings had to be isolated from contact with viable Earth organisms to prevent uncertainty about the results of experiments to determine whether any viable organisms existed on other planets. Sterilization and decontamination procedures imposed still another requirement on materials and components designed for space use. Materials could not deteriorate electrically or physically, within certain specified tolerances, after the sterilization process. To qualify for NASA approval, materials had to remain stable after ethylene oxide and thermal exposure the two key sterilization methods. Surface decontamination entailed subjecting the component to six separate cycles of ethylene oxide exposure. Each cycle entailed exposure to an ethylene oxide-gas mixture (12 percent ethylene oxide, 88 percent Freon 12) at 50 percent relative humidity for 30 hr at 50°C. Internal sterilization was accomplished by thermal exposure, an additional six cycles of 92 hr each at 135°C (Figure 7-4). Details of the dry-heat-sterilization and ethylene oxide-decontamination procedures are given by Kalfayan.[12]

It may be appropriate at this point to differentiate between "decontamination" and "sterilization," terms which are sometimes used interchangeably. *Decontamination* is reduction of the biological population to some number greater then zero, whereas *sterilization* is the attainment of a zero viable population. Sterilization is therefore defined as total destruction, and is not expressed in degree. However, since at the present time absolute sterilization cannot be demonstrated, the condition of sterility is actually a probability statement. A probability level of 10^{-4} was established as a requirement for the Mars mission. This means that the chance must be less than 1 in 10,000 that a single viable organism is still present.

Materials Problems

Plastics constitute the largest single class of insulating and protective materials used in electronic packaging. A typical electronic assembly may contain a wide

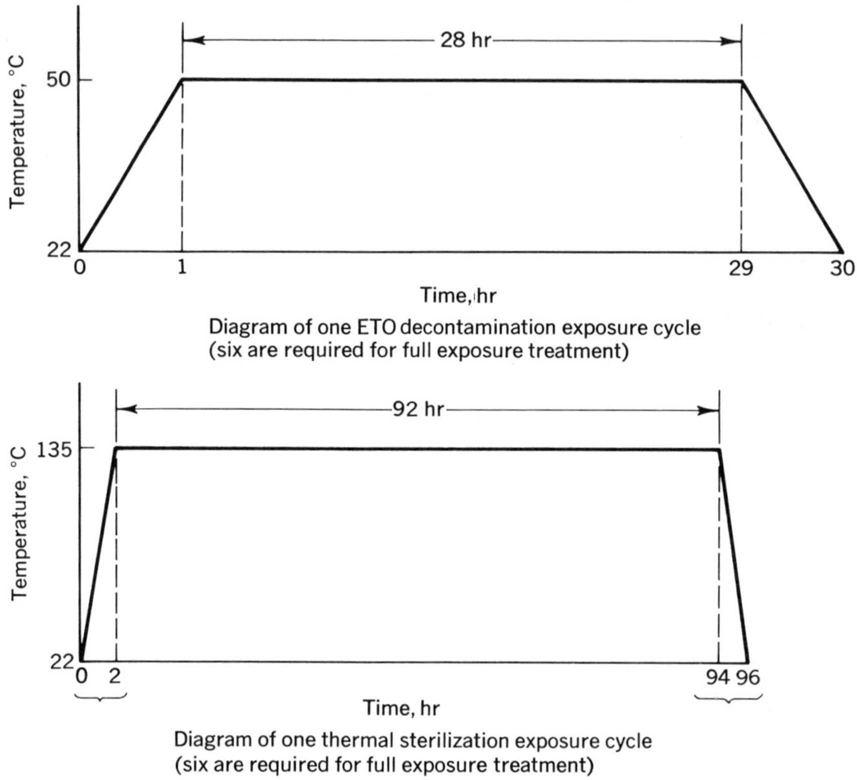

Figure 7-4: Ethylene oxide decontamination and thermal-sterilization cycles.

variety of polymeric materials in the form of coatings, adhesives, encapsulants and laminates. Although the polymer resin itself may be quite stable to the sterilization procedures, other ingredients of the formulation—plasticizers, catalysts, excess curing agents, ultraviolet stabilizers, pigments, colorants, fillers, and solvents—may not be. Although unaffected at ambient conditions, these additives may be subject to partial or complete decomposition at high temperatures, or may be reactive or soluble in the ethylene oxide-Freon mixture. A thorough understanding of the chemical-reaction mechanisms arising from exposure to the sterilization environments is therefore dependent on the conditions of exposure and on the polymeric structural properties. For example, the extent of reaction to heat or to ethylene oxide will depend on the degree of unsaturation in a polymer molecule, the temperature and time of exposure, and the presence of foreign substances in the plastic. Other materials which are thermally stable, such as silicones, may soften and swell because of their solubility in the Freon portion of the ethylene oxide mixture.

Thermal-sterilization Effects. Organic coatings deteriorate and discolor to varying degrees upon thermal sterilization. The thermal sterilization of several circuit-board assemblies (Figure 7-5) showed a number of failure modes, one of

which was the discoloration of the polyurethane conformal coating. Although this apparently did not affect the electrical functioning of the module (Table 7-3), it could affect the long-term reliability of the circuit board through corrosion or increased moisture permeation.[13] Coating types having low thermal stability, such as the polystyrenes, whose maximum service temperature is about 85°C, can also present serious decomposition and outgassing problems. Coatings such as the epoxies and polyimides, which have high thermal stability, have been found satisfactory. Kalfayan[12] evaluated several wire enamels and found that the polyimide enamels best withstood both ethylene oxide and thermal exposure.

Ethylene Oxide Effects. The effects of ethylene oxide on organic coatings are not as pronounced as the thermal effects. A few coating types, however, are seriously damaged by the Freon portion of the ethylene oxide mixture. These include some silicones, which swell considerably, and polysulfides, which will soften and flow.[14]

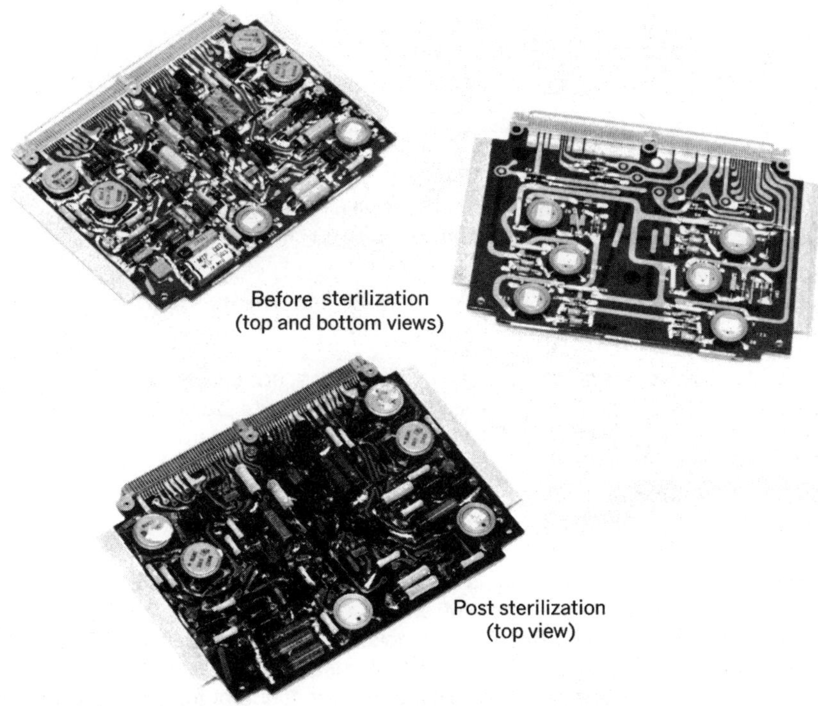

Figure 7-5: Effects of sterilization on circuit boards.

Table 7-3: Effects of Thermal Sterilization on Electronic Subassemblies[13]

Subassembly	System usage	Effects of thermal sterilization
Voltage regulator	Computer	Polyurethane conformal coating and filleting material discolored; tantalum wet electrolytic capacitors failed open; function within tolerance after six cycles; thermal conductive adhesive bond to power transistor separated.
Electronic power switch....	Flight control	Polyurethane conformal coating and filleting material discolored; insulation resistance increased; function within tolerance after 3,679-hr poststerilization operation.
Servo amplifier	Flight control	Polyurethane conformal coating and filleting material discolored; insulation resistance increased.
Power supply and regulator	Flight control	Polyurethane conformal coating and filleting material discolored; insulation resistance increased; 10 of 17 tantalum wet electrolytic capacitors failed open; one polystyrene dielectric capacitor charred; two zener diodes partially shorted.
Electronic switch and sync. signal demod.	Flight control	Visible cracking of potting material; intermittent electrical discontinuities; physical-stress failure of transistor; solder-lead separation; cracked printed-circuit connector.
Electronic switch-roll control	Flight control	No visible effects function within tolerance after six cycles.
Ref signal generator	Navigation and guidance	Discoloration of CPC chips and fiberglass circuit board; function within tolerance after six cycles.
400-hz amplifier..........	Navigation and guidance	Discoloration of CPC chips and fiberglass circuit boards; two of five tantalum wet electrolytic capacitors failed open, causing secondary failure of three transistors.

SPACE AND NUCLEAR RADIATION

In selecting coatings for space or military applications, such as planetary probes, manned orbiting laboratories, satellites, aircraft, and missiles, consideration should be given to the possible effects of solar, space and nuclear radiation. A comprehensive discussion of radiation effects on electronic systems is given by Messenger and Ash.[15] In particular, transient radiation effects on electronic systems is covered in this publication. Radiation effects on coatings are difficult to evaluate or predict because of the difficulty in reproducing the exact radiation environment, which consists of many types of radiation, fluxes, energy levels, and high vacuum (space) or high heat (nuclear), and because coating formulations are proprietary admixtures of many ingredients, each having a distinctive behavior. Accordingly, generalizations should not be made, except perhaps for pure unformulated polymer resins under controlled radiation test environments.

The Space-radiation Environment

A detailed description of the various elements comprising the space-radiation environment is given by Goetzel, Rittenhouse, and Singletary[16] and by McIlwain.[17] In general, space radiation may be classified as solar or penetrating. *Solar radiation* consists of x-rays, ultraviolet light, visible light, and infrared radiation, and *penetrating radiation* consists of cosmic, trapped, auroral, and solar-flare radiation. The penetrating portion consists largely of charged particles; for example, cosmic radiation consists primarily of protons (positively charged hydrogen nuclei), some alpha particles (positively charged helium nuclei), and other positively charged particles. Radiation consists of high-energy protons and electrons that have been trapped several hundred miles above the earth in its magnetic field. The most famous and thoroughly studied of these trapped radiation belts is the *Van Allen belt.* This belt contains ionizing radiation including protons with energies up to 800 Mev and electrons with energies up to 1 Mev. The Van Allen belt has further been described in terms of an inner and outer belt; estimates of proton and electron fluxes for each are given in Table 7-4.

Table 7-4: Estimates of Particle Flux in Van Allen Radiation Belts[18]

Electron flux	*Proton flux*
Inner belt	
99% between 10 and 600 kev	1% between 10 and 800 Mev
2×10^9 electrons/(cm^2)(sec), E > 20 kev	2×10^4 protons/(cm^2)(sec), $E > 40$ Mev
Outer belt	
$10^8 - 10^{11}$ electrons/(cm^2)(sec), $E > 20$ kev	
10^8 electrons/(cm^2)(sec), $E > 200$ kev	10^2 protons/(cm^2)(sec), $E > 60$ Mev
10^6 electrons/(cm^2)(sec), $E > 2.5$ Mev	

The Nuclear-radiation Environment

Explosion of a thermonuclear warhead releases energy in the form of x-rays, gamma rays, and neutrons. Gamma rays strip electrons from air molecules and generate an electromagnetic pulse (EMP). Also, intense heat (thermal radiation) and thermomechanical shock accompany a nuclear blast.[19] Each type of radiation produces several effects making it difficult to simulate the radiation environment in the laboratory.

It should be noted that for most space and military applications, sufficient shielding is designed into the equipment that coatings used in interior parts (memory coatings, circuit-board coatings, etc.) are not directly exposed to the full impact of radiation. Where shielding is used, secondary radiation becomes a more important factor. Charged particles, for example, may be stopped by the transfer of their energy by ionization, but the secondary gamma radiation that is generated, referred to as *Bremsstrahlung,* is much more difficult to stop. Uncharged particles such as x-rays and neutrons are also very penetrating and difficult to stop. They have no finite range, and a certain percentage is always transmitted through the material.

Effects of Radiation on Polymeric Materials

Many of the radiation types described do not possess sufficient energy per quantum to cause cleavage of chemical bonds, and in general, their interaction with organic polymers is negligible. In the case of visible and infrared radiation, the principal effect is an increase in thermal agitation, with little structural change. Cosmic radiation is also considered benign. The effective ionization dose rate due to primary cosmic radiation is very low, about 10^{-4} rad/hr, that due to secondary emission is about 10^{-3} rad/hr, and most organic materials do not deteriorate until well above 10^5 rads. However, both ultraviolet and x-rays are sufficiently energetic to rupture many chemical bonds, resulting in decomposition, molecular rearrangement, or reconstitution of the polymer structure.

The exposure of insulating coatings to high-energy ionizing radiation has both transient and permanent effects. *Transient effects* are those which eventually disappear after removal of the radiation field. *Permanent effects,* which are irreversible, and either degrade or improve the physical or electrical properties of the coating. There may also be *side effects,* in which decomposition products from the coating may corrode or damage other parts of the system. A knowledge of these effects is extremely important in microelectronics, where the construction materials are so intimately related to electronic functions and electrical performance.

Permanent Effects. The mechanism for the interaction of gamma rays and neutrons with atomic nuclei and their electron clouds is adequately described in the literature.[20,21] Irradiation with neutrons, electrons, gamma rays, or x-rays may have any one or more of the following effects on a plastic:

1. Free-radical formation
2. Ionization
3. Rupture of the C—C or C—H bonds in a plastic structure, giving rise to lower molecular-weight fractions
4. Further cross-linking and polymerization

Any modification in the structure resulting from such mechanisms is usually detrimental to the physical and electrical properties of the material. Following excitation and loss of a bonding electron, the molecule partially decomposes, giving rise to reactive species, either ions or radicals. Recoupling of fragments to form new polymer species and rearrangements of structure can result in very different physical and electrical properties.[22]

Because these mechanisms alter the carbon skeletal structure, they will have permanent, irreversible effects on the material. As an example, polyethylene becomes brittle or glassy on irradiation at 10^{10} rads. The mechanism is postulated to be as follows:

1. Free-radical formation with abstraction of a hydrogen atom
2. Abstraction of another hydrogen by the first hydrogen atom, forming hydrogen gas
3. Cross-linking by combination of carbon free-radical portions

Thus irradiation changes the flow properties of polyethylene to no flow above its melting point. Irradiation can introduce unsaturated or vinyl groups into the polyethylene structure, which then lowers its oxidation resistance.

Transient Effects. Besides the permanent effects of high radiation on organic insulating materials, transient effects can occur at both high and low radiation dosages. These effects disappear either immediately or slowly after the radiation source has been removed. Transient effects are usually the result of electron excitation and temporary ionization and will normally affect the electrical properties of the plastic. Electron excitation increases the number of conducting electrons and hence the electrical conductivity, a process similar to photoconductivity induced by light. However, generalizing from experimental data is difficult because the results are often nonreproducible. For example, results for polystyrene, the material most extensively studied, show variations of several orders of magnitude. This is not surprising, since variations in formulation, type and amount of filler or plasticizer, degree of cure, purity, and other factors have pronounced effects on the final properties.

Where coatings provide insulation between closely spaced conductors, as in flexible printed circuits, radiation-induced electrical changes can cause electrical shorting. An indication of how radiation can affect electrical properties is shown for Teflon 6 in Figure 7-6.[23] These data show that the dissipation factor increases sharply up to exposures of 1×10^6 rads. Figure 7-7 shows that the dissipation factor recovers, but in most cases this is a very slow process, requiring more than 10 days. Figure 7-8 shows the percentage increase in dielectric constant for Teflon as the absorbed radiation increases. It is apparent that the presence of oxygen influences the electrical properties of Teflon when irradiated above 4×10^4 rads. In the absence of oxygen, Teflon may be useful at dosages up to 10^7 rads. Low-frequency loss properties of Teflon are drastically affected by x-rays and it is anticipated that gamma radiation would show similar effects. Loss properties at high frequencies are less affected.

Side Effects. In addition to the radiation effects described above, there may be side effects on both components and materials caused by gases and other volatiles arising from the decomposition of the plastic. Gases commonly evolved from

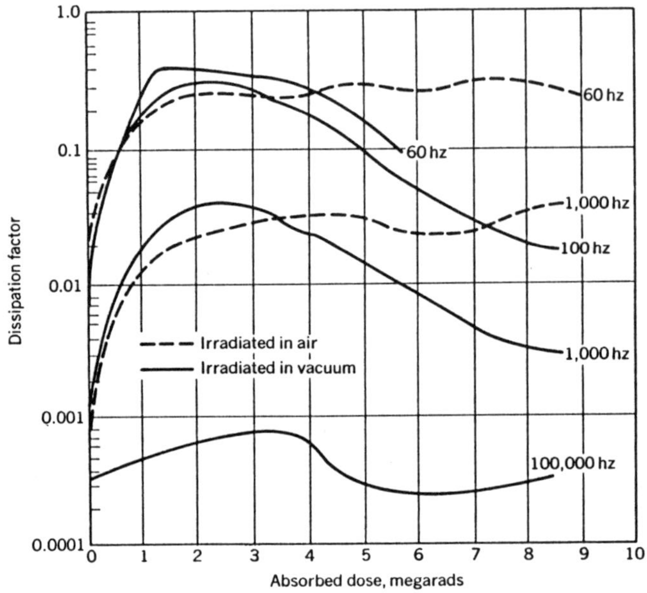

Figure 7-6: Effect of x-ray irradiation on Teflon TFE-6.[23]

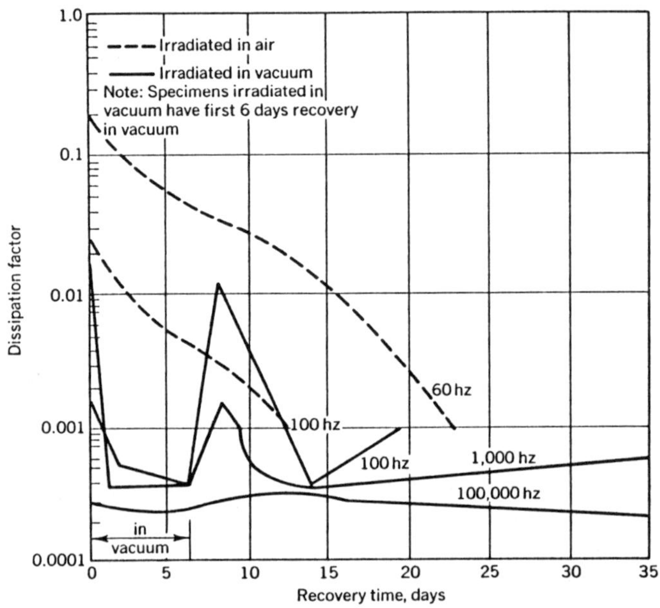

Figure 7-7: Recovery characteristics of Teflon TFE-6 after x-ray irradiation.[23]

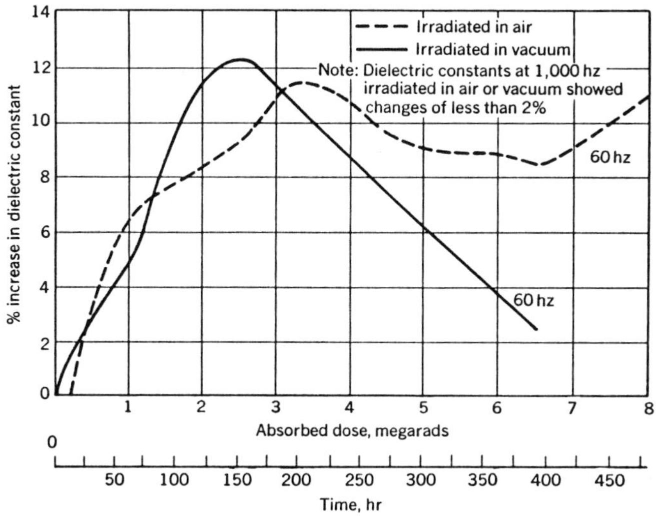

Figure 7-8: Effect of x-ray irradiation on dielectric constant of Teflon TFE-6.[23]

plastics include ammonia, water vapor, methane, low molecular weight hydrocarbons, carbon dioxide, hydrogen halide, and elemental halogen. Gas evolution can be particularly detrimental in the case of hermetically sealed electronic parts. A buildup of pressure within a sealed package can cause rupture of the seal or case.[23]

Outgassing becomes significant at radiation dosages of 10^6 rads and above. Very little data exist for values below 10^5 rads. However, it can be shown by extrapolation (Figure 7-9) that neoprene outgasses to the extent of about 7×10^{-4} ml/g at 10^4 rads, and Mylar film has very slight outgassing of 7×10^{-5} ml/g. Outgassing at dosages of less than 10^4 rads is therefore negligible and is not considered a problem. At higher dosages, the severity of outgassing of specific materials, such as those used in computers, must be determined experimentally.

Materials evaluated to date show a linear response in the amount of gas evolved to gamma dosage, as shown in Figure 7-9.[24] A comparison of gas evolution of various plastics at dosages of 3 to 7×10^8 rads is presented in Table 7-5. Several conclusions are evident from this table:

1. The more aromatic the plastic structure, the less gas is evolved. Aromatic structures of varying degrees are represented by items 2 to 5, 7, 9, and 10 in the table.
2. The nature of the filler has a pronounced effect on gas evolution. As indicated by comparison of the paper-filled and the asbestos- or graphite-filled phenol formaldehyde, paper is a poor material to use as a filler.
3. Polyvinyl formals, widely used as wire insulation, are among the worst gas evolvers and should be avoided for radiation-hardened systems. Polyimide Pyre M.L. wire insulation is reported to be very resistant because of its highly aromatic polymer structure.

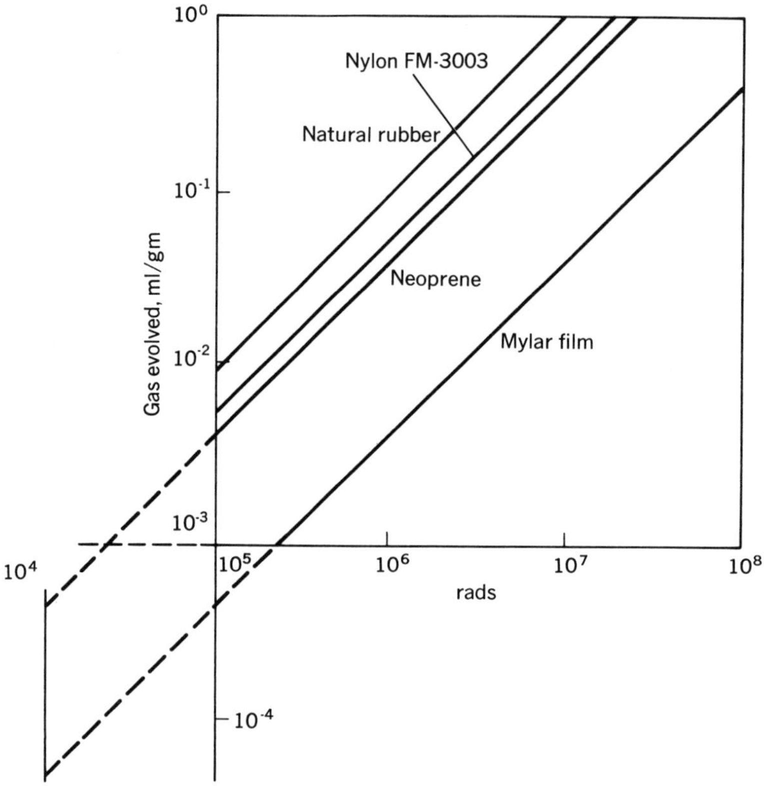

Figure 7-9: Outgassing of plastics resulting from irradiation.[24]

Table 7-5: Gas Evolution from Plastics Irradiated at 3 to 7 x 10^8 Rads

Plastic type	*Gas evolved, ml/g*
1. Methylmethacrylate	12.0
2. Phenol formaldehyde, unfilled	1.1
3. Phenol formaldehyde, asbestos filled	0.05
4. Phenol formaldehyde, graphite filled	0.01
5. Phenol formaldehyde, paper filled	6.0
6. Polyamide	11.0
7. Polyester	3.4
8. Polyethylene	38.0
9. Polyethylene terephthalate (Mylar)	1.2
10. Polystyrene	0.1
11. Polyvinyl formal	37.0

The evolved gas can cause other side effects, such as corrosion of adjacent metallic parts. This has been shown for polyvinyl chloride-insulated wires and cables, which outgassed chlorine or hydrogen chloride and caused green corrosion products on adjacent mumetal. Normally, polyvinyl chloride coatings should not be used on aluminum in any high-intensity-radiation environment. The failure—blistering of the coating and corrosion of the aluminum—occurs by a two-step mechanism: chemical decomposition of the polymer releasing the hydrogen chloride, and chemical reaction of hydrogen chloride with aluminum, resulting in aluminum chloride corrosion products.

Radiation Effects on Plasticizers and Fillers

Probably the most serious effect of radiation is on the stability of plasticizers in a coating formulation. Increased cross-linking caused by irradiation results in polymer shrinkage and squeezing out of plasticizer. The plasticizer may then agglomerate in droplets throughout the specimen or exude to the surface. The combination of cross-linking and separation of the plasticizer increases the rigidity and brittleness of the coating.[21]

Ingredients such as driers, pigments, and flame retardants used in small amounts are generally little affected by radiation. Whereas some finely powdered metallic fillers have been known to catalyze radiation damage in polymers, inorganic fillers such as silica and alumina suffer comparatively little deterioration and, in fact, can reduce the total damage to the plastic matrix. Terai[25] has shown very little effect due to pigmentation of paints at radiation dosages of less than 1.5 X 10^7 rads (gamma), but considerable differences at levels of 8 X 10^8 rads or more. At the higher levels, for example, titanium dioxide accelerated the damage of nitrocellulose lacquer, whereas carbon black inhibited damage. At lower levels no differences were noted with white lead, titanium dioxide, antimony oxide, calcium carbonate, barium sulfate, talc, iron oxide, chromium oxide, carbon black, Hansa yellow, and toluidine red.

Correlation of Organic Structures with Radiation Resistance

One factor in the selection or development of more radiation-resistant materials is that the aromatic or highly resonating structures such as benzene or carbazole are excellent radiation energy sinks.

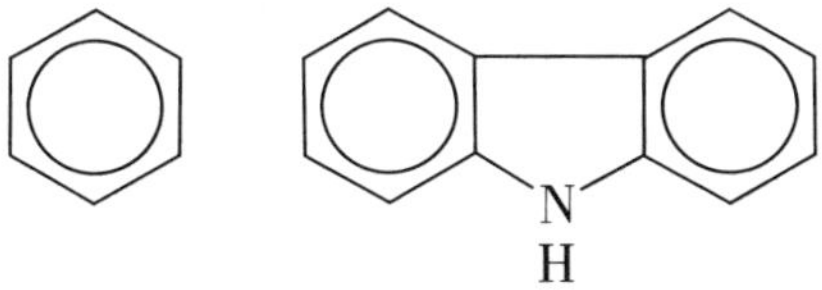

Benzene Carbazole

Polymers consisting of many repeating aromatic units will therefore have a high degree of radiation resistance. This principle has been derived from extensive data reported in the literature. For example, an epoxy cured with an aromatic amine retained 60 percent of its flexural strength after exposure to 10^9 rad, whereas the same epoxy cured with a low molecular weight aliphatic amine retained only 25 to

30 percent of its flexural strength after exposure to 4 X 10^8 rads.[26,27] The aliphatic-amine-cured epoxies also showed three times greater gas evolution than the aromatic-amine-cured epoxies.

Polystyrene, which has an aromatic structure (repeating benzene rings), is one of the most radiation-resistant plastics known.

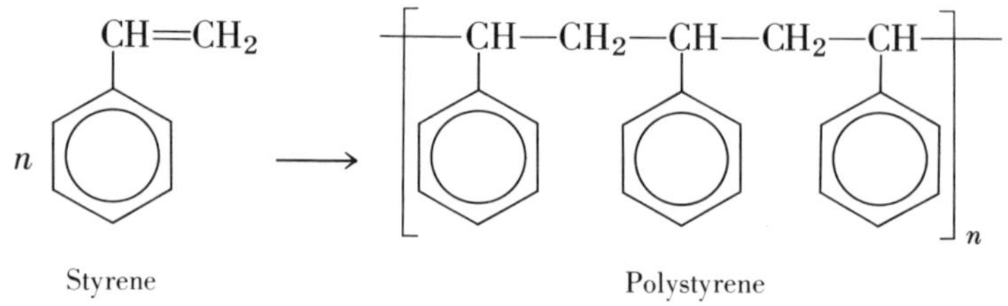

As can be seen from Figure 7-10, most physical properties did not change appreciably at electron dosages up to 1 X 10^8 rads. Tear strength was an exception, with about 75 percent retention at the same radiation dosage.

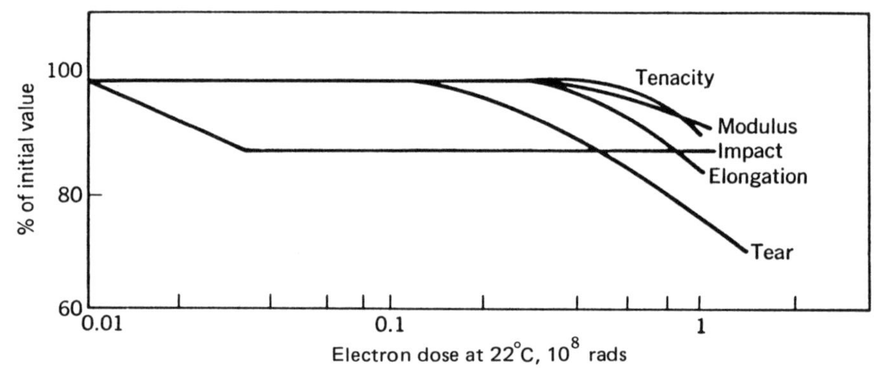

Figure 7-10: Effect of irradiation on the mechanical properties of polystyrene.[28]

Polystyrenes have not been used widely for electronic applications because of their flammability, high shrinkage rates, high exotherm, and generally poorer mechanical properties. However, copolymers and modifications of the presently used epoxies and polyurethanes with styrene should be explored, as they may yield all the desirable properties. Epoxies may be cured with vinyl-type monomers, but the resulting polymers have not been studied extensively. Although no specific mention is made of styrene, it is anticipated that it will behave similar to other vinyls tested. Other polymers that have highly aromatic structures include epoxies, phenolics, furanes, and silicone alkyds.

Another principle in formulating radiation-resistant resins is that aromatic additives such as aromatic plasticizers, flame-retardant agents, and dyes are preferable to aliphatic types. Moreover, inorganic or ceramic fillers are preferable to organic fillers.

Relative Radiation Resistance of Coatings

Although there is considerable information on radiation effects on molded and cast plastics, there is very little information on coatings. The data for molded plastics, however, may be used as a guide for estimating effects on coatings of the same polymer type. Most data for coatings have been obtained on ultraviolet-light stability because of the importance of this factor in commercial paints and finishes that must withstand outdoor sunlight exposure.[29]

The approximate gamma-radiation resistance of six coating types applied to AZ-31 magnesium with Dow 7 pretreatment is given in Table 7-6. Coatings were tested for adhesion, abrasion resistance, humidity resistance, and flexibility. Of those tested, the best performance was for the phenolic and the worst was for the nitrocellulose. Silicone alkyd, although excellent at 9 X 10^8 rads, became powdery and brittle at 44 X 10^8 rads.

Table 7-6: Relative Radiation Stabilities of Organic Coatings[30]

Coating type	*Maximum γ-radiation resistance, rads, × 10^8*
Phenolic (phenol formaldehyde)	44
Silicone-alkyd enamel	9–44
Alkyd enamels	
40 percent phthalic anhydride	9–44
32 percent phthalic anhydride	6–9
Epoxy	4–9
Fluorinated vinyl (trifluorochloroethylene–vinylidene chloride copolymer)	4–9
Nitrocellulose (white lacquer)	4–6

In other studies the effect of ionizing radiation on 45 plastics was examined.[31] High-energy radiation from a Van de Graaf accelerator at a total flux of 5.8 X 10^{16} electrons/cm^2 was used. This was estimated as equivalent to 18 years in a Van Allen radiation belt. The main effect noted was embrittlement and change in color (darkening). The polymers showing the greatest radiation stability were the polystyrenes and polyethylenes. Polystyrene and styrene-divinylbenzene copolymer showed very little change in dielectric constant when measured at 10^3 hz, but the dissipation factors showed large increases. Plastics such as diallylphthalates and epoxies, with initially high dielectric constants and dissipation factors, showed small changes after irradiation. Most of the polymers tested were molding plastics, but it is expected that the same polymer types formulated as coatings would behave similarly. The effects of ionizing radiation on other polymers have also been reported.[32,33] The relative resistance to gamma radiation of some organic insulating materials is shown in Figure 7-11.[34]

Ultraviolet Irradiation. More work has been performed on the effects of ultraviolet radiation on organic coatings than any other type of radiation. This is probably because of the availability and lower cost of ultraviolet sources, such as the mercury vapor lamps. Xenon- and carbon-arc sources, such as those em-

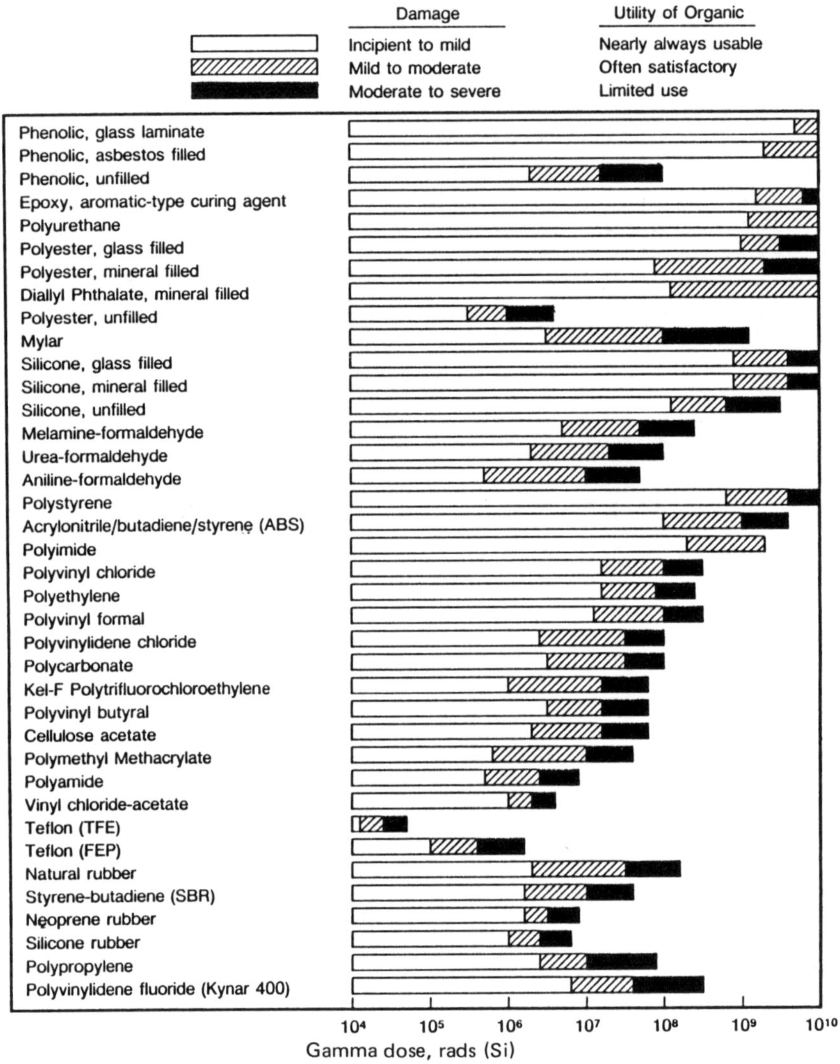

Figure 7-11: Relative radiation resistance of organic insulating materials. Based upon changes in physical properties.[34]

ployed in the Weather-Ometer (trademark of the Atlas Electric Devices Company), are often used, but they require more expensive equipment and yield less intense radiation in the low-frequency regions. A comparison of the spectral distribution of various light sources is given in Figure 7-12.

A variety of gaseous decomposition products may be obtained from the ultraviolet irradiation of coatings. Some of these, such as carbon dioxide and carbon monoxide, are common to many coatings, regardless of type. Other products are specific to the plastic structure. Nine decomposition products were obtained from an alkyd film after mercury-arc irradiation in air for 6 hr or xenon-arc exposure for

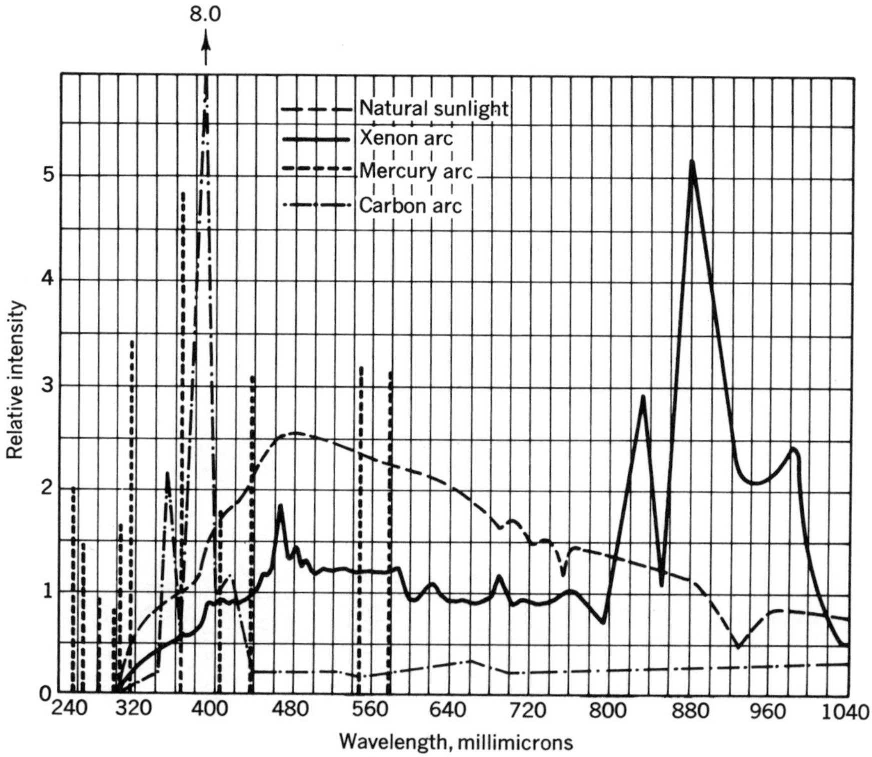

Figure 7-12: Spectral distribution of light sources.[29]

54 hr.[29] The products evolved included carbon dioxide, carbon monoxide, formic acid, methanol, hydrocarbons, ethylene, acetylene, and small amounts of acetone and vinyl compounds. Carbon dioxide was the major constituent evolved in this case, as it was in the case of vinyl, vinyl-alkyd, polyvinyl acetate, epoxy-amine, and epoxy-polyamide films. Polyvinyl chloride films released hydrogen chloride only with mercury-arc irradiation.

Combined Space-environment Effects. Some data are available on the combined effects of vacuum, radiation, and cryotemperature on the dielectric properties of plastics. Polyvinylidene fluoride, polycarbonate, melamine, and silicone plastics were exposed to a nuclear-radiation environment of 1×10^{10} ergs/g (carbon) of gamma radiation and 3×10^{15} neutrons/cm^2 of fast neutrons (E = 1.0 Mev) at room temperature and atmospheric pressure and also at temperatures as low as -195°C and 10^{-6} torr vacuum.[33,35] The data in Table 7-7 show that the combination of cryotemperatures and radiation had greater effects on the dielectric properties of polyvinylidine fluoride polymer (Kynar 400) than on the other materials tested.

Table 7-7: Electrical Values for Irradiated Plastics[35]

Material	Environment		Dielectric constant (normalized)			Dissipation factor, %			Volume resistivity, ohm-cm		
	Temp., °C	Vacuum, torrs	Pretest	Posttest	At power	Pretest	Posttest	At power	Pretest	Posttest	At power
Pennsalt Kynar 400 (vinylidene fluoride)	15–35	760	1.000	0.999	...	0.0042	0.0056	*	10^{15}	10^{10}	*
	−193	10^{-6}	0.576	0.584	...	0.0014	0.0075	...	10^{16}	10^{16}	
	−160	10^{-6}	...	...	0.615	...	...	0.025	...	...	10^{14}
G. E. Lexan (polycarbonate)	15–35	760	1.000	0.997	1.02	0.0015	0.0019	0.093	2×10^{15}	3×10^{12}	6×10^{10}
	−195	10^{-6}	0.958	0.961	...	0.0002	0.0019	...	10^{17}	10^{17}	
	−185	10^{-6}	...	...	0.974	...	...	0.016	...	...	2×10^{15}
Cyanamid Lamicoid 6038E (melamine and fiberglass)	15–35	760	1.000	0.999	1.03	0.0026	0.0032	0.005	6×10^{14}	2×10^{12}	4×10^{10}
	−190	10^{-6}	0.863	0.866	...	0.0013	0.0018	...	3×10^{15}	2×10^{15}	
	−170	10^{-6}	...	...	0.907	...	...	0.016	...	...	10^{15}
Dow Corning Sylgard 182 (silicone resin)	15–35	760	1.000	1.011	1.003	0.0015	0.0017	0.11	2×10^{14}	4×10^{12}	7×10^{10}
	−193	10^{-6}	1.000	0.973	...	0.0027	0.0017	...	10^{16}	10^{15}	
	−175	10^{-6}	...	...	1.048	...	...	0.057	...	...	10^{15}

* Losses too high to measure.

REFERENCES

1. Abstracted from L. Jaffe and J. B. Rittenhouse, *JPL Tech. Rept.*, 32–161. Oct. 30, 1961.
2. Fulk, M. M., and K. S. Horr: Sublimation of Materials Problem in Electronic Packaging for Spacecraft, *Advances Electron. Circuit Packaging,* vol. 4 (1964).
3. Compound, Printed Circuit Board, Conformal Coating, Elastomeric, Marshall Space Flight Center-SPEC-393, Oct. 9, 1964.
4. Markley, R. R. Roman, and R. Vosecek: Outgassing Data for Several Epoxy Resins and Rubbers for the Zero Gradient Synchrotron, *Trans. Eighth Vacuum Symp. Second Intern. Cong.,* 1961.
5. Baron, R. S., and R. P. Govier: A Mass Spectrometric Study of the Outgassing of Some Elastomers and Plastics, *United Kingdom Atom. Energy Authority Res. Group Rept.* CLM-R16, 1962.
6. Parts, Materials, Packaging, *IEEE Trans.,* December, 1965.
7. Thorne, J. A., C. L. Whipple, and A. B. Boehm: Space Environmental Effects on Silicone Insulating Materials, *Sixth Elec. Insulation Conf. New York, IEEE Publ.* 32C3-60, Sept. 12–16, 1965.
8. Fisher A. and B. Mermelstein: A Compilation of Low Outgassing Polymeric Materials Normally Recommended for GSFC Cognizant Spacecraft, NASA TMX-65705, July 1971.
9. Campbell W. A., R. S. Marriott, and J. J. Park: A Compilation of Outgassing Data for Spacecraft Materials, NASA TMD 7362, Sept. 1973.
10. Chou D. T. and R. D. Hermansen: Novel Thermally Conductive Elastomeric Adhesive for Electronic Component Assembly, *2nd Intl. SAMPE Electronics Conference, Proceedings, Vol. 2,* 1988.
11. Sax, N. I.: "Dangerous Properties of Industrial Materials," Sixth Edition, Van Nostrand-Reinhold, 1984.
12. Kalfayan, S. H., B. A. Campbell, and R. H. Silver: Effects of Ethylene Oxide-Freon 12 Decontamination and Dry Heat Sterilization Procedures on Polymeric Products, *JPL Tech. Rept.* 32-1160, Sept. 15, 1967.
13. Lee, S. M., J. J. Licari, and R. O. Fewell: Sterilization Effects on Microelectronics, *Natl. Electron. Packaging Prod. Conf., Long Beach, Calif.,* Jan. 31–Feb. 2, 1967.
14. Lee, S. M., and J. J. Licari: Effects of Decontamination and Sterilization on Spacecraft Polymeric Materials, *Rept. JPL Contract* 951566, *NASA* NAS7-100, January, 1968.
15. Messenger, G. C., and M. Ash,: "The Effects of Radiation on Electronic Systems," Van Nostrand Reinhold, New York, 1986.
16. Goetzel, C. G., J. B. Rittenhouse, and J. B. Singletary: "Space Materials Handbook," Addison-Wesley, Reading, Mass., 1965.
17. McIlwain, C. E.: The Radiation Belts, Natural and Artificial, *Science,* **142:** 355 (1963).
18. Chapin, W. E., D. J. Hamman, E. N. Wyler, and D. Jones: Space Radiation Damage to Electronic Components and Materials, Battelle Memorial Institute, *Radiation Effects Information Center Rept.* 32. Oct. 30, 1963.
19. Rudie, N. J.: "Principles and Techniques of Radiation Hardening, Volume 1," Western Periodicals Co., 1976.
20. Kircher, J., and R. E. Bowman: "Effects of Radiation on Materials and Components," Reinhold, New York, 1964.
21. Bolt, R. O., and James G. Carroll (eds.): "Radiation Effects on Organic Materials," Academic Press, New York, 1963.
22. Charlesby, A.: Radiation Effects in Polymers, Electrical Insulation Conference, 1966, *Natl. Acad. Sci. Natl. Res. Council Publ.* 1484, 1967.
23. Jolley, C. E., and J. C. Reed: "The Effect of Space Environments on Insulation of Teflon TFE and FEP Resins," E. I. du Pont de Nemours & Company.
24. Levine, J. H.: Radiation: Effects, Considerations, in G. A. Walker (ed.), "Advances in Electronic Circuit Packaging," vol. 2, Plenum Press, New York, 1962.

25. Terai, I.: Damage to Various Films by Gamma Irradiation, *Paint, Oil, Chem. Rev.,* **121** (2): 6 (January, 1958).
26. Aitken, I. D., and K. Ralph: Some Effects of Radiation on Cast Epoxy Resin Systems, *British Rept. AERE* R-3085, February, 1960.
27. Mixer, R. Y., and D. B. Parkinson: Nuclear Radiation Effects on Structural Plastics and Adhesives, *Stanford Res. Inst. Rept.* WADC-TR-56-534, part III, August, 1957.
28. LeClair, H. G., and W. H. Cobbs, Jr.: Effect of Radiation on Plastic Packaging Films, *Ind. Eng. Chem.,* **50** (3): 323 (March, 1958).
29. Hearst, P. J.: Volatile Degradation Products of Organic Coatings Irradiated in Air, *J. Paint Technol.,* **39** (March, 1967).
30. Horrocks, L. A.: How Radiation Affects Six Organic Coatings, *Matls. Des. Eng.,* **47** (1): 120 (January, 1958).
31. Pascale, J. V., D. B. Herrmann, and R. J. Miner: High Energy Electron Radiation Resistance of Plastics, *Mod. Plastics,* vol. 41 (October, 1963).
32. Bovey, F. A.: "Effects of Ionizing Radiation on Natural and Synthetic High Polymers," Interscience, New York, 1958.
33. (a) Swallow, A. J.: "Radiation Chemistry of Organic Compounds," Pergamon, New York, 1960. 33(b) Reichmanis, E., and J. H. O'Donnell: "The Effects of Radiation on High-Technology Polymers," ACS Symposium Series 381. American Chemical Society, Washington, D.C., 1989.
34. Hanks, C. L. and D. J. Hamman: The Effects of Radiation on Electrical Insulating Materials, *Radiation Effects Information Center Rept.* 46, June 16, 1969.
35. McMillan, W. D., et al.: The Measurement of Dielectric Properties of Materials under Vacuum, Reactor Radiation, and Cryogenic Conditions, *Trans. ASTM, Atlantic City, N.J.,* June 27, 1966.

8

Wire and Coil Coatings

Magnet wire is used principally in the electrical and electronics industries for coils, inductors, transformers, armatures, solenoids, and other windings. Wire may be insulated and protected with a wide variety of coating materials which may be classified according to the manner of application. Wires may be dipped in a liquid bath of the resin solution, or coated by extrusion of solid thermoplastic resins.[1] Extruded insulation, such as polyvinyl chloride, polyethylene, or polytetrafluoroethylene, is relatively thick in relation to the wire diameter and is therefore used chiefly for interconnect and lead wires. Dip coatings may be applied in very thin layers and consequently are employed for the insulation of thin magnet wire and for the impregnation of coils and windings. This chapter covers only the liquid-resin coatings such as the varnishes, enamels, and impregnating coatings. For purposes of comparison, however, some properties of the extruded types will also be given.

MAGNET-WIRE CLASSIFICATIONS

Magnet wire may be classified as coated wire, coated wire with fibrous wrappings, and wire with impregnated fibrous wrapping. In the third type the fibrous wrapping may be cotton, polyester, or glass impregnated with resinous materials. Wire coated with only an organic coating is frequently referred to as *enameled wire* or simply coated wire; the organic coating is applied directly to the conductor and is the only dielectric material used.

Magnet wire may also be classified according to National Electrical Manufacturers Association (NEMA) thermal ratings (Table 8-1). In order to qualify for an NEMA thermal classification, a wire must meet two criteria: its extrapolated life must be 20,000 hr or more at or above the thermal class rating, and the tested life

Table 8-1: Magnet-Wire Classifications[2]

Type of insulation	Type	NEMA designation	Class, C
Oleoresinous-enamel-coated, round	E	MW1-C	105
Polyurethane-coated, round	U	MW2-C	105
Nylon-coated, round	N	MW6-C	105
Polyvinyl, formal-coated, round	T	MW15-C	105
Polyvinyl, formal-coated, rectangular	T	MW18-C	105
Polyvinyl, formal-nylon-coated, selfbonding, round	TN	MW17-C	105
Polyurethane-nylon-coated, round	UN	MW28-C	130
Polyester-coated, round	L	MW5-C	155
Polyester-nylon-coated, round	LN	MW24-C	155
Polyester-imide or polyester-amide-imide-coated, round	H	MW72-C	180
Modified polyester or polyester-imide or polyester-amide-imide overcoated with polyamide-imide, round	K	MW35-C	200
Modified polyester or polyester-imide or polyester-amide-imide overcoated with polyamide-imide, rectangular	K	MW36-C	200
Aromatic polyimide-coated, round	M	MW16-C	220
Aromatic polyimide-coated, rectangular	M	MW20-C	220
Polyester-glass-fiber-covered, round	DgV	MW45-C	155
Polyester-glass-fiber-covered, rectangular	DgV	MW46-C	155
Polyester-glass-fiber-covered, round	DgH	MW47-C	180
Polyester-glass-fiber-covered, rectangular	DgH	MW48-C	180
Polyester-glass-fiber-covered, rectangular	M_2DgGM	none	220

of the specimen must not be less than 5,000 hr at 20°C above the thermal class rating. Another specification, Federal Specification J-W-1177, defines the same thermal classes as the NEMA, but the type designations are different.[2] Performance requirements for each of these classes and types, including elongation, solvent resistance, and abrasion resistance, are given in J-W-1177.

Finally, magnet wire may be classified according to the thickness of the coating. Enameled wire is furnished in various coating thicknesses as Single, Heavy, Triple, etc., again according to NEMA standards.[3] Classifications for wire coated with thin polyvinyl formal may be found in Tables 8-2 and 8-3.

WIRE-COATING TYPES

Most of the coating types discussed in Chapters 1 to 3 may be used as wire coatings and as coil-impregnating resins. Although they are formulated differently, the basic properties of each type are essentially the same. Accordingly, silicones, polyimides, and fluorocarbons are best suited for very-high-temperature applications, polyurethanes for ease of removal, and epoxies for solvent and chemical resistance. Several additional polymer types are also used for wire coatings. Among these are the oleoresinous varnishes and the polyolefins (polyethylenes, polypropylenes, and irradiated polyolefins).

In general, wire and impregnation coatings consist of high molecular weight polymers, either of the thermoplastic (linear) or thermosetting (cross-linked) types. The thermoplastics have definite softening or melting points, above which the

Table 8-2: Dimensions of AWG Wire Sizes 31 to 44

AWG size	Bare-wire diameter, in.			Single		Heavy	
	Min.	Nominal	Max.	Min. increase in diam., in.	Max. overall diam., in.	Min. increase in diam., in.	Max. overall diam., in.
31	0.0088	0.0089	0.0090	0.0006	0.0100	0.0013	0.0108
32	0.0079	0.0080	0.0081	0.0006	0.0091	0.0012	0.0098
33	0.0070	0.0071	0.0072	0.0005	0.0081	0.0011	0.0088
34	0.0062	0.0063	0.0064	0.0005	0.0072	0.0010	0.0078
35	0.0055	0.0056	0.0057	0.0004	0.0064	0.0009	0.0070
36	0.0049	0.0050	0.0051	0.0004	0.0058	0.0008	0.0063
37	0.0044	0.0045	0.0046	0.0003	0.0052	0.0008	0.0057
38	0.0039	0.0040	0.0041	0.0003	0.0047	0.0007	0.0051
39	0.0034	0.0035	0.0036	0.0002	0.0041	0.0006	0.0045
40	0.0030	0.0031	0.0032	0.0002	0.0037	0.0006	0.0040
41	0.0027	0.0028	0.0029	0.0002	0.0033	0.0005	0.0036
42	0.0024	0.0025	0.0026	0.0002	0.0030	0.0004	0.0032
43	0.0021	0.0022	0.0023	0.0002	0.0026	0.0004	0.0029
44	0.0019	0.0020	0.0021	0.0001	0.0024	0.0004	0.0027

Table 8-3: Characteristics of AWG Wire Sizes 45 to 56

AWG size	Theoretical* nominal bare-wire diameter, in.	Conductor resistance at 20°C, ohms/ft†			Single		Heavy	
		Min.	Nominal	Max.	Min. increase in diam., in.	Max. overall diam., in.	Min. increase in diam., in.	Max. overall diam., in.
45	0.00176	3.080	3.348	3.616	0.00010	0.00205	0.00030	0.00230
46	0.00157	3.870	4.207	4.544	0.00010	0.00185	0.00030	0.00210
47	0.00140	4.868	5.291	5.714	0.00010	0.00170	0.00030	0.00190
48	0.00124	6.205	6.745	7.285	0.00010	0.00150	0.00020	0.00170
49	0.00111	7.744	8.417	9.090	0.00010	0.00130	0.00020	0.00150
50	0.00099	9.734	10.58	11.43	0.00010	0.00120	0.00020	0.00140
51	0.00088	12.32	13.39	14.46				
52	0.00078	15.69	17.05	18.41				
53	0.00070	19.48	21.17	22.86				
54	0.00062	24.82	26.98	29.14				
55	0.00055	31.54	34.28	37.02				
56	0.00049	39.73	43.19	46.64				

* Theoretical nominal bare-wire diameters are in accordance with the Copper Wire Tables, *Natl. Bur. Stds. Handbook* 100.

† Conductor diameter tolerances are shown as resistance values and shall be determined by measuring the resistance of the wire in accordance with the USA Standard "Method of Test for Resistivity of Electrical Conductor Materials," C7.24-1961, where applicable. A specimen at least 5 ft long shall be used.

materials are not usable, especially where local stresses are to be applied. Examples of thermoplastic coatings include Teflons and nylons. The thermosetting coatings are more resistant to "cut through" and have improved solvent resistance. Examples include silicones, polyesters, and polyvinyl formals. Some wire-coating types, trade names, and manufacturers are given in Table 8-4, and a summary of advantages and limitations of wire coatings is given in Table 8-5.

As indicated in Table 8-1, many wire coatings consist of a combination of two, or even three, different polymers. In this way characteristics are achieved which could not be obtained with any one polymer alone. Thus polyvinyl formal may be coated with nylon to provide the lubricating properties of nylon for greater ease in winding. Polyvinyl formal is also frequently overcoated with polyvinyl butyral to render the coating self-bonding either by heat or solvent action. This property is especially important in the fabrication of formed coils.

Plain-enamel or Oleoresinous Coatings

The terms "plain enamel" or "oleoresinous varnish" refer to the natural resinous coatings that have been used as standard wire insulation for more than 60 years. They are derived from natural drying oils such as linseed oil or tung oil and are cured through the air oxidation of double bonds. The coatings may be augmented by "cooking in" fossil or synthetic resins to improve their hardness and resistance to solvents. Because of the dark color of the finish wire, such coatings are sometimes referred to as "black enamels." Oleoresinous coatings are rapidly being replaced by the higher-performance synthetic coatings.

Table 8-4: Wire Coating Types[4]

Manufacturer	Trade name	Chemical type	Temperature rating, C	Applications
Anamag	Anaclad-A (HAC-A)	Polyester/polyimide-amide	200(Cu)/220(Al)	Hermetic refrigeration, generators, integral motors, ballast transformers, microwave motors.
	Nylac (SNL)	Polyurethane/polyamide	130 to 155	Electronic coils, motor coils, lighting transformers.
	Formvar (HF)	Polyvinyl formal	105	Oil-filled transformers, pole mount and pad mount high voltage circuit breakers.
	Anatherm (HAT-NC)	Polyester/polyamide	180	Fractional and subfractional open motors, portable generators, telephone equipment.
	HML	Aromatic polyester	220	Hermetic refrigeration motors, sealed relays, encapsulated windings, dry-type, oil-filled transformers, large hp motors.
Arcor	Hot-Melt (Nyleze)	Nylon over polyurethane	130 to 200	Transformers, solenoids, coils, motors, ballasts
Belden/EWP	Celenamel	Cellulose acetate	130	Transformers, ignition coils, relays
	Celemid	Polyester	180	Transformer coils, automotive coils, relays.
	Beldtherm 200	Cross-linked, modified polyester	200	Coils, relays; constant temperature of 200 C; low outgassing.
	Belden ML	Polyimide	220	Coils, relays, transformers.
				transformers.
Bridgeport	Polyurethane	Polyurethane	155	Small motors, relays, transformers.
	Formvar	Polyvinyl formal	105	Small motors, relays,
	Poly-Bond	Polyurethane with polyvinyl butanol topcoat	105	Small motors, relays, transformers.
	Poly-Nylon	Polyurethane with nylon topcoat	130	Small motors, relays, transformers.

(continued)

Table 8-4: (continued)

Manufacturer	Trade name	Chemical type	Temperature rating, C	Applications
	Terasod	Solderable polyester	155	Small motors, relays, transformers.
	Isonel 200	Modified polyester	180	Small motors, relays, transformers.
	Isomid	Modified polyester-imide	180	Small motors, relays, transformers.
Canada Wire	Polyvinyl formal Polythane (Westinghouse)	Polyvinyl formal	105	Oil-filled transformers.
	Formel Epoxy	Self-bonding polyvinyl formal	105	Oil-filled transformers.
	Nypol	Polyurethane-nylon	130	Automotive generators and alternators, ballasts, open motors.
	Magnesol	Solderable modified polyester	155	Automotive generators and alternators, ballasts, open motors.
	Armored Magnetemp	Modified polyester over-coated with polyamide-imide	200	Mercury ballasts, motors, dry-type transformers, coils, hermetic motors.
	ML	Aromatic polyimide	220	Dry-type and distribution transformers, heavy duty motors, traction motors.
	Kapton	Kapton	220	Traction motors, coils under strong shock.
Essex Group	Thermetex GP-200	Polyester/polyamide-imide	200	Motors, dry-type transformers, large coil applications.
	Allex	Aromatic polyimide	220	Fhp* and integral hp** motors; high temperature continuous duty coils and relays; hermetic and sealed units; heavy duty hand tool motors, encapsulated coils.
	Nytherm	Polyester/polyamide	180	Fhp and integral hp motors; relays and coils; control and dry-type transformers; encapsulated coils; d-c field coils.

(continued)

Table 8-4: (continued)

Manufacturer	Trade name	Chemical type	Temperature rating, C	Applications
	MR/GP200	Polyester/polyamide-imide	200	Fhp and integral hp motors; dry-type transformers; automotive and electric tool armatures.
	Solidex	Modified polyester-imide	180	Special transformer coils, shaded pole motor coils, automotive coils, electronic coils.
	Soderex	Polyurethane	130	Small motors, relays, bell ringer coils, encapsulated coils.
	Nomex	Polyamide tape	200/220	Dry-type or oil-filled transformers, lifting magnets, form-wound coils.
Elektrisola	Polysol 140	Modified polyurethane	140	Molded and encapsulated coils; automotive coils; audio RF, instrument, and small power transformers; small motors, armatures.
	Polysol-N	Polyurethane-nylon	140	Molded and encapsulated coils; automotive coils; audio RF, instrument, and small power transformers; toroidal coils.
	Ester-N	Polyimide-amide-ester	155	Specialty power transformers; subfractional motors; control coils; automotive coils.
	Estersol 180	Polyimide-amide-ester	180	Specialty power transformers; subfractional motors; control coils; automotive coils.
	Amidester 200	Polyimide-amide-ester	200	Hermetic motors; d-c motors; dry-type transformers; control transformers; electronic coils.

(continued)

Table 8-4: (continued)

Manufacturer	Trade name	Chemical type	Temperature rating, C	Applications
Phelps Dodge	A.P. Bondeze (AP-B)	Thermosetting Polyester with linear amide-imide overcoat, self-bonding topcoat	130	Yoke coils; clutch and brake coils; bobbinless solenoids.
	Armored Poly-Thermaleze 2000 (APTZ)	Thermosetting Polyester with linear amide-imide overcoat, self-bonding topcoat	200(Cu)/220(Al)	Dry-type transformers; hermetic motors; tool motors; television fly-back transformers; toroidal TV yokes; automotive alternator stators; solenoids.
	Bondeze-T (BDZ-T)	Tinnable polyester with self-bonding topcoat	150	Fhp motors; solenoids; clutch and brake coils.
	Formareze (F)	Polyvinyl Formal	105	Oil-filled transformers.
	Imideze	Aromatic polyimide	220	Dry-type transformers, traction motors, d-c field coils, submersible pump motors.
	M/R Thermaleze (M/R TZ)	Thermosetting polyester with modified aramid overcoat	200	Motors used in high moisture and hermetic application.
	Nyleze (NYLZ)	Polyurethane with nylon (polyamide) overcoat	130	Coils, universal motors.
	Sy-Bondeze (S-Y-B)	Polyurethane with nylon	105	Voice coils, encapsulated solenoids, toy motors.
	Sodereze (SDZ)	Polyurethane	105/155	Ignition coils, low voltage transformers, relays, solenoids.
Rea	Formvar	Polyvinyl formal	105	Oil-filled distribution transformers.
	Solvar	Modified polyurethane	105	Coils.
	Nysol	Modified polyurethane with nylon overcoat	130	Coils.
	Hy-Silk Amid-Therm	Modified polyester with proprietary overcoat	180	Motors.
	Thersol	Modified, solderable polyester	180	Coils, motors.
	THERM-ID	Modified polyester	200	Coils.

(continued)

Table 8-4: (continued)

Manufacturer	Trade name	Chemical type	Temperature rating, C	Applications
Westinghouse	THERM-AMID	Modified polyester with polyamide-imide overcoat	200	Hermetic motors.
	Super Hy-Silk 200	Modified polyester with polyamide-imide overcoat with proprietary lubricant	200	Motors.
	Polythane	Modified polyurethane	105	Small coils.
	Nythane	Modified polyurethane with polyamide overcoat	130	Coils.
	Omega-Klad	Modified polyester with amide-imide overcoat	200	Transformers, motors.
	Hydroshield	Modified polyester with amide-imide overcoat	200	Transformers, motors.
	Omega-Stix	Bondable modified polyester with amide-imide overcoat	200	Coils.
	ATOML	Aromatic polyimide	220	Motor manufacturing and repair.

*Fractional horsepower.
**Horsepower.

Table 8-5: Summary of Properties of Various Coated Wires

Coating	Thermal rating, °C	Advantages	Limitations*
Polyvinyl formal	105	Toughness, dielectric strength; compatible with other coatings; heat-shock resistant.	Crazes in polar solvents.
Polyurethane	105	Dielectric strength, chemical resistance, moisture and corona resistance; compatible with solvents and chemicals; solderable without stripping.	Low thermal resistance.
Polyamide (nylon)	105	Toughness, dielectric strength, solvent resistance; solderable; good windability.	High moisture absorption; high electrical loss at all frequencies.
Polyvinyl formal, nylon . . .	105	Toughness, solvent resistance; good windability; heat-shock resistant.	Nylon portion subject to same limitations as nylon.

(continued)

Table 8-5: (continued)

Coating	Thermal rating, °C	Advantages	Limitations*
Polyvinyl formal, polyvinyl-butyral	105	Bondability, dielectric strength; heat-shock resistant.	Vibration; high mechanical stress.
Polyurethane, nylon	105	Solvent resistance; solderable.	High moisture absorption; high-frequency losses.
Polyester	155	Toughness, dielectric strength, chemical resistance, cut-through resistance	Hydrolyzes in moist sealed atmosphere.
Polytetrafluoroethylene (Teflon)	200	Thermal stability, chemical stability, dielectric strength; low dielectric constant.	High abrasion; high gas permeability; cold flow; poor adhesion.
Polyimide	220	High overload resistance, thermal resistance, chemical stability, radiation resistance; high cut-through resistance.	Stripping difficulty; crazes in some solvents.

* A prebake at 125°C for 2 to 4 hr after winding relieves film stresses so that crazing does not occur during subsequent varnish or resin encapsulation.

Polyvinyl Formal and Modified-Polyvinyl Formal Coatings

Coatings based on polyvinyl formal (Formvar), first introduced in 1938 as magnet-wire insulation, are still widely used magnet-wire coatings. The base polymer is synthesized by partially hydrolyzing polyvinyl acetate and then reacting the free hydroxyl groups of the resulting product with formaldehyde to form the six-membered formal structure. The polymer will therefore contain three basic functional units—vinyl formal, vinyl alcohol, and vinyl acetate, as follows:

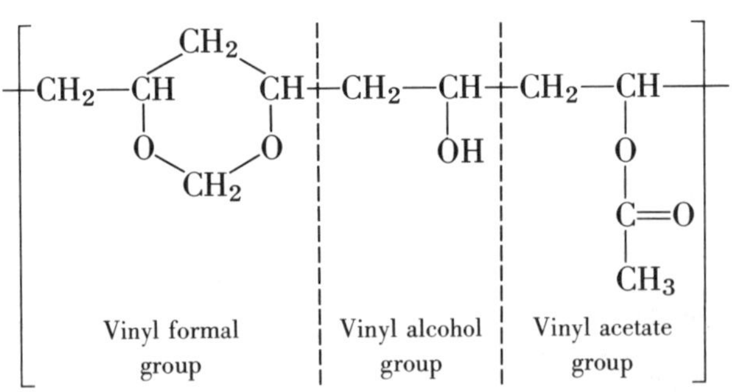

Molecular Structure of Polyvinyl Formal Resin

Polyvinyl formal used alone has only marginal solvent resistance, hardness, and other properties. Its properties are greatly improved by modifying the basic resin through reaction with phenol formaldehyde or cresol formaldehyde resins. Both resins condense with the free hydroxyl groups of the polyvinyl formal to produce coatings which are extremely tough and flexible and which possess good abrasion and solvent resistance. Polyvinyl formal can also be modified with resins such as melamine, urea, epoxy, and urethanes.[5]

The modifications with urethane are especially favored because of their solder-through properties. Dip-tinning or soldering at high temperatures (up to 350°C) may easily be performed without stripping the insulation. Three-component coatings consisting of polyvinyl formal, polyurethane, and polyester are also easy to solder through and, in addition, have good flow, flexibility, and heat-shock properties, as well as good resistance to abrasion and solvents.[6] In fact, practically all the solderable magnet-wire coatings on the market are urethane modifications.

Polyurethane Coatings

Polyurethanes used as wire enamels are generally the reaction products of hydroxylated polyesters or hydroxylated polyethers with tolylene diisocyanate. Admixing these components, however, results in very rapid polymerization—too rapid for any practical application. For this reason, polyesters containing excess isocyanate groups are prereacted with phenol to yield adducts which are relatively stable at room temperature. After application to the wire and subsequent high-temperature baking, the phenol portion of the adduct is released. In this manner the free isocyanate groups which are formed can react with hydroxyl groups of the polyester to form the desired polyurethane coating. This type of polyurethane is the ASTM Type 3 described in Chapter 1. Other polyurethane wire enamels consist of reaction products of isocyanates with formvar and, as already mentioned, offer good solderability. In fact, the largest share of the solderable insulation market utilizes polyurethane systems.[7]

Polyester Varnishes and Enamels

Polyester varnishes are produced by the condensation of terephthalic acid, a glycol, and a polyol such as glycerine. When the proper proportions are mixed and heated, partial polymerization occurs, leaving the product soluble in solvents. The general structure for the polyester polymer is

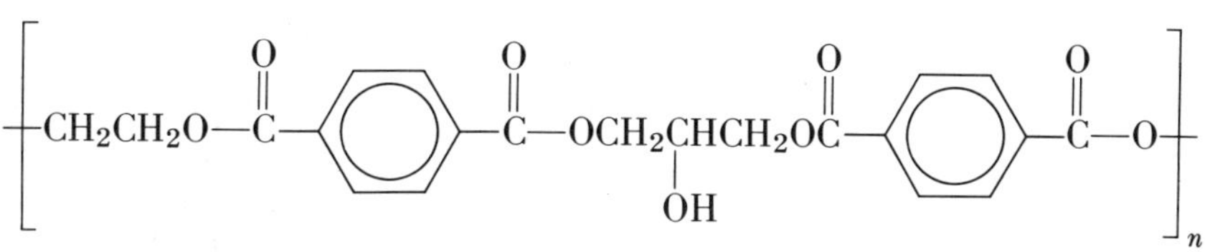

Polyester coatings may be applied to wire by either an automated or a manual dipping operation and then baked at 150 to 205°C. They may be used alone or in conjunction with other compatible wire enamels such as nylons, formvars, polyimides, and polyurethanes. Polyester wire coatings are especially noted for their excellent thermal properties, which derive from the stability of the ester linkages,

high cross-linking density, and the symmetry of the para-substituted benzene rings. They qualify for Class A (105°C) through Class 200 (200°C) applications. Thermal endurances, measured according to ASTM D 2307, for typical polyester and polyester-polyimide wire are given in Figure 8-1. Polyester varnishes also provide good chemical and solvent resistance, except to strong alkalis and mineral acids. Both alkalis and acids react slowly at room temperature or more rapidly at elevated temperature, resulting in hydrolysis of the ester linkages and cleavage of the polymer chain at these sites (Table 8-6).

Polyimide Coatings

Polyimide varnishes and enamels consist of polyamic acid resins dissolved in highly polar solvents such as dimethylformamide. After evaporation of the solvents, the coatings are cured by high-temperature baking, which effects ring closures through the elimination of water. The chemistry and properties of polyimides were discussed in Chapter 3.

Polyimide coatings excel in high-temperature and radiation resistance. They belong to Class 220 thermal insulation, i.e., they are capable of withstanding at least 20,000 hr at 220°C, according to IEEE No. 57 test procedures. Electrical properties, such as dielectric strength and insulation resistance, are not impaired even at temperatures of 240°C. The exceptional thermal stability of polyimide

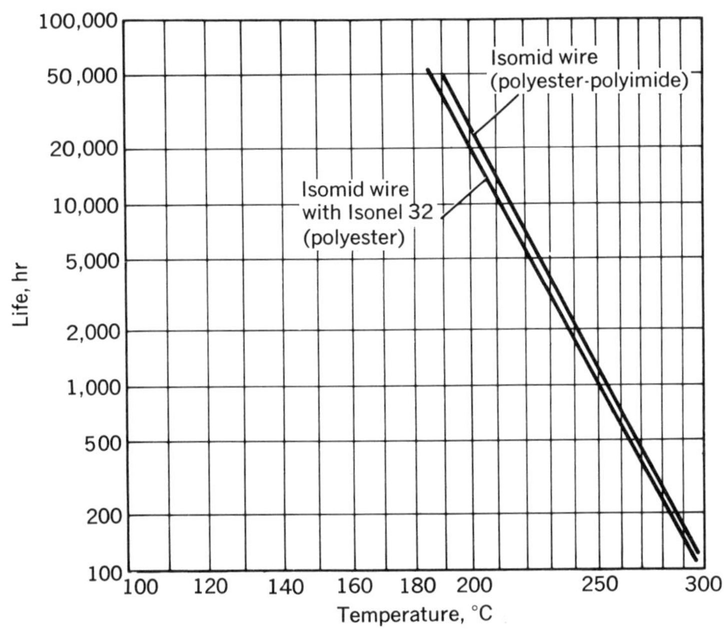

Figure 8-1: Thermal endurance of polyester-polyimide varnished wire (ASTM D 2307, 1,000-volt proof test). (Schenectady Chemicals, Inc.)

Table 8-6: Effects of Chemicals and Solvents on Isonel 51 Polyester Varnish

Chemical	Exposure	Observed effect	Dielectric strength, volts/mil
Control, varnished sample	None	...	3,150
Transformer oils............	48 hr at 100°C	Unaffected	2,920
Acetone	24 hr at 25°C	Slight softening	3,080
Xylol	24 hr at 25°C	Unaffected	2,980
5 percent sulfuric acid	24 hr at 25°C	Slight attack	2,690
2 percent sodium hydroxide...	24 hr at 25°C	Slight swelling	2,980

By permission of Schenectady Chemicals, Inc.

varnishes permits the manufacture of smaller motors, transformers, and generators. Because of their low weight loss at elevated temperatures and their high chemical resistance, polyimide varnishes are useful in encapsulated windings and hermetically sealed components. Polyimide wire coatings are also compatible with numerous phenolic, polyester, and epoxy varnishes and encapsulants and are resistant to most solvents and chemicals. Disadvantages include its hydrolytic instability at temperatures above 105°C in a sealed system containing moisture and the difficulty in stripping it off wire. There are very few good methods of removing polyimide from wire.

IMPREGNATING VARNISHES

After coils and their cores are assembled, the electrical part (for example, a transformer or toroidal coil) is impregnated with varnish and encapsulated or otherwise embedded in one of a wide variety of materials (see Figure 8-2). Complete impregnation serves four main functions: it insulates the wires, allows better dissipation of heat during operation, provides protection against moisture and contaminants, and provides better shock and vibration resistance. The exclusion of air and oxygen by thorough impregnation with organic coatings results in substantial improvements in thermal life. Epoxy impregnation, for example, may convert a Class A insulation to a Class B. The cooler operation of coils and transformers has also permitted a degree of miniaturization never before possible.[6] Transformers for household electronic equipment, which are not subject to severe environments, may be coated with low-cost solvent varnishes, treated with asphalt, or encapsulated with potting resin. Transformers subject to severe environmental conditions, as in military or aerospace equipment, may be coated with high-performance synthetic polymers (Figure 8-3), potted in plastic cups, molded or cast in plastic resins designed to meet the more severe conditions.

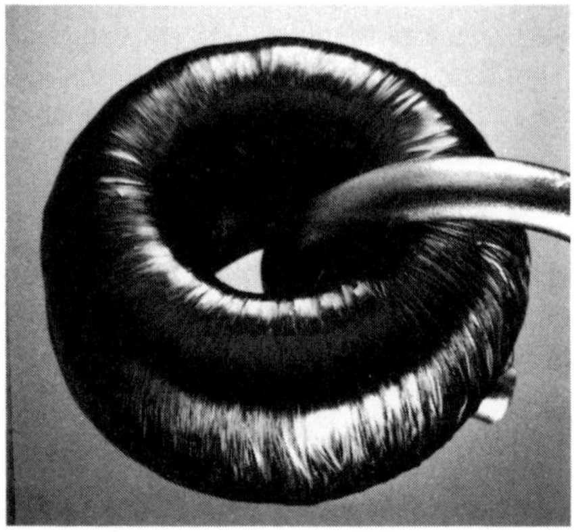

Figure 8-2: Silicone-coated toroidal coil. (Dow Corning Corp.)

Figure 8-3: Insulated transformers: (left) silicone-coated; (right) Westinghouse Fosterite, polyester-coated. (Westinghouse Electric Corporation)

APPLICATION AND WINDING METHODS

Wire coatings are normally applied from solvent solutions. The wire is passed through a varnish tank, then through a stripper die for build regulation, and finally through an oven or heating tower to effect solvent removal and cure. Most wire coatings are applied in multiple thin coats, because single coats invariably contain pinholes or other breaks in dielectric continuity. Each succeeding coat reduces the probability of pinholes.

After the coating is applied and cured, the wire is wound into coils for various electrical uses. The ease with which wire may be wound is a direct function of the lubricating properties of the coating. Coatings, such as nylons and Teflons, which have low coefficients of friction possess inherent lubrication. These coatings are preferred for automated high-speed winding operations. In other cases lubricating oils may be applied to the wire insulation to facilitate winding, but it is reported that these oils outgas and contaminate adjacent surfaces. Proprietary surface treatments have been introduced which are capable of reducing the coefficients of friction by 20 to 90 percent over those of untreated surfaces. One process, developed by Essex Group, entails grafting polymers to the surface of the wire coating. The coefficient of friction for nylon-coated insulation is 0.17 but for other commonly used coatings it may be as high as 0.27.

The reliability of magnet-wire coatings depends not only on their inherent material properties and method of application, but also on the winding process used. The winding method must be carefully chosen to avoid stresses and mechanical damage to the wire and insulation and to allow for close packing. Coil winding may take any one of several forms, depending on the design of the part. The most common form is one in which the wire is wound in layers, one wire thick, with each turn adjacent to the next. This type of coil is generally wound on a rectangular or square coil form, which may be plastic, ceramic, or, in the case of commercial equipment, even paper. The wire enamel normally provides sufficient insulation for resisting turn-to-turn electrical stress, but for added reliability paper or other insulating materials are frequently used between layers, where the electrical stress is higher.

In another winding process the coil is wound on molded plastic bobbins, and as with coils wound on coil forms, sheet material may be inserted between layers. The finished coil may also be covered by a coil wrapper of sheet material, such as paper, held in place by pressure-sensitive tape.

In a third type of winding, referred to as *precision winding,* the wire is passed over an adhesive applicator and then wrapped on a precisely grooved arbor. Each turn of wire rests in the groove between turns in the layer below. Since each layer has a relationship to the layer beneath of a left-hand thread to a right-hand thread, a crossover occurs with each turn; the arbors are so designed that all the crossovers are on one of the four sides of the coil. With fast-drying coatings, the coil is removed by collapsing the arbor on which it is wound. Precision-wound coils have the advantage of possessing very high wire density.

A fourth type of winding is necessary where the transformer or inductor core is toroidal. Intricate winding machines thread the wire through the toroid, forming the coils directly over the core. Other coils are assembled by inserting the core through the coil form or bobbin and completing the magnetic circuit by cementing the pieces of the core or interleaving the individual magnetic-iron punchings.

TESTING WIRE COATINGS

Coated wire is most often tested according to procedures described in NEMA MW 1000 or, in the case of military contracts, J-W-1177. Many of the tests in these two standards are identical. Some of the more commonly used tests are described in the following discussion.

Film Thickness

Film thickness is determined by measuring the total diameter of the coated wire with a micrometer, removing the coating in a manner that is not destructive to the wire, and then measuring the diameter of the bare wire. The difference between these diameters, divided by 2, gives the thickness of the coating. One method of removing the coating from copper wire is to burn it off by heating the wire and then quickly immersing the hot wire in ethyl alcohol. The alcohol dip minimizes the formation of copper oxide.

Adhesion

Adhesion is measured by elongating a 10-in. length of wire at 12 in./min. for Size 13 AWG (American Wire Gauge) and larger or by rapidly pulling wire Sizes 14 and finer. Examination of the wire for separation or cracking at specified magnification for different wire sizes determines failure.

Flexibility

Flexibility is determined by winding the wire around a mandrel of specified size and then examining it at a magnification dependent on the size of the wire. Flexibility is frequently expressed as 1×, 2×, 3×, etc., signifying that the particular wire coating will withstand winding around a mandrel 1, 2, or 3 times the AWG size of the wire. Adhesion and flexibility tests are frequently followed by elongation and by a mandrel-wind test.

Heat Shock

Heat shock is determined by elongating or stretching the wire, winding it around a mandrel of a size dependent on the wire size, and exposing it to oven heat. The specimen is then examined under magnification for cracking or separation from the metal.

Elongation

Elongation is measured by stretching a 10-inch piece of wire at a rate of 12 inches per minute until the wire breaks. The results are reported as a percentage.

Scrape Resistance

Scrape resistance is measured by either repeated scrape or unidirectional scrape. In the repeated-scrape test a machine rubs a 0.016-in.-diameter needle or wire parallel to the axis of the wire under test, with provision for a specified load at a rate of 60 strokes/min. Each stroke consists of a 360° rotation of an eccentric drive. The machine is equipped with an electrical circuit which stops the action and a counter which records the number of strokes at the time of breakthrough of the coating.

The unidirectional-scrape-resistance test, as the name implies, consists of a scraping action in one direction only. The machine rubs a 0.009-in.-diameter needle on the coated wire at 16-in./min at right angles to the wire, with a constantly increasing load. The machine is equipped with an electrical circuit for detecting the time to failure. Failure is recorded as "grams to fail."

Springback

Springback is a measure of the ability of the wire to unwind. A wire sample is wound three turns around a mandrel of a size dependent on the wire size at a rate of 5 to 10 rpm. After winding, one end of the coil is clamped and the other end allowed to unwind. The softer the wire the less the tendency to spring back. This test is used for wire sizes 14 to 30 AWG inclusive.

Thermoplastic Flow

Thermoplastic flow, also referred to as *cut-through,* is the temperature at which the coating softens and flows under defined conditions of pressure. According to the NEMA test procedure, two pieces of 18 AWG insulated wire are placed at right angles to each other and pressed by a load of 2,000 g. or 36 AWG wire is used with a 1000 g load. The wire conductors, pressed against a steel plate, are connected to a 110-volt ac power supply with an indicating device in the circuit (Figure 8-4). The unit is then placed in an oven, and the temperature is increased at a rate of less than 5°C/min. Deterioration of the coating and cut-through are registered when the indicating device shows short circuiting. At this point the temperature of the plate supporting the wires is measured and recorded.

Dielectric Strength

The dielectric strength of most wire coatings is determined by employing specimens of twisted wire pairs. The specimens are formed by twisting two wires under

Figure 8-4: Thermoplastic flow test setup, per J-W-1177. (Rockwell International Corporation)

a specified tension for a length of 4.75 in. The total number of twists in this length will, of course, depend on the wire size. The conductors of each pair are connected to a 60-cycle ac power source, and the voltage is increased at a rate of 500 volts/sec until breakdown occurs.

Continuity

The continuity of insulation is determined by passing 100 ft of wire through a mercury bath at a rate of 100 fpm and measuring dc resistance. A potential of 20 to 75 volts dc is applied between the wire and a mercury bath, and any discontinuities in the insulation result in a closed electrical circuit (Figure 8-5). A discontinuity device operates when the resistance between the mercury bath and wire conductor is less than 5,000 ohms but not more than 10,000 ohms. The test circuit is arranged so that the total number of dielectric breaks can be recorded. Results are reported as the number of breaks per 100 ft of wire.

High-voltage continuity is checked by passing 100 ft of wire between grooved metal rollers, with the wire grounded to the takeoff drum. The test circuit permits a 60-cycle voltage to be imposed between the takeoff drum and the roller and has a sensitivity of 1 megohm at the test voltage. The voltages used depend on the thickness of the coating and the wire size. The test circuit indicates when dielectric breaks occur during the test.

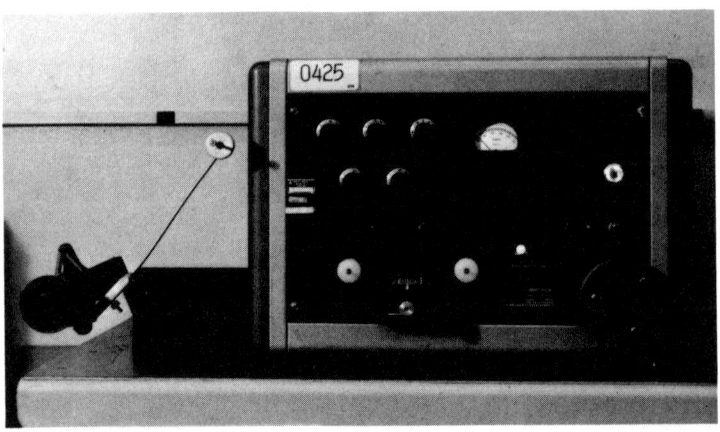

Figure 8-5: Continuity of insulation tester, per J-W-1177. (Rockwell International Corporation)

Solvent Resistance and Completeness of Cure

Deterioration of wire coatings owing to solvent exposure may be determined by an accelerated test in which the wire specimens are exposed to various solvents and their scrape resistance is measured. The scrape-resistance value may then be compared with that of an unexposed control wire.

The weight loss of the insulating coating after extraction with solvents is another

method of assessing both solvent resistance and the degree of cure of the coating. Solvent can be extracted by the Soxhlet method[9] with toluene, methanol, or other solvents. Extractables are calculated from the original weight of coated wire, from the weight of the extracted solution after the solvent is evaporated, or from the weight of wire after all the coating has been removed. For most wires, the extractable content also provides a good index of the degree of cure. The degree of cure is particularly important for wire that is later to be exposed to the action of various liquids, such as the oils or fluids used in transformers.

Still a third method of determining solvent resistance entails microscopic examination of the coated wire after it has been exposed to solvents. Wire coated with polyvinyl formal, for example, is exposed for 5 min to a mixture of 30 percent toluene and 70 percent denatured alcohol (by volume). Extraction of self-bonding resin from the coating or swelling within 0.5 in. from the end is considered a failure criterion. Besides solvent-resistance tests, the completeness of cure may also be determined by dissipation-factor and other electrical-measurements (see Chapter 4).

Solderability

Solderability, principally of nylon- and polyurethane-insulated wire, is measured by immersing the wire in rosin-alcohol flux, then in molten 50:50 tin-lead solder at temperatures dependent on the size of the wire, and visually inspecting the insulation. The sample is prepared by twisting the ends of a 12-in. length of wire for a distance of ¾ to 1 in. and cutting off the end of the twist, leaving 5 to 10 turns.

Electrical Overload

A pair of twisted wires is used to measure electrical overload. Five successive steps of current, determined by wire type and size, are passed through the wires for 3 minutes each. Temperatures from 345 to 570°C are obtained while a 115 V source is applied between arms of the specimen. This is used to measure burn-out time.

Bond Strength of Self-bonding Wire

Twisted-pair specimens, prepared like those for the dielectric-strength test, are bonded by heating for 30 min at 125°C. The bond strength is then measured using a tensile-testing machine, by clamping the wire ends to the jaws of the machine, pulling at a rate of 12 in./min, and recording the load that causes one wire to slide along the other.

Alcohol Tack

Self-bonding wire is used for the alcohol tack test. The wire is passed over a felt pad saturated with ethyl alcohol and then wound tightly. The coil is subsequently heated at approximately 80°C until all the alcohol has evaporated from the coil (minimum one-half hour).

Thermal Rating and Thermal Stability

Thermal ratings are based on heat-aging tests of twisted-pair specimens of 18 AWG wire with a heavy-film coating, in accordance with ASTM D 2307. According

to this rather complex but widely used test, sets of 10 or more twisted-pair specimens are exposed to temperatures at least 20°C apart for specified cycle periods and are tested by proof voltages which depend on the coating thickness. The proof voltages (60-cycle ac) are selected to produce a voltage stress of approximately 300 volts/mil. The results are extrapolated by statistical procedures outlined in ASTM D 2307 for a predicted life of 20,000 hr (this method serves as the basis for the thermal ratings in Tables 8-4 and 8-5. The thermal-life curves from which extrapolations are made are given in Figures 8-6 to 8-9 for four coatings.) Another, equally lengthy method of evaluating the thermal resistance of magnet-wire insulation may be found in IEEE No. 57.

The thermal stability of wire coatings may be determined more rapidly and accurately by using instrumentation specifically made for thermal analysis. There are three primary techniques for thermal analysis: (1) *Differential Scanning Calorimetry,* (2) *Thermogravimetric Analysis,* and (3) *Thermomechanical Analysis.*

Differential Scanning Calorimetry (DSC). This procedure continuously monitors the heat input (endotherm) or heat generation (exotherm) of a material while it is subjected to a precisely controlled temperature rise. These heat changes correspond to phase changes at the indicated temperature; for example, they may consist of glass transitions, softening or melting, oxidation, sublimation, or decomposition—all characteristics of a given material.[10,11] Applications include determination of the degree of cure of encapsulating compounds and thermosetting resins such as epoxy printed circuit boards.[10]

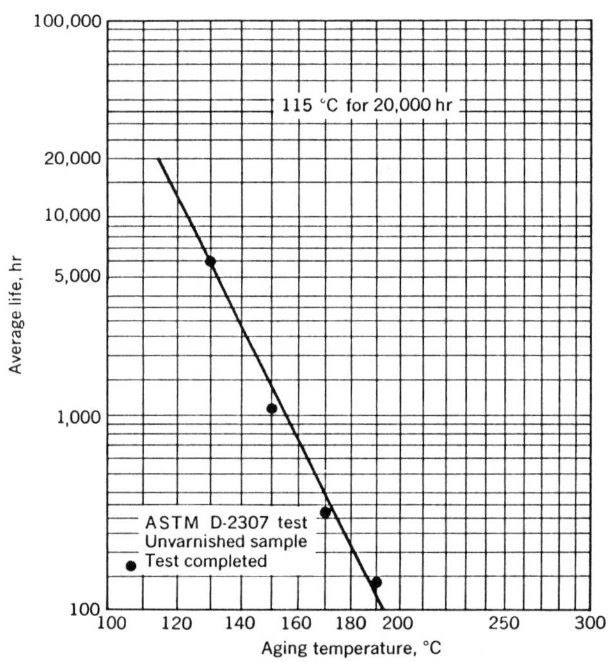

Figure 8-6: Thermal-life curve for Formvar coated wire (polyvinyl formal). (Anaconda Wire & Cable Co.)

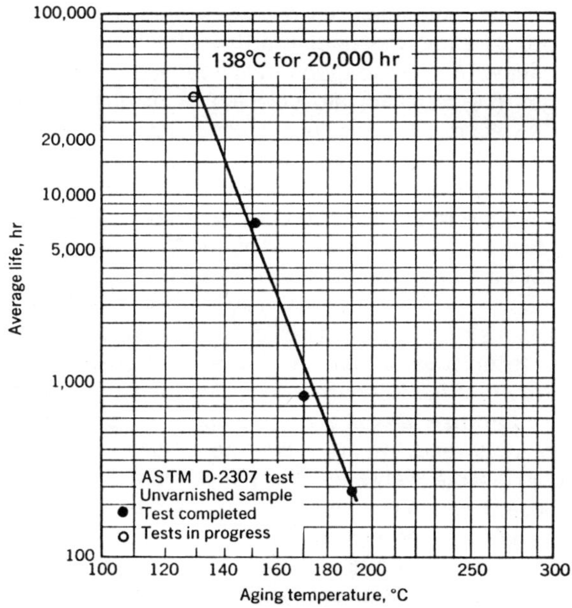

Figure 8-7: Thermal-life curve for Analac-coated wire (polyurethane). (Anaconda Wire & Cable Co.)

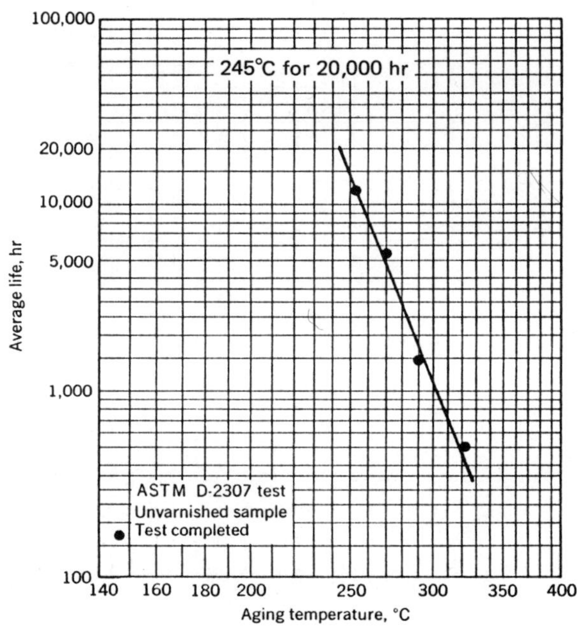

Figure 8-8: Thermal life curve for Pyre-M.L. coated wire (polyimide). (Anaconda Wire & Cable Co.)

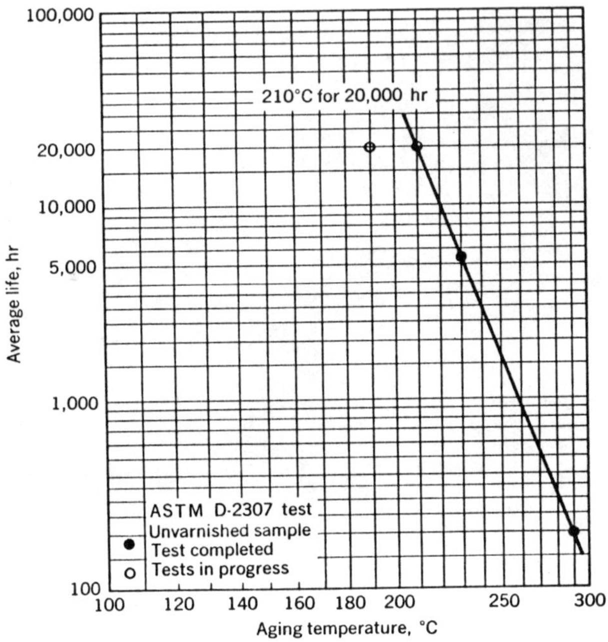

Figure 8-9: Thermal-life curve for Anamid-M coated wire (polyester-imide polyamide-imide). (Anaconda Wire & Cable Co.)

Thermogravimetric Analysis (TGA). This procedure is used to measure weight changes in a sample while subjecting the material to a programmed temperature rise (Figure 8–10). TGA is usually used to determine residual solvents, filler content, and thermal stability of a material.[10]

Thermomechanical Analysis (TMA). This procedure is used to determine a material's expansion coefficient or deflection temperature under load. TMA continuously monitors the expansion or contraction of a sample as a function of loading and temperature. This method is typically used to measure expansion characteristics of heat shrinking wire insulation[11,12] and glob-top encapsulants.[10] Further details and applications of these techniques are discussed elsewhere.[13–18]

EFFECTS OF RADIATION

The modes and mechanisms of radiation damage to wire insulation are essentially the same as those described in Chapter 7; in brief, radiation exposure may result in the breakdown of polymer chains, in further cross-linking, or in outgassing. In the case of wire coatings, radiation has a more pronounced effect on mechanical properties than on electrical properties. Whether the mechanical damage is serious or not depends on the service conditions the wire must meet. For example, embrittlement and loss of flexibility after irradiation are critical only if the wire must remain flexible while it is in service. If the wire is wound on a coil form and

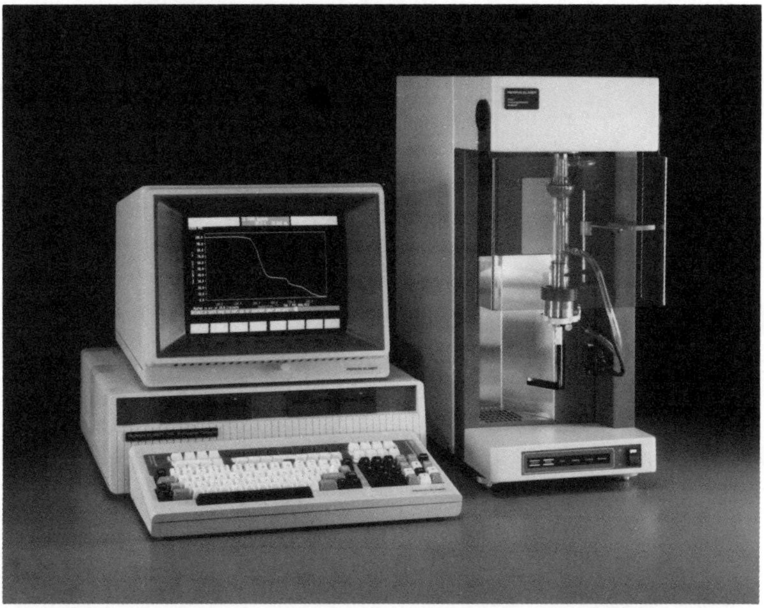

Figure 8-10: Thermogravimetric Analyzer (TGA) with computer. (Courtesy of Perkin-Elmer)

subsequently varnish coated or encapsulated, it may be able to tolerate very high radiation doses because of the added insulation. Even though the wire insulation loses much of its mechanical strength, it is held in place by the encapsulant.

Relative radiation stabilities for various wire coatings are given in Table 8-7. One of the most radiation-resistant coating types is polyimide, and one of the poorest is polytetrafluoroethylene. Only minor changes in the electrical and physical properties of polyimides occur on exposure to 3×10^9 rads,[19] consisting of either 1.33- and 1.17-Mev gamma radiation or 2-Mev electron bombardment for about 500 hr.[20] Table 8-8 shows the very small changes in electrical properties of polyimide-coated glass fabric which occurred at various levels of ionizing radiation up to 3×10^9 rads.[21]

A study was done comparing the radiation resistance of three popular families of insulating varnishes: solventborne oil-modified polyester, solventless epoxy (bisphenol A epoxy based), and unsaturated solventless polyester.[23] A radiation dosage of 10^8 rads was applied, after which the mechanical and electrical properties of the varnishes were measured. Only the epoxy showed a degradation in properties, while the solventborne oil-modified polyester was basically unchanged and the unsaturated solventless polyester had enhanced electrical properties and no mechanical degradation.

The physical properties and life expectancies of wire coatings exposed to radiation alone may differ considerably from those exposed to combined environments, such as heat and radiation or gaseous ambients and radiation.[24] The temperature

Table 8-7: Relative Radiation Stability* of Magnet-Wire Insulations

Insulation	*Stability*
Formex	Good
Alkenex	Good
Polyurethane	Unknown, probably fair
Nylon	Fair
Teflon	Poor
Formex nylon	Fair to Good
Formex Butvar	Good
Alkenex Butvar	Good
Polyimide enamel or varnish	Very good

* Ratings are based on general tests on the insulating materials or on an estimate of radiation tolerance based on the chemical structure of the materials.

Table 8-8: Radiation Resistance of Polyimide Varnish[21,22].

Electrical property*	Initial value	Value after various dosages of 2-Mev electrons		
		1,000 Mrads	2,000 Mrads	3,000 Mrads
Dissipation factor (10^3 hz)	0.0062	0.031	0.0259	0.0388
Dielectric constant (10^3 hz)	3.5	3.4	3.9	4.2
Volume resistivity, ohm-cm $\times 10^{14}$	6.5	5.8	5.3	3.1
Dielectric strength, volts/mil	1,700	1,610	1,720	1,695

* Tests conducted on 4-mil coated product.

effect especially has been shown to be a critical factor, with some materials being improved on irradiation at one temperature and deteriorated at another temperature. Some quantitative data for magnet wires coated with a variety of enamels and varnishes have been obtained by Campbell[25] and are given in Tables 8-9 and 8-10. The average life of wire coated with polyvinyl formal (Formvar) improved on combined thermal-irradiation exposure as compared to thermal aging alone, while polytetrafluoroethylene (Teflon) coatings behaved in entirely the opposite manner. Because of such unpredictable behavior, it is always a safe practice to evaluate and select coatings on the basis of the actual application conditions, and not on the expected radiation dose alone.

STRIPPING OF WIRE COATINGS

To complete an electrical circuit, magnet wire coil leads must be joined to corresponding lead wires. If the wire insulation is not solderable it must be removed prior to connection. Four methods are available for coating removal: mechanical, plasma, thermal, and chemical stripping.

Table 8-9: Effects of Gamma Radiation on Thermal Life of Magnet-Wire Insulations[24,25]

Insulation material	(*a*) Thermal aging, no irradiation		(*b*) Thermal aging following irradiation, aging temperature as in (*a*)				(*c*) Combined environment: aging temperature as in (*a*), dose rate as in (*b*)		
	Aging temp., °C	Av. life, hr	Dose rate, Mrads/hr	Total dose, Mrads	Av. life, hr	Percent of thermal life	Total dose, Mrads	Average life, hr	Percent of thermal life
Polypyromellitimide	300	940	0.37	...	...	...	2,900	7,750	820
Modified polyester with oil-modified phenolic varnish	200	660	0.2	125	720	110	1,020	5,125	780
Polyester with oil-modified phenolic varnish	200	3,160	0.02	125	2,640	84	25	1,260	40
	200	3,160	0.4	...	...	...	2,540	6,350	200
Polyvinyl formal	180	280	0.5	210	240	86	280	560	200
Polyvinyl formal	200	90	0.4	...	...	...	120	300	330
Polyvinyl formal	160	630	0.4	...	...	...	2,200	5,510	870
Polyvinyl formal	180	280	0.09	...	...	...	900	10,000*	>3500

* No failures; exposure discontinued.

Table 8-10: Effects of Gamma Radiation on Thermal Life of Magnet-Wire Insulations[24,25]

Insulation material	(*a*) Thermal aging, no irradiation		(*b*) Thermal aging following irradiation, aging temperature as in (*a*)				(*c*) Combined environment: aging temperature as in (*a*), dose rate as in (*b*)		
	Aging temp., °C	Average life, hr	Dose rate, Mrads/hr	Total dose, Mrads	Av. life, hr	Percent of thermal life	Total dose, Mrads	Average life, hr	Percent of thermal life
Silicone with silicone varnish	240	350	0.5	125	320	90	290	570	160
Modified silicone	240	500	0.4	...	...	...	124	310	60
Polytetrafluoroethylene with silicone varnish	270	>10,000	0.015	25	12	<0.1	1.2	60	<0.6
Polytetrafluoroethylene	270	>10,000	0.015	...	...	...	1.0	50	<0.5
Polytetrafluoroethylene	180	>10,000	0.014	...	...	...	1.5	105	<1
Polytetrafluoroethylene	13	∞	0.015	...	...	...	34	2,280	
Polytetrafluoroethylene	13	∞	0.32	...	...	...	240	760	

Mechanical stripping is the preferred process. Unlike the other three methods, no chemicals are required, therefore no hazardous waste must be handled or disposed. Mechanical strippers utilize cutters or abrasive wheels to remove the insulation. Equipment can be automated and programmed to strip a variety of coatings and wire types and gauges.

For small-gauge wire, plasma stripping works well. This is the same technique described in Chapter 6 for removal of conformal coatings from printed circuit boards. Generally, an oxygen or an oxygen/fluorocarbon plasma is used. This process requires masking (with foil or heavy tape) the portion of the wire or coil that is to remain coated—often Teflon. Plasmas are only efficient when removing thin coatings of a few mils or less.

Thermal stripping requires high temperatures that may damage the wire. Some coatings can be removed simply by dipping in molten solder while others require immersion in molten inorganic salts to burn off the wire insulation. However the high temperatures and oxidizing nature of these salts can corrode the wire.

Chemical strippers, popular in the sixties and seventies, have lost favor due to recent regulations concerning toxic and hazardous substances. Also, the insulators most often used, polyimides and polyurethanes, are very resistant to chemical removal. The chemical strippers used are corrosive (usually alkaline) and can damage the wire or cause electrical failures if residues are not rinsed off completely after stripping.

REFERENCES

1. "Modern Plastics Encyclopedia," vol. 45, no. 1A, pp. 764–772, McGraw-Hill, New York, 1968.
2. Wire, Magnet, Electrical," J-W-1177A; September 1976.
3. Magnet Wire, NEMA MW 1000.
4. *Electri-onics Electrical Edition,* January 1987, pp 14–27.
5. Cohen, S. M., R. E. Kass, and E. Lavin: Chemical Interactions in the Poly(vinylformal)-phenolic Resin System, *Ind. Eng. Chem.,* **50:** 229 (1958).
6. Lavin, E., A. H. Markhart, and R. W. Ross: Recent Developments in Magnet Wire Based on Formvar, *Insulation,* April, 1962.
7. Sollner G. N., et al: Review of Solderable Magnet Wire Insulation, *Proceedings of the 17th EEIC,* 1983.
8. Wahlgren, W. W.: Use of Epoxy Resins in High Reliability Transformers, *Electron, Equipt. Eng.,* January 1959.
9. Fieser, L. F.: "Experiments in Organic Chemistry," pp. 40–42, Heath, New York, 1941.
10. Brennan, W. P. and R. B. Cassel: Thermal Analysis Application Study 25, 1978 Pittsburgh Conference Paper No. 569, April 1978.
11. Brennan, W. P.: Thermal Analysis Application Study 24, March 1978.
12. Earnest, C. M.: Modern Thermogravimetry, *Analytical Chemistry,* **56,** 1471A, 1984.
13. Hunt, C. F., and A. H. Markhart: *Insulation,* vol. 6 (November, 1960).
14. Saito, Y., and T. Hino: *Proc. AIEE,* 1960.
15. Anderson, D. A., and E. S. Freeman: *Anal. Chem.,* **31:** 1697 (1959).
16. Ke, B., and A. W. Sisko: *J. Polymer Sci.,* **50:** 87 (1961).
17. Murphy, C. B.: Differential Thermal Analysis, *Anal. Chem.,* **30:** 867–872 (1958).

18. Murphy, C. B.: Differential Thermal Analysis, *Anal. Chem.,* **32:** 168R-171 (1960); Murphy, C. B.: Differential Thermal Analysis, *Anal. Chem.,* **34:** 298R-330R (1962).
19. Pyre-M.L. Varnish (RK-692), *Du Pont Bull.* 1, April, 1966.
20. Milek, J.: Polyimide Plastics: A State-of-the-Art Report, *Hughes Rept.* S-8, October, 1965.
21. Leppla, R. R., and R. R. Carryer: Polyimide Insulation System for Higher Operating Temperatures, More Compact Units, *Insulation,* June, 1963.
22. Schweitzer, F. E., et al.: An Evaluation of Insulation Systems Based on Polymer ML, Fourth Electrical Insulation Conference, *NEMA Publ.,* 1962.
23. Johnson, D. S. and E. A. Thomas: The Radiation Resistance of Some Common Insulating Varnishes, *Proceedings of the 17th EEIC,* 1983
24. Brancato, E. L., and L. M. Johnson, Thermal Aging Problems in Electrical Insulation: Past, Present, and Future, Eighth Electrical Insulation Conf., *IEEE Pub.* 68C6-EI, December, 1968.
25. Campbell, F. J.: Combined Environments Versus Consecutive Exposures for Insulation Life Studies, *IEEE Trans.,* vol. NS-11, no. 8 (November 1964).

9

Coatings for Electronic Components and Devices

Organic coatings, either alone or in combination with transfer-molded or cast resins, provide an extremely important and economical means of packaging and protecting electronic components. Organic coatings are also gaining popularity as passivation coatings, interlayer dielectrics, and alpha-particle barrier coatings, replacing traditional inorganic materials. Plastic-encapsulated devices are rapidly replacing those in hermetically sealed cans for many applications. Resistors, capacitors, transistors, diodes, and integrated circuits are all now available in plastic packages. So great has been the impact of plastics that production of encapsulated transistors alone now numbers in the tens of millions of devices per month. Indeed, the sealing of devices in plastics appears to be the dominant trend, especially in the commercial semiconductor industry.

The popularity of plastics stems from the simplicity and rapidity with which they can be processed and applied, resulting in substantial economics.[1,2] Dramatic cost reductions have been achieved, especially for transistors and integrated circuits. Plastics may also be molded in an almost infinite variety of configurations, although most plastic containers are designed to be interchangeable with their metal-can counterparts (Figure 9-1). In this chapter coatings, glob-tops, and transfer-molding compounds for discrete integrated circuits will be discussed. Also, thin-film coatings used as passivation, interlayer dielectrics, and barrier coatings for integrated circuits will be described. Finally, coatings for hybrid microcircuits will be covered. The qualities which plastics and coatings must have to yield reliable devices, the limitations of plastics for some military and NASA environments, examples of available materials and suppliers, and application methods are also discussed.

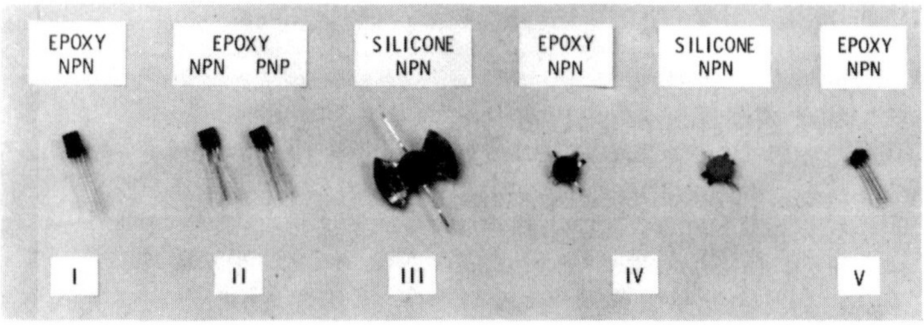

Figure 9-1: Plastic-packaged transistors[2]: (I) small-signal npn in epoxy; (II) small-signal npn and pnp in epoxy; (III) high-power high-frequency npn in silicone; (IV) high-power audio npn in silicone and epoxy; (V) small-signal npn switch in epoxy.

FUNCTIONS OF PLASTIC PACKAGING MATERIALS FOR DISCRETE DEVICES

The combination of a coating and a molding plastic provides greater rigidity, dimensional stability, impact resistance, and moisture protection than a coating alone. The main functions of the coating, also referred to as the *glob-top coating, barrier coating,* or *junction coating,* are:

1. To hold delicate components and fine interconnection wires together, allowing them to be handled, dropped, or vibrated without electrical or mechanical damage. This function is sometimes called "ruggedization."
2. To impart stress relief. Stresses arising from the transfer-molding operation, resin shrinkage, differences in coefficients of expansion between the plastics and substrates, thermal cycling, and thermal shock may be equalized and dissipated by using barrier coatings.
3. To provide some degree of moisture, gas, and dust protection, thus preventing corrosion and electrical malfunctioning.
4. To provide an insulation barrier, thus preventing electrical shorting.

The main functions of the transfer-molding or casting compounds are:

1. To provide a well-defined structure with exact dimensional tolerances comparable to those for hermetic cans. This permits standardization and interchangeability of parts.
2. To provide high impact and mechanical-shock resistance.
3. To provide a longer path for moisture, reducing moisture penetration. The use of the coating alone does not afford a sufficient moisture seal.

PLASTIC PACKAGING MATERIALS

Early experimental work on coatings for the protection of semiconductor junctions was performed between 1955 and 1960. Early compositions consisted of waxes combined with plastics.[3,4] Later, silicone resins filled with lead tetroxide or mercuric oxide were used to coat transistors and germanium rectifying diodes.[5] In 1962 the first epoxy-packaged transistors were mass produced for the consumer and automotive markets.[6] Finally, plastic-packaged devices were introduced in 1964 for industrial and computer applications, and in 1966–1967 for some military applications.

Several types of coatings and molding compounds are presently employed for plastic-packaged semiconductor devices and other electronic components. These types are predominantly epoxies and silicones, with polyimides and polyurethanes being used to a lesser extent (Tables 9-1 and 9-2). The type and specific formulation are dictated by the function of the component and its operating environment.

Table 9-1: Molding and Casting Compounds for Electronic-Component Packaging

Trade name	Manufacturer	Chemical type	Characteristics
DK18-05	Hysol	One-part epoxy powder	Low temperature fusing powder for temperature sensitive electronic components; for all types of capacitors.
DK36	Hysol	One-part epoxy powder	For high volume coating of electronic components; good moisture protection; suitable for all capacitors and certain semiconductor devices.
136	Insulcast	Two-part epoxy	For casting of electronic components; good flow; conforms to MIL-I-16923.
140	Insulcast	Two-part epoxy	High thermal conductivity and low viscosity for casting of electronic components.
70-C.C.	Insulcast	Two-part epoxy	Semiflexible, low stress and shrinkage for encapsulating electronic components.
4128	Crest	Two-part epoxy	Soft potting compound that can removed for repair of components.

(continued)

Table 9-1: (continued)

Trade name	Manufacturer	Chemical type	Characteristics
Uniset 925-13	Emerson & Cuming	One-part epoxy	Highly filled epoxy with good resistance to heat, water, Freon; low expansion coefficient and high thermal conductivity; service temperature above 180 C.
Uniset G-508-1	Emerson & Cuming	One-part epoxy	Pourable, flame retardant epoxy with good resistance to heat, water, Freon; low expansion coefficient; service temperature above 180 C.
Stycast 2850FT	Emerson & Cuming	One-part epoxy	Thermally conductive; low grit; low thermal expansion.
Stycast 1365	Emerson & Cuming	Two-part epoxy gel	Repairable gel; optically clear.
Stycast 1090	Emerson & Cuming	One-part epoxy	Lightweight, thermally insulating encapsulant with low dielectric constant.
4757	APCO	Two-part epoxy	High performance, room temperature setting; very low exotherm.
4936	APCO	Two-part epoxy	Used for potting transformers, electronic components and products that require large masses with a low heat rise.
Aratronic 5001	Ciba-Geigy	Bisphenol A epoxy resin	High purity, low chlorine content; for potting, encapsulating and glob top coatings
Aratronic 5011 5012, 5014	Ciba-Geigy	Bisphenol A epoxy resins	High purity, low chlorine content; different melting points; for molding and fluidized bed coatings.
Aratronic 5057	Ciba-Geigy	Epoxy novolac resin	High purity, low viscosity; designed for high temperature applications.
Epolite 4004	HEXCEL	Two-part epoxy	Used for capacitors, micro-electronics and other modular components; low viscosity and low surface tension.
Epolite 4007	HEXCEL	Two-part flame-retardant epoxy potting compound	Used for potting electronic components.

(continued)

Table 9-1: (continued)

Trade name	Manufacturer	Chemical type	Characteristics
Epolite 6518	HEXCEL	One-part epoxy	Used for small mass potting, encapsulating and vacuum impregnating of enameled wire wound components and taped coils.
Epolite 6523	HEXCEL	One-part self-leveling epoxy	Used for sealing, potting and encapsulating electronic components.
Epolite 6526	HEXCEL	One-part epoxy	Used for capping and sealing pin grid arrays and integrated circuits for moderately high temperature use.
Epolite 6534	HEXCEL	One-part semi-rigid epoxy	Used for potting and encapsulating solenoids and coils.
Polybac 131	Polygem	Two-part epoxy	Low viscosity, filled, excellent thermal shock; for potting, casting, dipping and impregnation encapsulation.
Polybac 416	Polygem	Two-part epoxy	Medium viscosity, service temperature 150 C; for potting, casting, dipping and impregnation encapsulation.
Polybac 426	Polygem	Two-part epoxy	Low exotherm; excellent thermal shock; for potting, casting, dipping and impregnation encapsulation.
Polybac 199	Polygem	One-part epoxy	Low viscosity; UL approved; for potting, casting, dipping and impregnation encapsulation.
Polybac 200	Polygem	One-part epoxy	Low viscosity; filled system; for potting, casting, dipping and impregnation encapsulation.
Polybac 1201	Polygem	One-part epoxy	Low shrinkage; for potting, casting, dipping and impregnation encapsulation.
Polybac 1221	Polygem	One-part epoxy	Thixotropic for dipping.

(continued)

Table 9-1: (continued)

Trade name	Manufacturer	Chemical type	Characteristics
ES4322	Hysol	Two-part epoxy	High purity epoxy for chip-on-board, hybrid and TAB applications.
ES4322	Hysol	Two-part epoxy	High purity epoxy encapsulant for chip-on-board, hybrid and TAB applications.
ES4323	Hysol	Two-part epoxy	High purity epoxy glob top for chip-on-board, hybrid and TAB applications.
FP4400	Hysol	Two-part epoxy	High purity, low stress, premixed and frozen epoxy encapsulant for potting and coating applications.
FP4401	Hysol	Two-part epoxy	High purity, low stress, premixed and frozen epoxy glob top encapsulant for potting and coating applications.
Stycast 2651	Emerson & Cuming	One-part epoxy	Meets MIL-I-16923.
Conacure RN-1000	Conap	Two-part epoxy	Low viscosity casting system for encapsulating and potting resistors, connectors, solenoids, transformers, coils and other electrical devices.
Conacure FR-1046	Conap	Two-part epoxy	Filled, nonabrasive, low viscosity, low shrinkage, low exotherm system; resistant to thermal shock; good electrical insulation.
Conacure RN-1200	Conap	Two-part epoxy	Low viscosity potting and encapsulating system with excellent impact and thermal shock resistance.
Conacure RN-1600	Conap	Two-part epoxy	Flexible, unfilled, low viscosity system for potting, encapsulating and impregnation.
Conathane EN-2527	Conap	Polyurethane	For potting and encapsulating applications where excellent thermal conductivity is necessary; can tolerate temperature extremes between -55 to 130 C.

(continued)

Table 9-1: (continued)

Trade name	Manufacturer	Chemical type	Characteristics
Conathane EN-2522	Conap	Polyurethane	Unfilled, low viscosity system; can tolerate temperature extremes between -55 to 130 C.
Conathane EN-4 to -12 series	Conap	Polybutadiene-based urethane	Elastomers that are hydrolytically stable; excellent electrical properties over a wide temperature range; low exotherm with minimal shrinkage and low toxicity.
Eccosil 4952	Emerson & Cuming	Condensation cure Silicone	Thermally conductive encapsulant; working temperature range -75 to 315 C.
306	Dow Corning	Glass-filled silicone	A glass-filled silicone that can be used for encapsulating integrated circuits, transistors rated at 25 W.
307	Dow Corning	Silicone with short glass fiber and silica-glass filler	Designed specifically for transfer molding of wire-wound power resistors and high-power semiconductor devices; molds at 150 to 180 C and 500 to 1500 psi.
308	Dow Corning	Silicone transfer-molding compound	Used in packaging integrated circuits, RF power transistors and other transistors, varistors, capacitors, and potentiometers; molds at 160 to 180 C and 400 psi in 30 secs; service temperatures range from -65 to 175 C.
480	Dow Corning	Silicone transfer-molding compound	Molding diodes, rectifiers, small signal and power transistors, integrated circuits, connectors and connector inserts; molds at 160 to 180 C and 500 psi in 1.25 min.
1-5021	Dow Corning	Silicone transfer-molding compound	High speed molding of power transistors and integrated circuits; molds at 160 to 180 C and 500 psi in 1 to 2 min.; low, nearly isotropic coefficient of thermal expansion.

(continued)

Table 9-1: (continued)

Trade name	Manufacturer	Chemical type	Characteristics
Q1-5036	Dow Corning	Silicone transfer-molding compound	High speed molding of small semiconductor devices such as rectifiers and transistors.
631	Dow Corning	Silicone-epoxy hybrid molding compound	Molding integrated circuits where a high degree of reliability is desired; superior moisture, salt atmosphere resistance; excellent lead seal and strength.
Eccosil 4122	Emerson & Cuming	Condensation cure Silicone	Low viscosity encapsulant; working temperature range -75 to 315 C.
Eccosil 4640	Emerson & Cuming	Condensation cure Silicone	Low density encapsulant; working temperature range -75 to 315 C.
Eccosil 5089	Emerson & Cuming	Addition cure Silicone	Repairable gel; cures in 20 minutes at elevated temperatures; working temperature range -75 to 315 C.
RTV 11	G.E.	RTV Silicone	Pourable; for general purpose electrical potting application.
RTV 31	G.E.	RTV Silicone	Pourable; for electrical potting and mechanical applications; service temperature is 260 C range.
RTV 8111	G.E.	RTV Silicone	Pourable, fast cure; meets MIL-S-23586; for military and aerospace electrical potting applications.
RTV 511	G.E.	RTV Silicone	Pourable, methyl-phenyl product for extreme low temperature (-115 C); for potting airborne electronic assemblies.
R.T.V.S. 11	Insulcast	RTV Silicone	Low viscosity; low durometer potting compound.
R.T.V.S. 13	Insulcast	RTV Silicone	Stabilized for high working temperatures of 260 C continuous, 315 C short term.
R.T.V.S. 65	Insulcast	RTV Silicone	Clear, low viscosity for optical clarity, easy repair.
R.T.V.S. 49	Insulcast	RTV Silicone	Thermally conductive and high temperature resistant.

Table 9-2: Silicone Barrier and Junction Coatings (Dow Corning*)

Trade name	Chemical type	Characteristics
HIPEC 643	Rigid, one-part, silicone coating, solvent solution	Semiconductor junction passivation; high temperature, high voltage applications; alpha particle barrier.
HIPEC 648	Rigid, one-part, silicone coating, solvent solution	Semiconductor junction passivation; high temperature, high voltage applications; alpha particle barrier.
HIPEC 649	Rigid, one-part, silicone coating, solvent solution	Semiconductor junction passivation; high temperature, high voltage applications; alpha particle barrier.
HIPEC R-6101	Clear, elastomeric, 100-percent solids, one-part, silicone coating; low viscosity	Semiconductor junction passivation for integrated and hybrid circuit protection, LED displays and optoelectronics; alpha particle barrier.
HIPEC R-6102	Black, elastomeric, 100-percent solids, one-part, silicone coating; low viscosity	Semiconductor junction passivation for integrated and hybrid circuit protection; alpha particle barrier.
HIPEC 1-4935	White, elastomeric, 100-percent solids, one-part, filled silicone coating; high viscosity	Protection and passivation of high voltage diodes and rectifiers.
HIPEC 3-6626	White, elastomeric, 100-percent solids, one-part, filled silicone coating; medium viscosity	Protection and passivation of high voltage diodes and rectifiers.
HIPEC 3-6550 RTV dispersion	Clear, elastomeric, RTV solvent dispersion, silicone coating	Protection of integrated and hybrid circuits; excellent moisture resistance; alpha particle barrier.
HIPEC R-6100	Black, elastomeric, 100-percent solids, two-part, silicone coating; high viscosity	Semiconductor junction passivation; alpha particle barrier.

(continued)

Table 9-2: (continued)

Trade name	Chemical type	Characteristics
HIPEC R-6103	Clear, elastomeric, 100-percent solids, two-part, silicone coating; low viscosity	Semiconductor junction passivation for optoelectronic applications; alpha particle barrier.
HIPEC R-6104	Clear, elastomeric, 100-percent solids, two-part, silicone coating; medium viscosity	Semiconductor junction passivation for optoelectronic applications; alpha particle barrier.
HIPEC Q1-4939	Clear, elastomeric, 100-percent solids, two-part, silicone coating; low viscosity	Semiconductor junction passivation for optoelectronic applications and integrated circuits; alpha particle barrier; provides stress relief and improved low temperature performance.

*Dow Corning Product Bulletins, Form No. 10-008C-86 and 10-120-86.

High-temperature resistance and stable electrical parameters at elevated temperatures are required for materials used to protect power transitors because of the appreciable amounts for heat generated by these components. The most critical plastic requirement for semiconductors is high purity, the absence of ionic and other contaminants. Even among transistor devices, the degree of chemical impurities that can be tolerated differs with the function of the devices. Low-current switches and metal-oxide semiconductor devices, for example, are more sensitive to chemical impurities than are amplifiers or silicon rectifiers. A different plastic formulation may be required for each of these transistor devices. The behavior of plastics on npn junctions has also been found to differ from that on pnp junctions, and again a specific plastic may have to be used for each type.

A number of so-called "semiconductor-grade" silicone and polyimide coatings are on the market. These combine purity with flexibility and high temperature stability. Some firms, however, purify their coatings prior to use by molecular distillation, solvent extraction, or other means. In addition to silicones and polyimides, the parylene coatings have been investigated and found to be useful barrier coatings (see Chapter 2). However, polysulfides, epoxy polysulfides, and polyurethanes, although they are very flexible, are not recommended for surface-active devices because of potential contamination, corrosion, poor electrical properties, or poor thermal stability. Some commonly used silicone barrier and junction coatings and their characteristics are given in Table 9-2.

Glob-Top Coatings

Glob-top gel coatings are possible alternatives to the traditional hermetic enclosures (metal or ceramic packages) used in military and space microelectronics.[7,8] They are generally very soft, high-purity silicones which are reported to protect delicate electronics from moisture, salt spray, vibration, handling, and alpha-particles.[9,10] Though not hermetic in the true sense of the word, since they are permeable to moisture to varying degrees, gels can provide a tightly adhering barrier, preventing moisture from condensing at the interface and on the surfaces of active and passive devices. Although plastic encapsulants and packages have been used for years in the protection and packaging of commercial electronic devices, they have not been considered reliable enough for military and space requirements. This point of view, however, is changing as new and improved coatings and encapsulants are developed and as improvements are made in the integrity of inorganic passivation coatings used in device fabrication. In fact, many integrated circuit chips, because of the excellent integrity of the top passivation layer, are considered hermetic in themselves. The rapid developments in the design and fabrication of hybrid wafer-scale integrated circuits make the use of gel coatings or other localized encapsulants ideal.[8-10] The weight and cost gains realized by wafer scale miniaturization can easily be forfeited by having to hermetically seal these wafers in large, heavy and costly metal or ceramic packages.

Yet another benefit of the silicone gels is that most of them are optically clear. This is particularly important for devices such as EPROMs (Erasable, Programmable, Read-Only Memories) that are programmed using visible or ultraviolet light.[11] Silicone gels have also been used inside hermetic packages to getter any particles trapped within the package.[11] A compilation of some commercially available silicone gels and their characteristics is given in Table 9-3.

THIN-FILM COATINGS

In recent years the use of polymeric materials in fabricating semiconductor devices has greatly expanded. This has been largely due to the development of new high-temperature stable polymers and improvements in the purification of these products. In particular, polyimides are now used extensively as fabrication aids, interlayer insulation, passivation coatings and alpha particle barriers.[12]

Fabrication Aids

As electronic device dimensions shrink, it is more difficult to maintain tolerances for conductors and interconnect vias and obtain optimum resolution of the desired pattern. To achieve this, low-molecular-weight polyimides are used in photolithography as a first layer to planarize the surface of a substrate onto which photoresist can be applied to define the circuit pattern. After the conductor pattern is defined in the polyimide, the photoresist is stripped off and metal is deposited over the entire substrate by sputtering or evaporation. Then the substrate is submerged in a solvent that dissolves the polyimide and "lifts off" the metal in all areas except the circuit pattern. This is called a bilayer lift-off process and permits high resolution photolithography.[13,14] In a trilayer lift-off process a third layer is used to improve the adhesion between the polyimide and the photoresists.[15,16]

Table 9-3: Silicone Gel Coatings

Trade name	Manufacturer	Chemical type	Characteristics
SC-17	Amicon	Clear, 2-component gel, 100 percent solids	Very soft dielectric gel for encapsulation/protection of devices
SC-2600	Amicon	High purity 2-component gel	Compatible with MOS, other active devices; adhers to die and hybrids; stress relief, alpha particle barrier
S-1286 A&B	Castall	Transparent, 2-component gel, RTV cure	For shock absorption needed to encapsulate delicate devices
HIPEC Q3-6550	Dow Corning	One-part RTV silicone	Designed to protect IC's and hybrids from moisture; alpha particle barrier
HIPEC X3-6646	Dow Corning	Transparent, 2-component, self-healing gel	Designed for sealing, protecting, preserving, cushioning; provides protection at low temperatures (-80°C)
Sylgard 527	Dow Corning	Transparent 2-component gel	Designed to seal, protect, and preserve microelectronics
Petrarch OE	Dynamit Nobel	Transparent, solventless, 2-component elastomer	Flexible, optically clear molding, encapsulating and coating compound
Petrarch 015	Dynamit Nobel	2-component RTV;addition cure	Viscoelastic encapsulating and coating compound
Petrarch 060	Dynamit Nobel	2-component RTV;addition cure	More viscous version of 015
Eccosil 5089	Emerson & Cuming	Transparent RTV; addition cure	Encapsulant for cushioning, vibration dampening, and environmental protection
Eccosil Gel	Emerson & Cuming	Very soft RTV; addition cure	Encapsulant for cushioning, vibration dampening, and environmental protection
Metasil 5016	Mereco Products	Transparent, solventless, 2-component RTV gel	Encapsulant for microelectronics
Metasil 5016	Mereco Products	Transparent, solventless, 2-component RTV gel	Encapsulant for microelectronics
RTV 132	Rhone Poulenc	2-component RTV addition cure	For potting delicate electronic assemblies; vibration suppression
KE 1052 A/B	Shin-Etsu Silicones	Transparent, 2-component liquid; RTV cure	Packaging and sealing electronic assemblies; moisture barrier; vibration suppression
KE 110 A/B	Shin-Etsu Silicones	Transparent, 2-component, deep section curing (150C)	Sealing, packaging, vibration dampening, moisture barrier
SC-139	Thermoset Plastics	Clear, 2-component, addition cure	Seal and protect microelectronics

Planarization with polyimide is especially important for multilayer circuitry where underlying structures (conductor lines and vias) can affect the resolution and profile of the next layer of circuitry. These processes make it possible to define extremely fine lines with good profiles. However, lift-off processes are limited to a maximum metal layer thickness on the order of 1 micron. Addition of an aluminum mask over the polyimide coating and reactive-ion-etching allows the polyimides to serve as a high-resolution ion-implant mask.[17]

Photosensitive polyimides. The increasing popularity of polyimides as coatings for electronics led to the development of photoimageable polyimide systems. By rendering the dielectric coating inherently photosensitive the secondary use of photoresist is eliminated. This reduces the number of steps required to fabricate multilayer structures thereby increasing efficiency of the process and improving the yield (Figure 9–2).

Photosensitive polyimides were pioneered at Siemens Company in the early 1980s.[18] Since then, new systems have been developed that have longer shelf lives, shorter photospeeds and can be applied thicker.[19] These coatings are solvent-soluble and act as negative photoresists (see Chapter 5 for discussion).

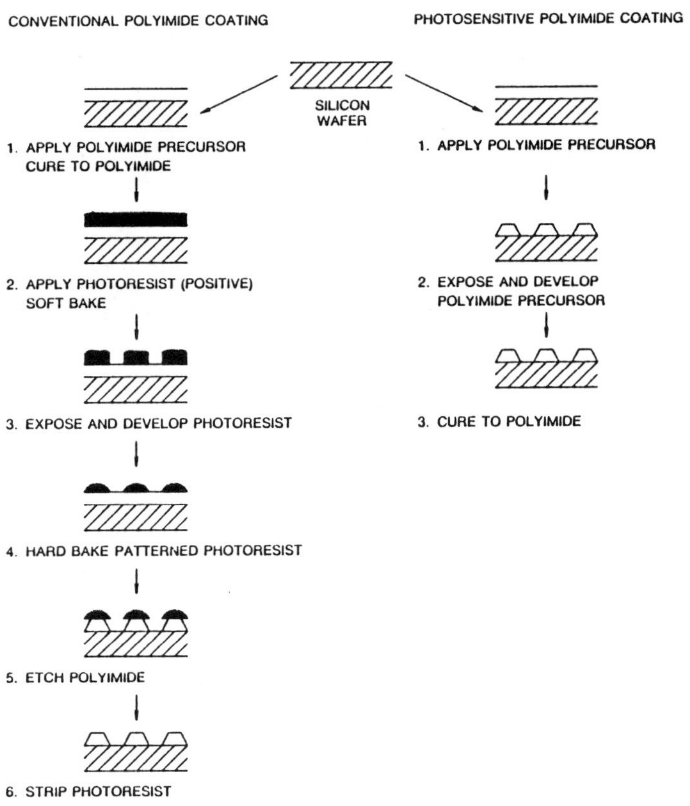

Figure 9-2: Comparison of processes for structuring conventional and photosensitive polyimides.

The photosensitivity is obtained by using a polyamic acid polymer bearing a photoreactive side group (examples in Chapter 2). After application and a soft bake, the polyimide is exposed to ultraviolet light which causes the polymer to cross-link. The masked areas of the coating remain solvent soluble and are removed during development. Finally, the coating is heat cured at 200 to 350°C resulting in a fully imidized polyimide.

Interlayer Insulation and Passivation

A permanent coating on an electronic device must be ultrapure (particularly, free of ionic impurities), have a low dielectric constant, good adhesion to the substrate and act as a barrier to moisture and other contaminants. In addition, the coating must be chemically resistant, mechanically tough, and withstand wafer processing temperatures up to 400°C. Often these coatings must serve as alpha particle barriers as well.

Both the solvent-soluble and photoimageable polyimides as well as some high-purity silicones are used for these purposes. Since the late seventies a great deal of work has been done on the use of polyimides as interlayer insulators[20–24] and passivation coatings.[25–27] Examples of polyimide passivation coatings are described in Table 9-4; silicones are described in Tables 9-2 and 9-3.

Table 9-4: Polyimide Interlayer Insulators and Passivation Coatings

Trade name	Manufacturer	Chemical type	Characteristics
Pyralin PI 2545	Du Pont	14% solids polyamic acid solution in NMP/ aromatic hydrocarbon	Low viscosity, high purity; for passivation of semiconductors; spin coating yields thickness of 1 micron.
Pyralin PI 2555	Du Pont	19% solids polyamic acid solution in NMP/ aromatic hydrocarbon	Low viscosity, high purity; for passivation of semiconductors; spin coating yields thickness of 1 micron.
Pyralin SP PI-2570-D	Du Pont	Self-priming polyamic acid; excellent adhesion to aluminum, silicon oxide/ nitride	Low viscosity, high purity; for passivation and mechanical protection of wafers; die and junction coating.
Pyralin PD PI-2701-D	Du Pont	Negative type, photodefinable polyimide solution in NMP/MIBK	Low viscosity, high purity; for interlayer dielectric and passivation coating or as a high temperature resist; can be applied by spinning or spraying.
Pyralin PD PI-2702-D	Du Pont	Negative type, photodefinable polyimide solution in butyrolactone/acetone	Low viscosity, high purity; for passivation coating, thick film interlayer dielectric or as a high temperature resist; applied by spinning.

(continued)

Table 9-4: (continued)

Trade name	Manufacturer	Chemical type	Characteristics
Probimide 284	Ciba-Geigy	Fully imidized solvent-soluble polyimide	For use as a passivation coating, interlayer dielectric, lift-off processes; spins to 1 to 3 microns; applied directly on circuitry or over glass or nitride.
Probimide 285	Ciba-Geigy	Fully imidized solvent-soluble polyimide	For use as interlayer dielectric; high planarization; spins to 0.5 to 1.2 microns.
Probimide 286	Ciba-Geigy	Fully imidized solvent-soluble polyimide	For use as x-ray mask.
Probimide 287	Ciba-Geigy	Fully imidized solvent-soluble polyimide	For thicker coating applications; can be spun or dispensed.
Probimide 293	Ciba-Geigy	Fully imidized solvent-soluble polyimide	For thicker coating applications and alpha particle protection; can be spun or dispensed.
Probimide 348	Ciba-Geigy	Fully imidized solvent-soluble, photodefinable polyimide	High solids content for thicker coatings; alpha particle barrier, general passivation; high resolution, high photospeed.
Probimide 337	Ciba-Geigy	Fully imidized solvent-soluble, photodefinable polyimide	Interlayer dielectric, general passivation, ion implant mask; high resolution, high contrast.
PIQ-13	Hitachi	14.5% Polyamic acid solution	Interlayer dielectric, general passivation and insulation
PIX-1400	Hitachi	14.5% Polyamic acid solution	Interlayer dielectric, general passivation and insulation; no adhesion promoter needed
PIQ-3200	Hitachi	14.5% Polyamic acid solution	Alpha particle barrier; wide range of thicknesses
PIX-8100	Hitachi	20.0% Polyamic acid solution	Alpha particle barrier; wide range of thicknesses; no adhesion promoter needed
PL-1100	Hitachi	14.5% Polyamic acid solution; photodefinable	Interlayer insulation, passivation
PL-1200	Hitachi	13.7% Polyamic acid solution; photodefinable	Interlayer insulation, passivation

FUNCTIONS OF COATING MATERIALS FOR HYBRID MICROCIRCUITS[28]

Particle Immobilizing Coatings

Though organic coatings are not generally used in hermetically sealed hybrid circuits, some NASA and military programs have required their use to assure that any particles that remain on the circuit after cleaning or that slough from the circuit after sealing do not migrate and cause electrical shorts. The function of the coating is to freeze in place all loose particles and to protect the circuit from any particles that may subsequently be generated during vibration, constant acceleration testing, or actual use.

Parylene Coatings. Parylene coatings have received the greatest attention as particle immobilizing coatings for hybrid circuits. Discussion of the characteristics and properties of these coatings is in Chapter 2. The application method and their use as particle immobilizers are described in Chapter 6.

Solvent-Soluble Coatings. High-purity coatings have been investigated as alternates to parylene because of the difficulties that have been encountered in removing parylene from circuits that need to be reworked. The most useful of these have been the block copolymers of silicones and styrenes. These styrene-modified silicone coatings are readily and completely soluble in toluene, xylene, the Freons, and chlorinated solvents. The coatings can be removed quickly and easily from the entire circuit by vapor degreasing with Freon TF solvent. In the original studies performed by Licari and Weigand, Dow Corning's DCX9-6326, a block copolymer of dimethylsiloxane and α-methyl styrene, was characterized for electrical, physical, and chemical properties, then applied to both hybrid microcircuits and printed wiring boards.[29-31] These studies were later corroborated and expanded by David.[32] In the studies, extensive long term reliability data were collected on the effects of the coating on sensitive chip devices and on gold and aluminum wire bonds to thick-film gold conductor pads. Wire bond pull strengths and electrical resistance of a series of wire bonds were measured after 1000 hrs at 150°C and after 1000 temperature cycles from −55 to +125°C. In all cases the performance of the coated wire bonds was almost identical to the uncoated bonds and no electrical failures of devices were noted. Though these coatings are more economical to process from the standpoint of ease of removal for rework, ease of masking, and the lower costs of spray application and equipment, they are still experimental and limited in availability.

Particle Getters. Plastic resins that cure to a tacky state may be used as alternates to coatings in immobilizing particles. These particle getters are soft silicone gels and are applied to the center of the inside of the lid, prior to sealing. The effectiveness of silicone getters has been demonstrated by several companies through controlled experiments in which known quantities of metal particles of various sizes were purposely added (seeded) to a circuit containing the getter, then vibrated and evaluated for loose particles by particle impact noise detection (PIND), x-ray, and electrical testing.[33] The types of particles used were representative of those that might be found in a hybrid package. They included gold, aluminum, solder balls, and pieces of eutectic alloy. It is important to apply the getter to the center of the lid so that the seal area does not become contaminated with organic "bleed-out" from the silicone which could degrade the seal. A distance of 0.1 inch minimum from the seal perimeter area should be kept free of all organic

contaminants. Another consideration in the use of silicone getters is the amount of outgassing. As with other organic materials, the extent of outgassing is a function of the material used, its degree of cure, and the vacuum bake schedule. Silicones and their properties are described in Chapter 1.

APPLICATION METHODS

Barrier coatings may be applied to components by dipping, spraying, spinning, screen-printing fluidized-bed techniques, or automated dispensing. Equipment is available that automatically dispenses coatings through needles or apertures of various sizes depending on the desired thickness. This equipment can apply controlled volumes of coatings to devices consistently over a large number of parts (Figure 9-3). Glob-top, barrier or stress-relief coatings applied to semiconductor devices must be localized so that they cover only the active surface, metallization,

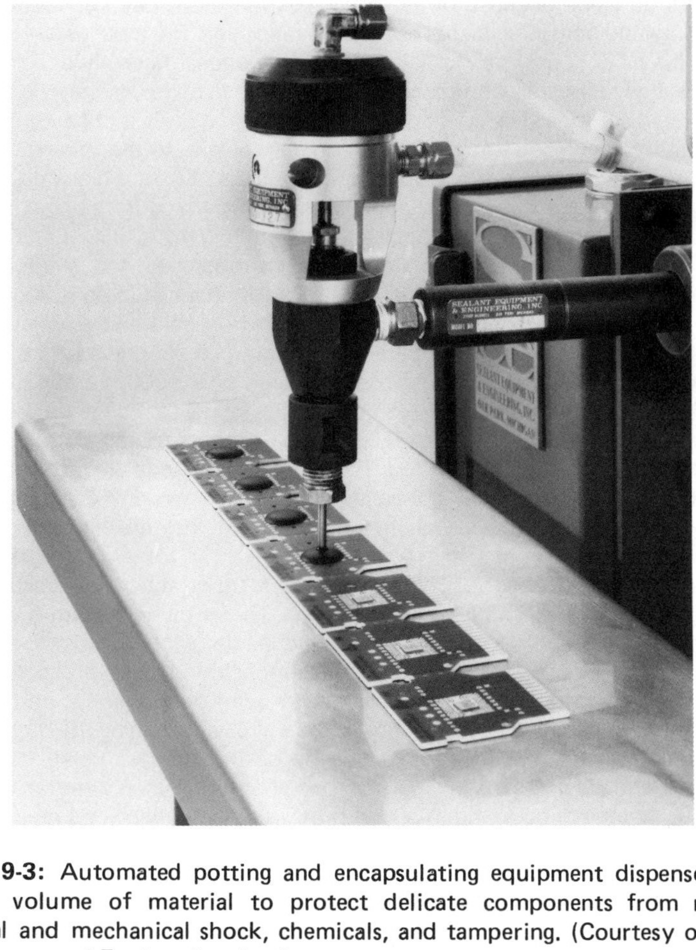

Figure 9-3: Automated potting and encapsulating equipment dispenses a controlled volume of material to protect delicate components from moisture, thermal and mechanical shock, chemicals, and tampering. (Courtesy of Sealant Equipment and Engineering, Inc.)

and a portion of the internal leads. The coating should not be allowed to spread to the edges of the device or onto the external leads, where it could affect the adhesion and moisture permeability of a subsequently applied molding compound or the solderability of the leads. Because of these restrictions, the only practical method of applying the coating is by dispensing it in small drops from an eyedropper or a microsyringe.

The structural plastic is subsequently applied by transfer molding, casting, or dipping. These techniques will be described more fully in the following sections. The relative position of the barrier coating with respect to the molding compound is shown in Figure 9-4 for a widely used 14-lead integrated circuit package.

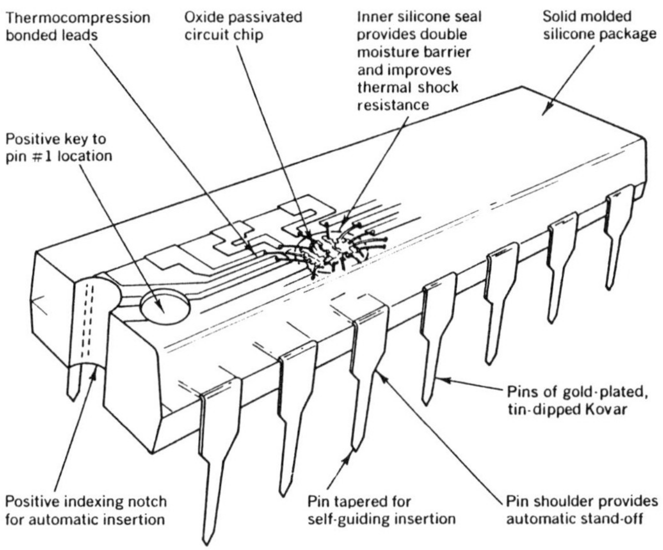

Figure 9-4: Plastic-packaged integrated circuit, showing position of barrier coating. (Signetics, a subsidiary of Corning Glass)

Transfer Molding

Transfer molding is successfully being used for the encapsulation of thin-film resistors, wire-wound resistors, diodes, bipolar and metal-oxide field-effect transistors, integrated circuits, coils, and many other components. Plastic powder is loaded into the hot chamber of a transfer-molding press; the plastic liquefies momentarily and is forced by a ram into a closed multicavity mold (Figures 9-5 and 9-6). In the mold, which is also preheated, the plastic quickly solidifies (cures) around the parts being encapsulated. The entire process requires only 3 to 5 min, at the end of which time the mold is opened and the parts are ejected. Multicavity molds of 64 to 100 cavities (Figure 9-7) are routinely being used for the encapsulation of integrated circuits in mass-production lines.

Although transfer molding is a fairly old process, it has only recently been extended to the packaging of semiconductor devices because of developments in

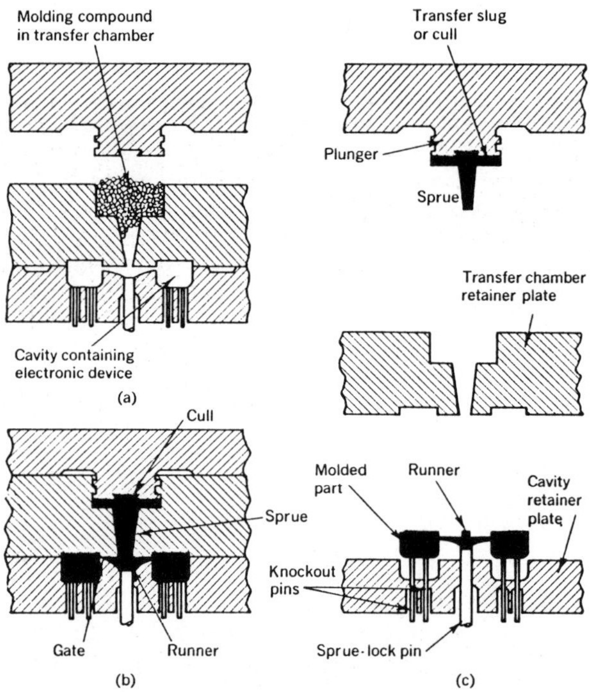

Figure 9-5: Transfer-molding process: (a) mold open; (b) mold closed; (c) part removal.

low-pressure and low-temperature plastic molding compounds. Previously available molding compounds required pressure of many thousands of pounds per square inch and temperatures above 180°C, conditions which are unacceptable for the packaging of pressure- and temperature-sensitive devices.[34] Newly developed molding compounds can be processed at pressures of 150 to 1,000 psi and temperatures of 130 to 180°C.[35] The molding compounds employed are glass- or mineral-filled resins in which the resin portion has been partially polymerized (B-staged). Other resins used are solids initially and need not be B-staged, but require blending only. Polymerization and cross-linking are completed through heat and pressure from the transfer-molding press, yielding a thermosetting infusable plastic. The main polymer types employed for transfer molding of electronic components are epoxies, silicones, phenolics, and polyesters.

As in other processes, certain limitations and problems are inherent in transfer molding and should be taken into consideration for successful results. First of all, the process is economically feasible only for mass production, because the initial cost of equipment (transfer-molding press, preheater, pelletizers, deburring equipment) and molds is high and can be amortized only over the production of large quantities of parts. In addition, since the process utilizes pressures which may vary from a few hundred to several thousand pounds per square inch and temperatures above 150°C, special attention must be paid to the component design and innate sensitivity of parts being molded to mechanical and thermal stress.[36] Once these

Figure 9-6: Transfer-molding press with microprocessor controls. Model T-90, 90 ton clamping. (Courtesy of the Hull Corporation)

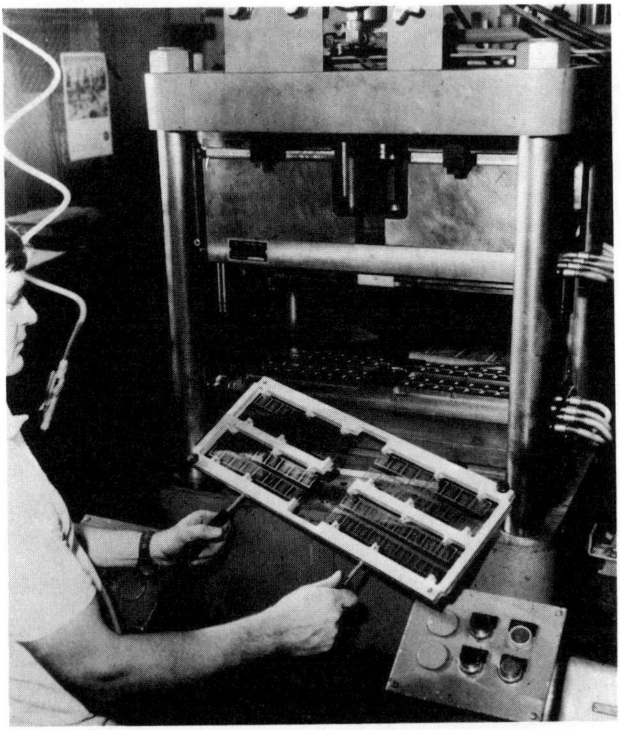

Figure 9-7: A 64-cavity integrated circuit mold showing unmolded strips in workloading frame about to be placed in press. (Hull Corp.)

component limitations are recognized, the mold design and molding parameters can be controlled so that mechanical stresses are minimized consistent with the production of dense void-free moldings.

Casting, Dipping, Dropping, and Spinning Processes

The plastic casting of devices consists of pouring liquid resin into mold cavities containing the devices and heat-curing (Figure 9-8*a*). The encapsulated components are then removed from the mold by means of ejector pins. The casting process is normally a manual operation and is ideal for small production runs. The process can be somewhat automated by using automatic metering and dispensing equipment. A second method of encapsulating components consists in dipping them into a liquid resin. A buildup of plastic occurs (Figure 9-8*b*), and the coated devices are then baked in an oven to harden the plastic. Finally, a dropping method may be employed (Figure 9-8*c*), in which the exact amount of resin required is dropped onto the device. For the application of junction coatings, microsyringes that dispense as little as 1 μl of resin can be used. This process may be automated by moving the components on conveyors past automatic dispensers.

Interlayer insulators and passivation coatings such as the polyimides are spun onto wafers like photoresist to give the desired thickness. In general, interlayer

insulators are thin films of a few microns thick. The top passivation coating can be between 10 and 250 microns thick.

Plastic resins employed in the casting, dipping, or dropping processes differ from those used in transfer molding, as shown in Table 9-5.

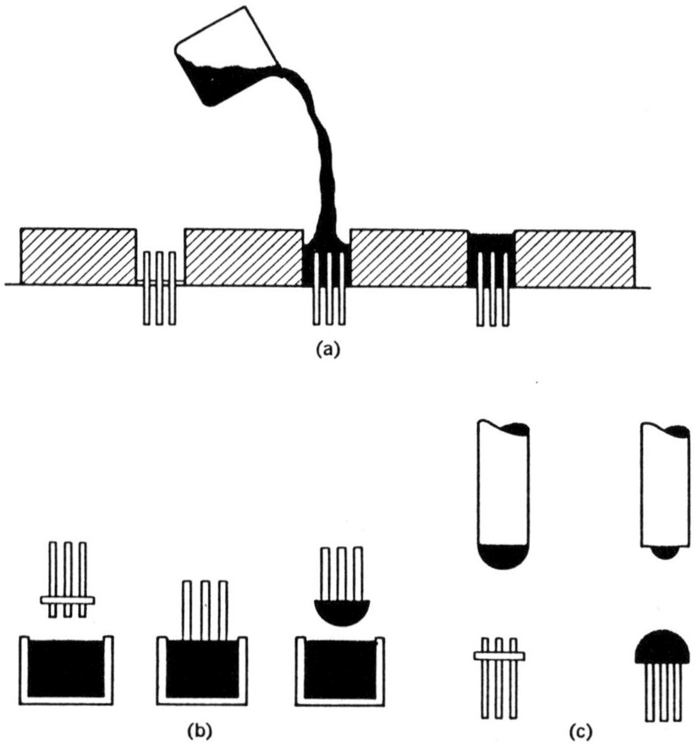

Figure 9-8: Device encapsulation with plastic resins: (a) casting; (b) dipping; (c) dropping.

Table 9-5: Comparison of Casting Resins with Transfer-Molding Resins

Casting resin	*Transfer-molding resin*
Usually a two-part system.	One-part system.
Requires weighing, preheating, mixing, and deaerating.	No weighing, mixing, or deaerating necessary.
Usually a liquid.	Usually a solid powder.
Does not require high pressure.	Requires temperature and pressure to flow into mold.
Not B-staged, requires many hours at elevated temperature to cure.	Generally B-staged, requires only minutes to cure.

Fluidized-bed Coating

In the fluidized-bed process the component to be coated is heated to a temperature above the melting point of the plastic powder and immersed into the power, which has been fluidized by forcing air through it. Details of this procedure were given in Chapter 5. The fluidized-bed process has been successfully used in coating and sealing wire-wound resistors, thin-film resistors, and capacitors. It has not, however, been found reliable for coating bare transistors and other active semiconductor devices, primarily because of the high temperatures required for adhesion and fusion of the coating and because of the limitations on the types of coatings which can be used. Many thermoplastic powders that are ideal from a processing standpoint lack the electrical and thermal properties required for microelectronic devices.

PARAMETERS AFFECTING ELECTRONIC DEVICES

The main parameters of a plastic that can affect the reliability of a device are permeability, purity, corrosion resistance, stress resistance, and adhesion. Although these parameters are treated separately in the following discussion, it should be emphasized that they are all closely interrelated. For example, lack of adhesion will increase moisture penetration, impurities will cause corrosion, and corrosion in turn can affect both adhesion and penetration.

Stresses and Stress Relief

Relieving Stress by Means of Barrier Coatings (Glob-top). Coatings can be used to minimize or eliminate the stresses imposed by subsequent transfer-molding or casting compounds and to protect electronic components from vibration and shock. The stress-relief coatings selected must, however, be flexible and possess a low modulus of elasticity. Silicones meet these requirements and have found many applications in component packaging. For example, one firm coats transistor components with 10 to 20 mils of silicone and then encapsulates with an *m*-phenylenediamine-cured epoxy. Another semiconductor manufacturer employs small amounts of silicone barrier coatings and then transfer-molds with silicone or epoxy.

The value of flexible dip coatings in relieving stresses is apparent from the data in Figure 9-9. The further improvement afforded by filled flexible resins over both unfilled flexible and rigid plastics is shown in Figure 9-10. In other experiments total stresses of 5,000 psi at −40°C were measured for components embedded in a silica-filled epoxy resin, but it was found that the stresses were greatly reduced when the components were precoated with room-temperature vulcanizing elastomeric silicone. At −40°C, a 1-mil silicone coating reduced the total pressure to 3,500 psi, and an 8-mil coating further reduced the pressure to less than 400 psi.[37,38]

Mechanical Protection with Rigid Encapsulants. Rigid encapsulation compounds that are transfer molded or cast onto a coated or uncoated device impart further mechanical protection to the device. The efficiency of plastics in damping vibration within components is evidenced by the fact that plastic-packaged devices can generally withstand vibration and impact better than their metal-packaged

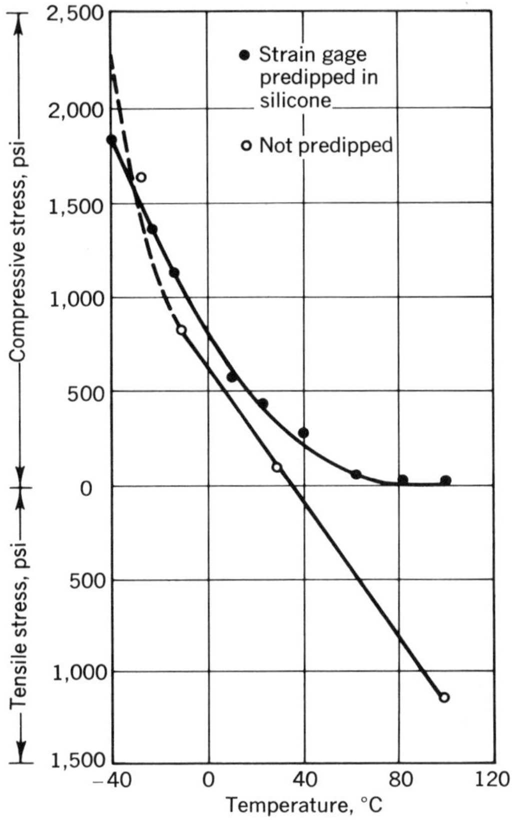

Figure 9-9: Effect of flexible precoat on stress of polyamide-cured epoxy.[39]

counterparts. The reason is that the plastic grips and secures all connections, wires, posts, and other projections. Loads to components are frequently applied through the wire leads; plastics can absorb these stresses and minimize their transfer to the component element (Figure 9-11).

The excellent shock and vibration protection provided by epoxy encapsulation has been demonstrated in tests on more than 500 transistors. Epoxy-encased transistors that were subjected to 100,000*g* acceleration forces for 1 min and to vertical-recovery gun shock showed no significant changes in electrical characteristics.[41]

Design Considerations in Minimizing Stresses. Although plastics play an important role in protecting components from severe mechanical stresses, they may also contribute stresses of their own. High stresses can be imparted to electronic components by shrinkage of the plastic during its cure or by mismatches in coefficients of thermal expansion between the plastic and the various metal and ceramic surfaces to which it adheres. Both shrinkages and expansion coefficients vary widely among plastics and depend on the type of polymer used and the

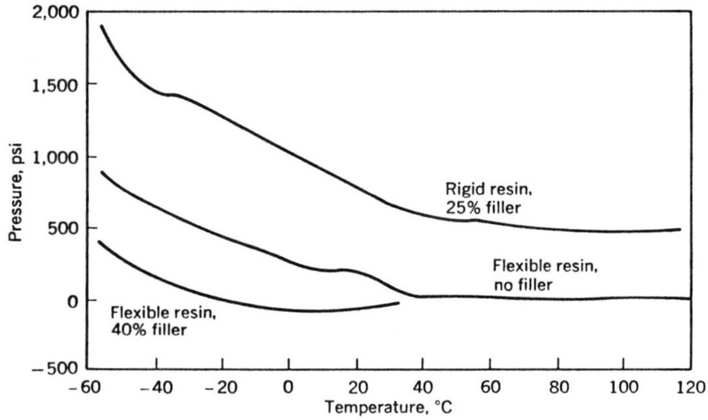

Figure 9-10: Pressure exerted by rigid and flexible epoxy-resin formulations during thermal cycling.[40]

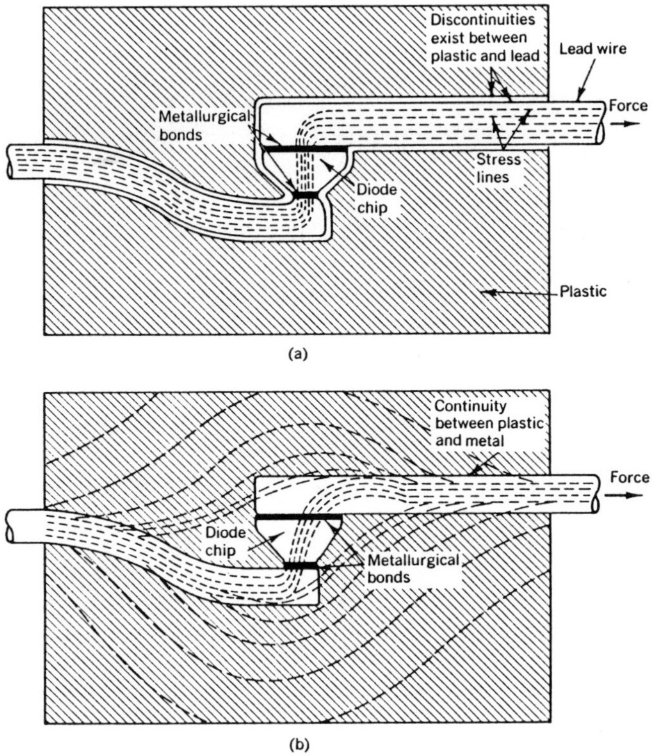

Figure 9-11: Effect of adhesion on dissipation of stresses in diode device: (a) poor adhesion of plastic to lead, poor stress dissipation; (b) good adhesion of plastic to lead, good stress dissipation.

volume percentage and type of filler employed. The percentage of shrinkage also depends on the temperature of cure, the geometry of the part, and the amount of plastic used.

A poor combination of material and process factors can result in stresses sufficiently high to cause fragile or thin sections of a component to reach ultimate strength and break. Figure 9-12 shows an integrated circuit whose flying lead wires were broken by the combined cure and thermal shrinkage of an electrical-insulating coating. The failure was caused primarily by applying too thick a layer of an unfilled solvent-containing coating. Unfilled plastics normally have much higher cure-shrinkage values and higher coefficients of expansion than filled plastics. A thinner application of the same coating and a gradual stepwise increase of the curing temperature would have minimized the stresses and avoided the wire breakage.

Figure 9-12: Integrated circuit showing lead breakage due to shrinkage of silicone varnish.

Plastic systems involving more than one coating or type of plastic may present problems if the inner coating has a higher coefficient of expansion than the outer one. An increase in temperature will obviously cause the inner coating to exert pressure against the confining outer shell. An example of this is the use of an inner unfilled silicone coating and an outer epoxy shell in some resistor and semiconductor devices. Stress problems can be avoided either by using a thinner inner silicone coating or by providing space between the inner coating and the outer shell into which the silicone can expand.

The transfer-molding or casting operation may impose still other kinds of stresses. Some of these arise from the molding operation and others are a function of the plastic material itself. For example, the combined pressure and heat during molding can result in momentary stress gradients of sufficient magnitude to cause stress failure in some materials. During the cure process the exothermic reaction

raises the temperature within the plastic mass even higher than the applied curing temperature; the cure shrinkage that occurs immediately thereafter imposes considerable compressive stress on the component.

Removal of the part from the mold also imparts stresses, depending on the mold design, the efficiency of the mold-release agent used, the type of molding material, and the care exercised by the operator. The hot strength of the molding compound is a factor in removing the components from the mold. Materials that are soft at molding temperatures (about 150 to 205°C) allow stresses to be transferred to the component. Epoxies, for example, are generally soft and pliable when hot, and greater care should be taken in handling them.

Cooling of the molded parts induces further stresses owing to differences in coefficients of expansion of the several materials used in the construction and encapsulation of the device. These internal stresses can change as the materials stress-relieve themselves upon standing at room temperature. In other words, there is a certain amount of creep which occurs with time. Internal stresses can often be relieved if the molded part is postcured or postbaked. Postbaking is usually performed anyway to complete the cure of the plastic or to stabilize the electrical characteristics of the device.

Stress Testing. Numerous methods exist for testing the mechanical integrity of plastic-packaged devices. Among these are vibration, centrifugal-force, thermal-shock, thermal-cycling, and impact tests. Most of these test procedures are detailed in military or federal specifications used for the procurement and environmental testing of equipment.

A quick method for measuring the stresses that a plastic may impart to a device consists in embedding resistance strain gauges or pressure transducers in the plastic and plotting the stresses as a function of temperature.[42,43] Temperature cycling accelerates any stress effects, especially those arising from large differences in coefficients of expansion between material interfaces.

Another measuring technique consists in recording the temperature of thermometers that have been calibrated to measure stress rather than temperature. It has been found that there is a linear relationship between temperature readings and pressure exerted on the bulb of the thermometer. In practice, the thermometers are embedded in plastic, and readings are taken at various temperatures.[38,44]

For the testing of rigid-molding or casting compounds, the Olyphant washer test is well established and is fairly standard in the industry.[45] According to this test, a complex metal washer (a machined-steel insert) is embedded in the plastic. High-shrinkage or high-stress plastics will crack around the washer immediately after curing or during thermal cycling.

Purity

The degree of purity is a critical factor for coatings and plastics used on any electronic assembly, but it is especially critical for materials that are to be applied directly to active microelectronic devices. Some devices appear to be less sensitive to surface contaminants than others and are therefore better suited for plastic encapsulation. For example, digital integrated circuits are less surface active than linear circuits. Metal-oxide semiconductors are more sensitive than pnp bipolar transistors, which in turn are more sensitive than npn bipolar transistors. However,

if the active devices are well passivated, purity of the plastic is not as critical. Indeed, the success of many plastic-packaged devices presently on the market has been due primarily to improvements in the reliability of inorganic passivation layers such as glass and silicon nitride.[46,47] Silicon nitride, in particular, has been shown to be one of the best barriers against sodium ions. It is virtually impervious to ion migration even at temperatures above 200°C with applied bias.[48]

Although plastics contain many types of impurities, the most critical ones are ions, both cations (positively charged ions) and anions (negatively charged ions). Ions can severely impair the insulating properties of a plastic, drastically change the electrical characteristics of a device, and under certain conditions cause corrosion of metal portions of a device. Other impurities consist of unreacted organic compounds such as amines in epoxy coatings, outgassing products, and additives such as plasticizers and flame retardants. The nature, sources, and potential effects of these impurities are discussed in the following sections.

Ionic Impurities. Insulation materials containing ionic impurities in contact with active or passive device surfaces can adversely affect the electrical parameters of the devices. It is fairly well known, for example, that sodium, potassium, and lithium ions cause inversion layers, high leakage currents, and decreases in breakdown voltages of diodes and transistors.[49–52] These effects and the decrease in electrical insulating properties of coatings are augmented by an increase in mobility of the charged carriers, brought about by moisture, electrical stress, or elevated temperature.

Ionic impurities can also result in electrochemical processes inducing corrosion and even the disappearance of metal films. With an applied potential, the usual faradaic electrolysis can take place, but at a much slower rate than is usual in electroplating baths. Electrical opens in wire-wound resistors and deterioration or actual disappearance of conductors and resistors in hybrid thin-film circuits have been known to occur by this mechanism. Figure 9-13 shows the corrosion and etching of the thin aluminum metallization of integrated circuits owing to penetration of moisture through the plastic-packaged device. In wire-wound or thin-metal-film resistors, voltage gradients, together with ionic conductivity, cause dissolution of metal from anodic areas, decreasing the cross-sectional areas of the resistive metal and increasing resistance values. Under more severe conditions, the etching of metal continues until it is completely removed and an electrical open occurs.

Most commercially available coatings contain ionic impurities. Ions are present as such or may be generated from inorganic compounds in the presence of moisture or an electrolyte.

The many reactants used in the synthesis of the resin, catalyst, or hardener may give rise to ions, and unless special steps are taken to purify during or after the synthesis of the resin, these impurities will be carried over into the final cured coating. For example, more than 12 steps are required to synthesize epoxies. Sodium hydroxide is widely used in the preparation of epoxies, and both sodium and chloride ions generate from the sodium chloride by-product. The presence of varying amounts of chloride ion has been verified in epoxies, as well as in silicones, phenolics,, and polyesters. Amounts ranging from 25 to 500 ppm and 5 to 100 ppm have been found in commercial-grade epoxies and silicones, respectively.

Ionic impurities may also arise from additives such as fillers, thixotropic agents, and flame retardants used in the formulation. Glass fillers, in particular, can be a major source of sodium ion contamination. A third source of ion contamination is

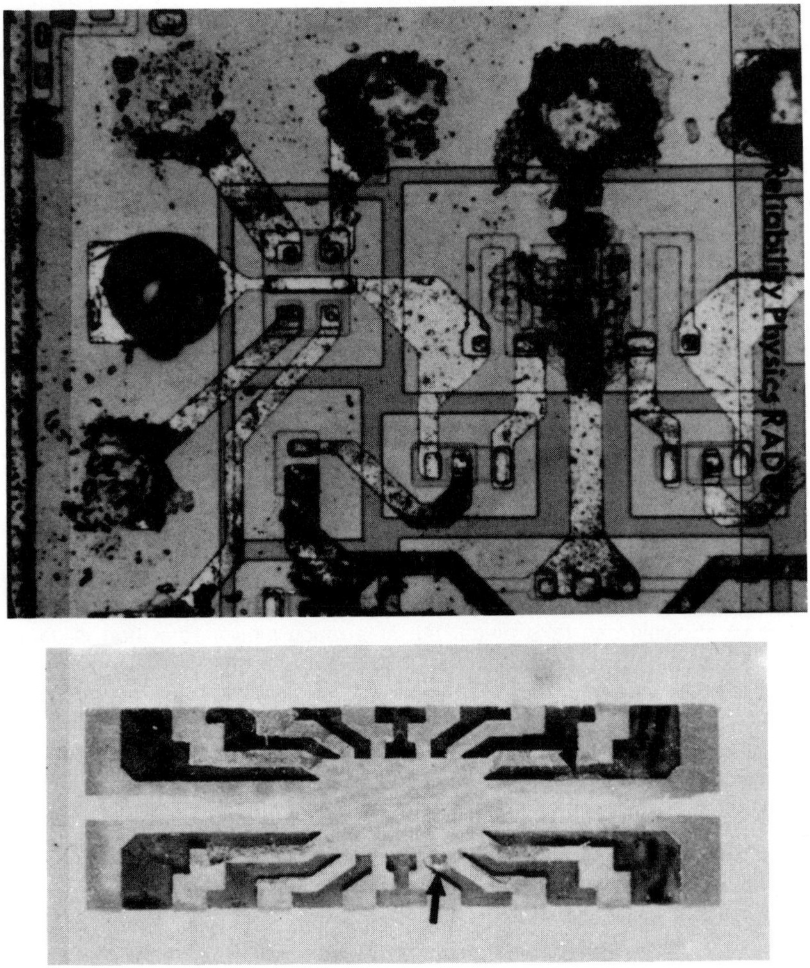

Figure 9-13: (Top) Corrosion of aluminum near ball bonds due to moisture penetration, resulting in open connections;[53] (bottom) residues on plastic package under lead-frame regions (arrows), indicating penetration of moisture and contamination.[53]

from metal or glass kettles, mixers, or other equipment used in processing. It has been shown by emission-spectrographic analysis that epoxy coatings mixed in metal containers acquire trace amounts of the metal.

Improved purification methods and controls now allow plastics to be used directly on active circuits. Semiconductor-grade products are purified, keeping sodium and potassium ions to less than 1-2 ppm, then passed through 0.1-0.2 micron filters to remove particulates. However, to maintain these low levels of particulates and ions the materials must be stored and used in an ultraclean environment (see Chapter 5 for discussion on clean rooms).

Other Impurities. In addition to ionic impurities, reactive outgassing products can evolve from plastics under ambient conditions, and even more so on thermal aging or irradiation. The outgassing of ammonia from certain phenolic compounds and its conversion to ammonium ions in the presence of moisture have been reported to cause inversion layers and an increase in the leakage current of microdiodes.[54] Katsev et al.[55] also investigated the effects of ammonia and other amine compounds on transistors. The detrimental effects of outgassing products, such as benzene and carbon dioxide, have also been shown[56] (Table 9-6). More

Table 9-6: Effect of Gas Ambient on Reverse-Current Leakage of Dual pnp Signal Transistors[56]

Condition	Collector-to-base current (I_{CBO}) at 20 volts reverse bias, nanoamp	
	Transistor *A*	Transistor *B*
Before vacuum bakeout	3.0	2.5
After exposure to benzene vapor and reverse bias on transistor *A* only	2,400	
After exposure to water vapor and reverse bias on transistor *A* only	10,000	
Vacuum bake at 300°C and no bias	2.5	2.5
Bias on transistor *B* in vacuum	...	3.0
After exposure to carbon dioxide and bias on transistor *B* only	...	650
Vacuum bake at 300°C	...	3.0

recent work with epoxies and silicones has shown the outgassing of numerous chemical species, some of which are considered to be active, that is, either deleterious to device performance or corrosive to metal surfaces. The specimens were sealed in an inert environment in glass ampules and aged at 150°C for 5 days. The gases evolved were then analyzed by gas-chromatographic and infrared-spectrographic techniques. Of the outgassing products detected, those considered active were the aldehydes, vinyl acetate, the amines, carbon dioxide, and moisture.

Optimum Resin-to-hardener Weight Ratios. A *stoichiometric mixture* of a two-component coating (resin and hardener) consists of the two components mixed in amounts proportional to their respective molecular weights according to a balanced molecular equation. If the mixture is not stoichiometric, some of the resin or some of the hardener is left unreacted. In practice, the optimum ratios may not coincide with the calculated stoichiometric ratios. Compositions which are slightly high in resin content are being used. However, too high or too low a resin content can render the plastic susceptible to moisture penetration and can degrade its electrical insulating properties. The unreacted constituents may contain active polar or ionic groups which can further affect the electrical parameters of the device. A plot of the electrical resistivities of water extracts for various resin-to-hardener ratios gives some indication of optimum ratios (Figure 9-14).

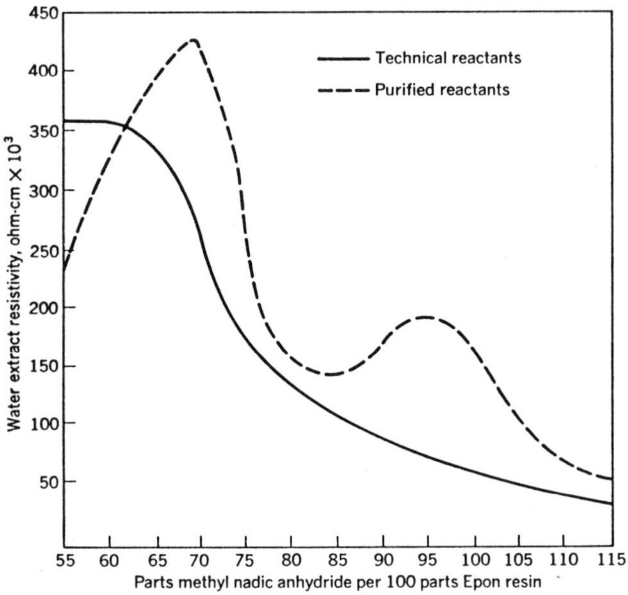

Figure 9-14: Optimum resin-hardener ratios by water-extract resistivity test.

Tests for Plastic Purity. A quantitative test for the total amount of ionized and ionizable constituents of a polymer consists in measuring the electrical conductivity of extracts of the plastic in either water or organic solvents. One test consists in digesting 1 g of powdered sample in 100 ml of deionized water at 70°C for 7 days.[54,57] The mixture is cooled to room temperature, and the electrical resistivity of the water extract is measured by means of a conductivity bridge. A modification of this test has been used by Dow Corning with similar results.[58] It has been found that plastics used in semiconductor packaging vary widely in this property (Table 9-7) and that there is some correlation between purity as defined by the water-extract resistivity and performance of the device (Table 9-8). Silicones having the highest water resistivities (lowest ionic impurities) provided the lowest number of failures on components after 1,000 hr of testing.

Table 9-7: Water-Extract Resistivity Data

Material	Sample size	Resistivity range, ohm-cm
Silicone junction coating	15	33,000–425,000
Silicone transfer-molding compound	5	30,000–280,000
Epoxy transfer-molding compound	6	4,900–50,000
Epoxy casting compound	6	32,500–250,000
Phenolic molding compound	1	4,700
Polyester molding compound	1	72,000
Deionized water (control)	. . .	600,000–900,000

Table 9-8: Typical Failure Data for Silicone-Coated Metal-Oxide Semiconductor Devices (Static Tested at 125°C/150 Milliwatts)[59]

Sample	Water-extract resistivity, ohm-cm	Sample size	Total failures*	
			After 1,000 hr	After 2,000 hr
Silicone *B*	80,000	7	3	
Silicone *E*	180,000	13	2	
Silicone *F*	190,000	7	3	
Silicone *G*	300,000	11	1	
Dow Corning DC-51	360,000	10	0	0
Dow Corning XR-62-044	660,000	10	0	0

* Failure indicators: threshold voltage, breakdown voltage, leakage current.

A highly sensitive test for organic ionic impurities is similar to the one just described.[60] A 1-g sample of plastic is extracted with 10 ml of highly purified trichloroethylene, and the electrical conductance of the extract is measured. The test is reported to be sensitive to 1 ppm. Other tests for assessing metal-ion impurities in plastics include emission-spectrographic analysis (Table 9-9), atomic-absorption analysis,[61] and flame-photometric analysis. The last two tests are particularly sensitive for sodium and potassium in concentrations as low as parts per billion.

Table 9-9: Emission Spectrographic Results*[59]

Sample	Cu	Al	Mg	Zn	Ca	Co	Fe	Si
Silicone *A*	<0.01%	<0.01%	<0.01%					
Silicone *B*	. . .	<0.01%	<0.01%	<0.5%	<0.01%			
Silicone *C*	<0.5%	. . .	<0.01%	<0.5%	<0.01%	<0.5%		
Silicone *D*	. . .	. . .	<0.01%	<0.5%				
Dow Corning XR-62-087	. . .	. . .	. . .	. . .	. . .	. . .	<0.01%	
Dow Corning XR-62-044	. . .	. . .	. . .	. . .	<0.01%			
Epon 828	<0.5%	<0.5%	<0.5%	. . .	<0.5%	. . .	. . .	<0.5%
Phenolic molding compound	<0.5%	<0.5%	<0.5%	. . .	10–100%	. . .	<0.5%	<0.5%

* Percentages are by weight of ash.

The elevated-temperature-reverse-bias test is particularly sensitive and provides an indirect method of establishing the presence of ions on an active semiconductor device. Semiconductor devices containing mobile ions form inversion layers and exhibit high reverse-leakage currents. The inversion layer is formed if ionic contaminants on the surface or in the oxide layer are mobilized at elevated temperatures and aligned by the application of an electrical field. Stress conditions generally used are 150°C and about 20 volts reverse bias.

The chloride-ion content of a plastic may be determined quantitatively by the potentiometric-titration procedure. The plastic sample is powdered and extracted with water, and the extract is titrated with a standard solution of silver nitrate.[62]

Moisture Resistance

All plastics absorb and transmit moisture to some extent. Hence they cannot be considered hermetic in the same sense that cavity packages or metal-to-glass seals are, nor do the same hermeticity and leakage tests apply. Nevertheless plastic-packaged devices have passed the stringent requirements of many military specifications, and devices have operated reliably after short exposures to high humidity and elevated temperatures. Their long-term resistance to humidity, salt spray, and warm, humid environments, however, is questionable.

Moisture can enter a plastic-encapsulated device by one or more of the following mechanisms:

1. Absorption in the plastic
2. Transmission through the plastic
3. Penetration along the device leads, through microcracks in the plastic, or through voids that may exist in the interior of the plastic

Water Absorption. The percentage of water that a plastic absorbs is a constant at a specified temperature after equilibrium has been reached. The percentage of water absorption can be simply determined by gravimetric methods. The plastic is immersed in water or exposed to constant humidity and temperature conditions and is weighed periodically. Measurements are continued until the plastic no longer absorbs water, i.e., until there is no further weight gain. Typical moisture-absorption curves for some silicones, epoxies, and phenolics are given in Figures 9-15 and 9-16.

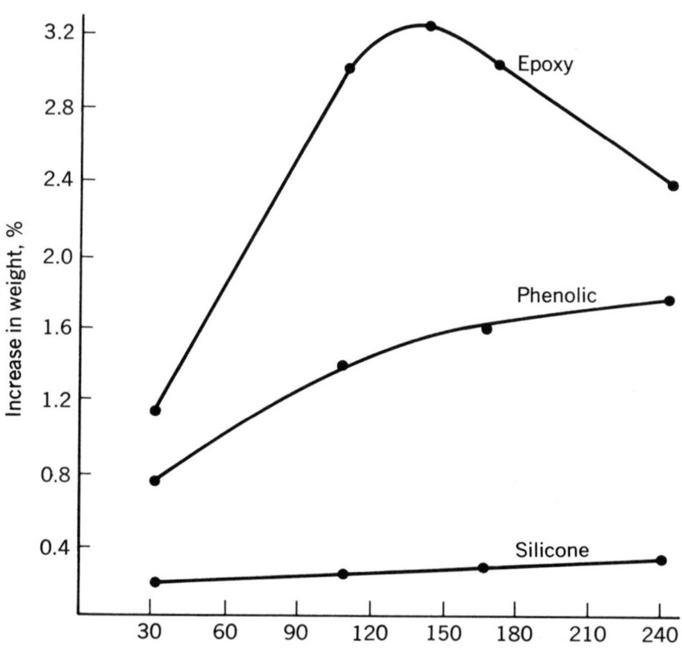

Figure 9-15: Stability of molding compounds to boiling water.[35]

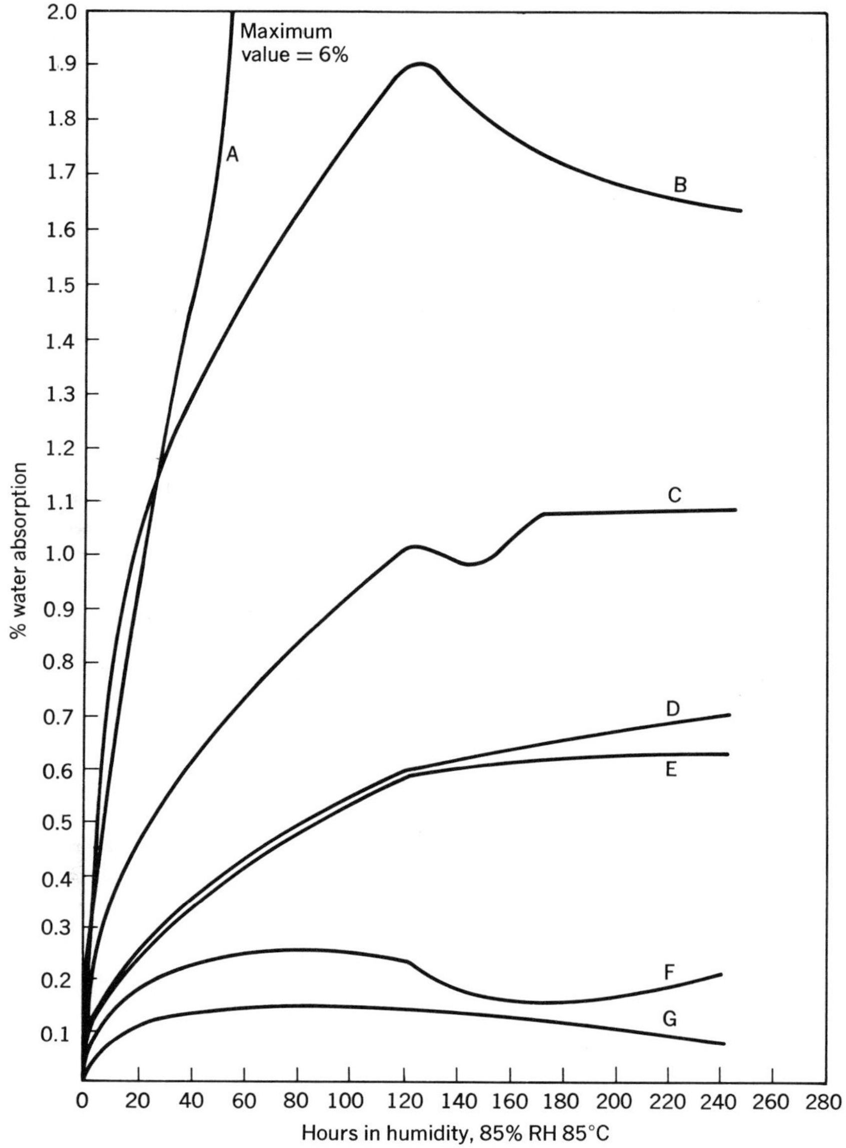

Figure 9-16: Moisture-absorption curves for integrated-circuit molding compounds (after transfer molding all samples postcured for 2 hr at 150°C): (A) epoxy, anhydride-cured with amine catalyst (vendor A); (B) Epoxy Novolac, amine-cured; (C) epoxy, amine-cured; (D) epoxy, anhydride-cured with amine catalyst (vendor B); (E) epoxy, anhydride-cured with amine catalyst (vendor C); (F) silicone with fused silica and short glass-fiber fillers; (G) silicone with fused silica filler.

Moisture-vapor Transmission. The amount of moisture vapor that will diffuse through a polymer coating depends on several factors as defined by Fick's laws of diffusion. Among these factors are the coating thickness, the exposed surface area, the duration of exposure, and the vapor-pressure differential. Integration of Fick's law of diffusion results in the following relationship:

$$Q = \frac{Dat\ (p_1 - p_2)}{l}$$

where Q, the amount of diffused moisture, is proportional to the cross-sectional area a of the plastic, the exposure time t, and the vapor-pressure difference $p_1 - p_2$, and inversely proportional to the thickness l of the plastic. The proportionality constant D represents the diffusion or permeability constant and is a function of the molecular structure of the plastic, its degree of cure, its polarity, and the interaction effects between the water and polymer molecules.

Many methods and elaborate test equipment have been designed to measure moisture-vapor transmission but as yet there has been no standardization throughout the industry. For the rapid screening of coatings, simple gravimetric methods are available.[63] One test consists in clamping a free or supported film of the coating over a flanged metal cup containing water and then accurately monitoring the weight loss of the cup with time. The cup is placed inside a desiccator to provide a consistent driving force for the water vapor to permeate the film. A description of the cup and details of the test procedure may be found in MIL-I-16923. Figure 9-17 shows the permeability cup and film arrangement. Results are reported in grams of water that will pass through a 1-mil-thick film 100 in.^{2} in area in a 24-hr period at a constant temperature. In some cases the rates must be calculated with a film area greater than the original area owing to bulging of the film after moisture equilibrium has been reached and the pressure inside the cup has increased. More precise areas can then be calculated with the assumption that the film surface is a segment of a sphere. This bulging phenomenon has been especially apparent in some silicone films.

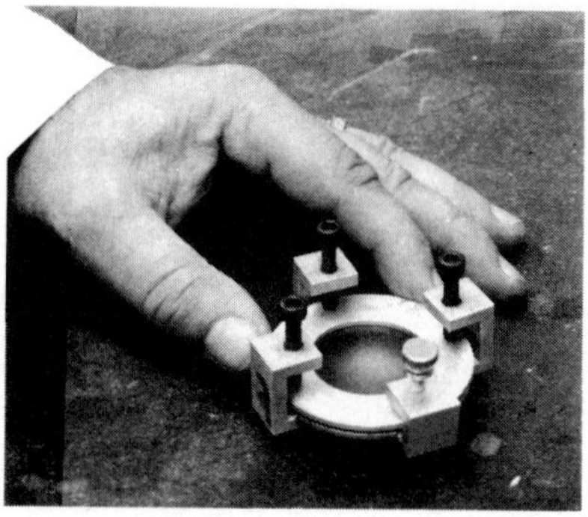

Figure 9-17: Cup and film arrangement for permeability test.

Another variation of the gravimetric method entails clamping or wax-sealing the film over an aluminum dish containing a desiccant and measuring the weight gain after placing the container in a humidity chamber.

The Pressure-cooker Test. This test, also referred to as the *steam-pressure test,* or HAST (Highly Accelerated Stress Test) is designed to force moisture through the plastic in a short period of time. The plastic devices are placed in a pressure cooker and subjected to saturated steam at pressures from 17 to 25 psig. The corresponding temperatures for saturated steam at these two pressures are 125 and 130 °C, respectively. This test will quickly disclose any component weaknesses associated with moisture-vapor penetration because the water-vapor pressure (the driving force for penetration) is appreciably higher than that encountered in even the most adverse hot, humid service conditions. The exposure time has not been standardized, but it can range from several hours to several days before the device deteriorates. After exposure the devices are checked visually and electrically for gross changes and failures. A more severe variation of this test entails thermal shocking the devices prior to steam exposure.

The Dye-penetration Test. The most critical area for moisture to penetrate to the interior of a device is along the leads at the lead-to-plastic interfaces. Poor adhesion or separation between the lead material and plastic due to differences in expansion coefficients will allow a rapid penetration of moisture. A fluorescent-dye test has been used to determine the extent of penetration. According to this test, the specimen or device is placed in a pressure vessel with fluorescent penetrant for 1 hr at 60 psi. The specimens are then cross-sectioned, and the depth of penetration is observed under an ultraviolet or black lamp. Some dye-penetration results for test specimens consisting of Kovar wire embedded in various plastics are shown in Figure 9-18. The photographs show complete, partial, or no penetration, depending on the plastic material used. Figure 9-19 shows the high penetration of dye solution along the leads in a plastic-packaged integrated circuit. Because the dye-penetration test is destructive in nature, it is best suited for quality control of devices on a sampling basis, for the selection and qualification of plastic encapsulants, and for failure analysis.

There is some uncertainty about the interpretation of this test. A positive result indicates the penetration of moisture. A negative result, however, is not unequivocal, because in some cases the plastic may filter out the dye. Experiments with tritiated water, such as those conducted by Texas Instruments, appear to be more meaningful.[64]

Corrosivity

The immediate or long-term effect of moisture may be the corrosion of metal portions of a device. The external leads, internal flying leads, interconnection bonds, and metallized conductors of a device are highly susceptible to corrosion. Ionic and other impurities on the surface of the device or within the plastic also play a key role in initiating and accelerating corrosion. Chloride ion contamination, for example, causes chemical corrosion of aluminum through the formation of aluminum chloride. Other ions, both cationic and anionic, contribute to electrolytic corrosion between external leads or closely spaced internal conductors. Corrosion of the aluminum metallization near ball bonds due to moisture penetration to the surface of a microcircuit device is shown in Figure 9-13 (top), and corrosion

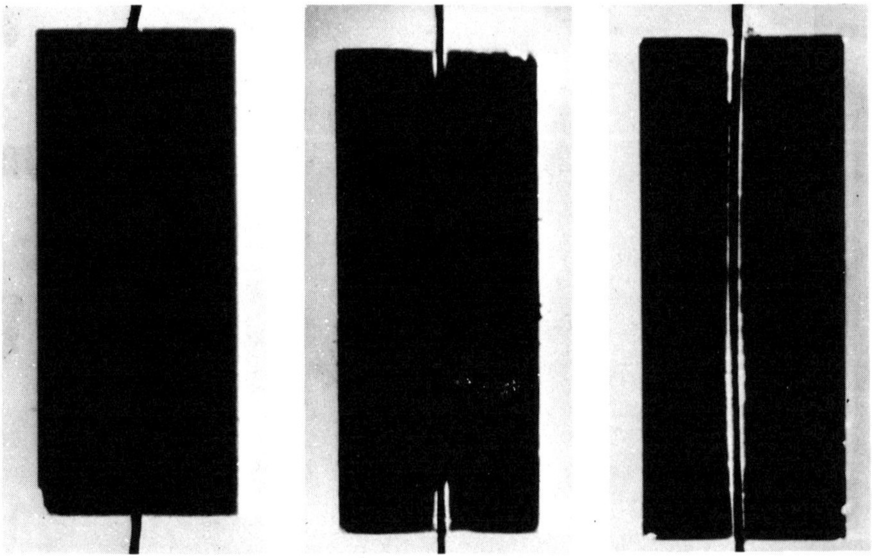

Figure 9-18: Dye penetration along wire leads: (left) no penetration, (center) partial penetration, (right) complete penetration. (Courtesy Rockwell International Corporation)

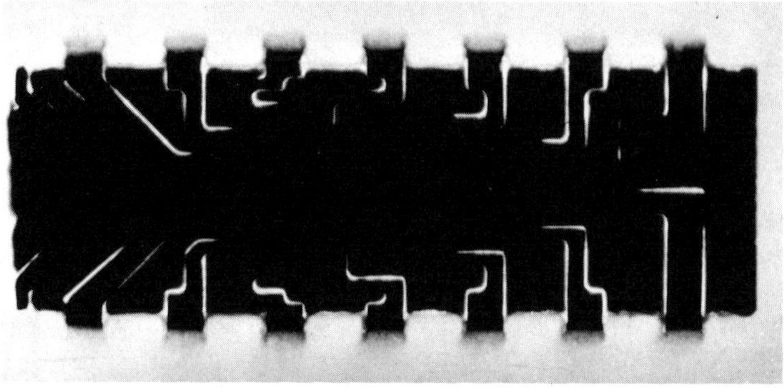

Figure 9-19: Dye-penetration test on integrated circuit, showing penetration along lead to plastic interface.

residues between conductors of the lead frame, indicating the penetration of moisture and the presence of contaminants, are shown in Figure 9-13 (bottom). A prevalent mode of corrosion in plastic-packaged devices is that occurring between external leads after exposure to humidity or salt spray. Heavy residues of corrosion products can form between and on the leads (Figure 9-20). Again, this type of corrosion can be minimized or avoided by employing extremely pure plastics. Some transfer-molding silicones with a very low ionic content have been shown not to display interlead corrosion.

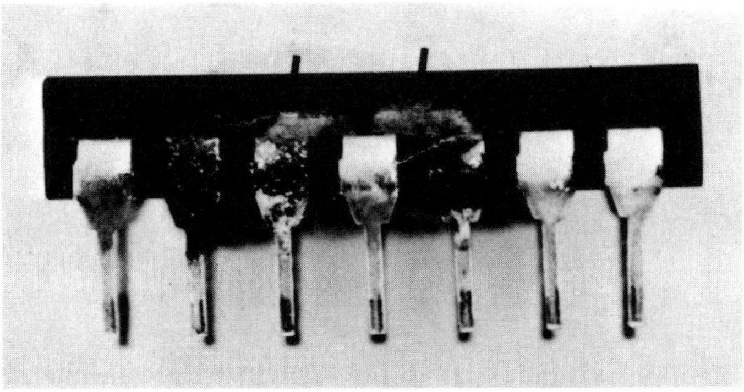

Figure 9-20: Plastic-packaged integrated circuit after humidity exposure, showing corrosion of leads and corrosion residues between leads, causing external leakage currents. (Rome Air Development Center)

In many instances catalysts used for curing resins have been known to induce corrosion of metal surfaces. In particular, peroxide catalysts used in curing acrylics and polyesters have been reported to attack copper (Figure 9-21). Even some aliphatic amines used in curing epoxies have been observed to attack copper. This attack occurs if the epoxy cure is carried out at room temperature; the same amine-epoxy system will have no discernible effect if the cure is performed at higher temperatures. A possible explanation of this phenomenon is that a copper-amine reaction competes with the amine-resin reaction, and at the higher temperatures the amine-resin reaction predominates. Still another contributing factor might be the higher moisture content that exists at the copper surface when cured at room temperature.

Corrosion Tests. The corrosive properties of plastics may be observed during the testing of devices under normal operating or accelerated conditions, or may be determined by several rapid screening tests. According to one of these tests, the plastic is contacted with strips of Mylar film that have been previously metallized. The test is accelerated by placing the metallized strips and plastic in test tubes containing small amounts of water, sealing them, and heat-aging in an oven. Any

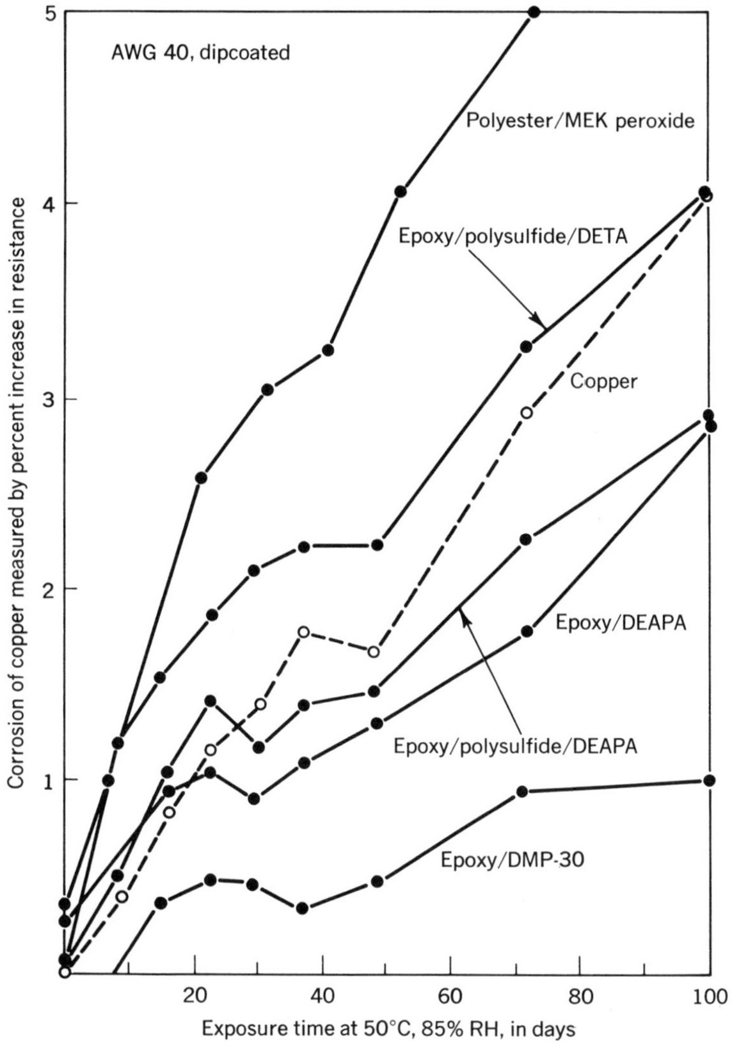

Figure 9-21: Corrosive effects of plastics on copper wire.[59]

corrosion effects should appear after several days of exposure, although the test may be extended to 1 month or until some corrosion is observed. The Mylar strips may be vacuum-metallized with aluminum, magnesium, copper, nichrome, or other metals or alloys to simulate material combinations used in the actual device. Metallized strips and corrosion testers are available commercially.

Another test, described by Doctor and Franklin,[65] entails embedding fine copper wire in the plastic or coating to be tested and measuring the percentage of increase in electrical resistance of the wire after various periods of temperature and humidity exposure. The extent of corrosion is a function of the increase in resistance of the wire. Although this method was developed for the evaluation of magnet wire and wire coatings, it is useful for evaluating plastics and coatings in general. A comparison of the corrosive properties of epoxies, epoxy polysulfides, and polyesters is shown in Figure 9-21. The effects of using polysulfides in epoxy formulations and of using peroxides in curing polyesters are evident from this figure.

Adhesion

Strong adhesion of plastics to components is essential both to effect a satisfactory seal against moisture and contaminants and to allow the encapsulating plastic to dissipate mechanical stresses to the leads during handling, installation, and service. Several factors affect adhesion:

1. Type of substrate material (metal or nonmetal)
2. Surface treatment prior to encapsulation
3. Surface roughness or porosity
4. Intrinsic adhesive properties of the plastic
5. Presence of internal release agents in transfer-molding plastics

The nature and preparation of the substrate are the most critical factors affecting adhesion. For example, the ease of bonding (bondability) varies greatly among metals, with the noble metals being more difficult to bond to. Examples of metals which are known to be difficult to bond to are gold, Kovar, nickel alloys, solder, solder plating, and cadmium.

A clean surface is essential for good adhesion. Cleaning methods vary from solvent cleaning to chemical or mechanical etching of surfaces. Primers are sometimes used with difficult-to-bond metals. For example, epoxy primers are used over gold before coating it with polyurethane. Special primers are frequently used to enhance the adhesion of silicones, particularly the room-temperature-vulcanizing types of polyimides.

The intrinsic bondability of polymers varies from type to type. Epoxies and phenolics, for example, are known for their inherently good adhesive properties. Polyurethanes exhibit excellent to fair bonds, while silicones exhibit adhesive properties which vary from good to very poor.

In the case of reactive metals, other critical adhesion factors are the type and mechanical strength of the oxide which forms rapidly on the surface under normal ambient conditions. This oxide layer may also form in the course of surface preparation. The nature of the hydrated oxide that forms on aluminum after etching depends on the temperature of the rinsing and drying operations. Rinsing and

drying above 70°C produce a hydrated oxide having poor bond strength to the metal, one which is likely to fail mechanically before the organic coating has a chance to adhere. In other cases the oxide layer may actually improve adhesion.

Tests for Adhesion. The adhesion of coatings to substrate surfaces can be measured easily by the classical tape test described in FED. STD. No. 141, method 6301. Panels simulating the surfaces of the actual device (such as gold plate, ceramic, or vapor-deposited aluminum) are coated and cured. Masking tape or Scotch Tape is applied to an area of the coating that has been scribed through to the substrate. The tape is then pulled off rapidly (Figure 9-22), and the degree of adhesion is reported in a semiquantitative way as the percentage of coating remaining within the scribed area. A more stringent version of this test entails exposure of the coated panels to humidity and thermal environments prior to tape testing.

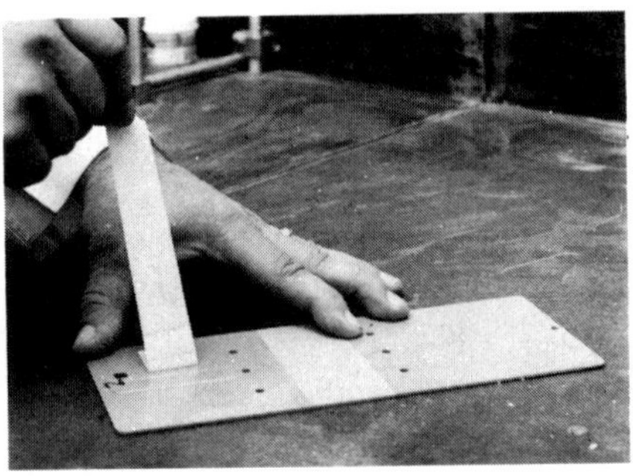

Figure 9-22: Tape adhesion test.

The adhesion of interconnection wires or lead-frame material to the encapsulant may be measured by embedding wire (for example, Kovar or gold-plated Kovar wire) in the plastic to be tested and then measuring the force necessary to pull the wire out with an Instron or other tensile tester. Table 9-10 gives some data on adhesion values obtained by pulling Kovar-wire inserts from 0.25-in.-diameter 0.62-in.-long cylindrical molded specimens. Initial adhesion values for three different epoxies tested were very consistent, but rapid loss of adhesion occurred after a 10-day pressure-cooker test. One silicone, however, retained most of its strength under the same conditions.

COMMERCIAL AND MILITARY USES OF PLASTIC PACKAGES

Plastic-packaged devices are used primarily for consumer and industrial products like television sets, radios, computers, and in-house test equipment. There is no question about their economy and performance for these applications. However, greater precautions and proof of long-term reliability are necessary for military

Table 9-10: Plastic-to-Wire Adhesion

Material	Initial adhesion, lb	Adhesion after 10-day pressure-cooker test, lb
Silicone transfer-molding compound *A*. .	11.3	11.1
Silicone transfer-molding compound *B*. .	4.6	Erratic results
Epoxy transfer-molding compound *A*. . .	10.6	8.5
Epoxy transfer-molding compound *B*. . .	10.4	4.4
Epoxy transfer-molding compound *C*. . .	10.6	2.6

Courtesy of Rockwell International Corporation.

systems because of the more severe environments that these systems are likely to encounter. The main drawback to more extensive use of plastics for military electronics has been their inability to act as adequate moisture barriers to meet stringent requirements such as those defined in MIL-E-5272, MIL-E-5400, MIL-E-16400, MIL-STD-202, and more recently, MIL-STD-883. In attempting to meet these requirements, some manufacturers have resorted either to the use of multiple coats of silicones or epoxies with high-temperature bakes after each application or to alternating several different coating types (for example, a silicone, followed by an epoxy). In the case of active devices and integrated circuits, improved protective layers of silicon nitride or special glasses can serve as the primary barriers for moisture and ionic contaminants, so that less reliance need be placed on the plastic for these functions. In fact, some firms report a very high level of reliability with plastic-packaged devices primarily because of the improved passivation layers which they have developed.

The lack of meaningful test methods has been another limiting factor in the wider use of plastics for military devices. It has been difficult to make a strict comparison between plastic-packaged devices and their hermetically sealed cavity-package counterparts because hermeticity tests applicable to one are not applicable to the other. For example, the standard helium-leak test used for metal-can-sealed devices cannot be used to test plastic-encapsulated devices because of the absence of a cavity. Hence new criteria applicable to plastic packages need to be developed.

REFERENCES

1. Robinson, W. M., and H. R. Lee: Pitfalls and Progress in Plastic Encapsulation of Semiconductors, *Insulation,* December, 1967.
2. Hakim, E. B.: Tests of Plastic Encapsulated Semiconductors, *IEEE Symp. Intern. Electron. Circuit Packaging,* vol. 9 (1968).
3. Pankove, J. I.: Method of Sealing a Semiconductor Device, U.S. Patent 2807558, April, 1954.
4. Rulison, R.: Surface Treatment of Germanium Circuit Elements, U.S. Patent 2768100, September, 1956.
5. John, H. F.: Protective Treatment for Semiconductor Devices, U.S. Patent 2937110, May, 1960.

6. Herr, E.: Reliability Assessment of Epoxy Transistors, *Inform, Microelectron, Navy Equipt., U-Notes* 25, U.S. Naval Ammunition Depot, Crane, Ind., August, 1968.
7. Kookootsedes, G.: Reliable Packaging Alternatives to Hermetic Sealing, *Hybrid Circuit Technology,* February 1987.
8. Otsuka, K. et al: Sealing Mechanism of Silicone Jelly Encapsulation with High Reliability, *IEEE Transactions,* Vol. CHMT-7, **3**, September 1984.
9. Kookootsedes, G. and R. Antonen: Selecting Protective Materials for Coating Hybrid Circuits, *ibid,* March 1986.
10. Kookootsedes, G. and R. Antonen: Specialty Silicone Elastomers Cope with Diversified Circuit Applications, *ibid,* September 1985.
11. Antonen, R. and G. Kookootsedes: Silicone Coatings for Integrated Circuit Protection, *Microelectronic Manufacturing and Testing,* April 1985.
12. Senturia, S. D.: Polyimides in Microelectronics, *Proceedings of the ACS Division of Polymeric Materials: Science and Engineering,* Vol 55, 1986.
13. Rothman, L. B.: Process for Forming Passivated Metal Interconnection System with a Planar Surface, *Journal of Electrochemical Society; Solid-State Science and Technology,* May 1983.
14. Ray, D. R., et al: "Polyimide Planarization in Integrated Circuits," MIT, June 1984, pp. 1–18.
15. Milgram, A. A.: Lift-off Process for Achieving Fine-Line Metallization, *J. Vac. Sci. Technol. B* 1 (2), Apr–June 1983.
16. Grabbe, P., E. L. Hu, and R. E. Howard: Trilevel Lift-off Process for Refractory Metals, *J. Vac. Sci. Technol.,* 21(1), May/June 1982.
17. Herndon, T. O., et al: "Polyimides for High Resolution Ion Implantation Masking," MIT Lincoln Lab.
18. Rubner, R., et al: A Photopolymer—The Direct Way to Polyimide Patterns, *Phot. Sci. Eng.* Vol 23(5), 1979.
19. Rohde, O., et al: Recent Advances in Photoimageable Polyimides, **SPIE** Vol 539, *Advances in Resist Technology and Processing II* (1985).
20. Bergeron, D. L., et al: Polyimide Interlevel Insulation Process Design Limitations, presented at IRPS Conference, pp. 1–5.
21. Herndon, T. O. and R. L. Burke: Inter-Metal Polyimide Insulation for VLSI, *Kodak Interface Conference,* October 1979.
22. Wilson, A. M.: Polyimide Insulators for Multilayer Interconnections, *IEEE Thin Solid Films,* 83, 1981.
23. Samuelson, G.: Polyimide for Multilevel VLSI and Alpha Protection, "Polyimide Coatings for Microelectronic Applications," August, 1981.
24. Senturia, S. D.: Polyimide as an Interlevel Insulator in Integrated Circuits, *Proc ANTEC 1985, Soc. Plastics Eng., Washington, D.C., April 29–May 2, 1985.*
25. *Rothman, L. B.: Properties of Thin Polyimide Films, Solid-State Science and Technology,* October 1980.
26. Kitade, et al: Effectiveness of Prevention of Soft Errors with Polyimide Resin Coating, *Hitachi, Ltd, Product Literature Bulletin* 63, 1982.
27. Economy, J.: High Temperature Polymers for Electronic Devices, *Science Engineering* **10,** 243, June 1974.
28. Licari, J. J. and L. R. Enlow: "Hybrid Microcircuit Technology Handbook," Noyes Publications, New Jersey, 1988.
29. Weigand, B. L. and J. J. Licari: Verification of Selected Solvent-soluble Coatings Using Production Hybrid Microcircuits, *AFWAL TR-80-4048,* 1980.
30. Weigand, B. L., J. J. Licari, and K. L. Perkins: Manufacturing Technology for Conformal Coatings, *AFWAL TR-80-4139,* 1980.
31. Licari, J. J., K. L. Perkins, and B. F. Barnett: Solder Mask/Conformal Coating Systems for Improved Maintainability of PWBs, *IPC-TP-304,* Institute of Printed Circuits, San Francisco, 1979.

32. David, R. F.: Block Co-polymer Coating of Hybrid Microcircuit, *Electronic Components Conf. Proc.,* Orlando, Florida 1983.
33. King, J.: Qualification of Dow Corning Q3-6527 as a Particle Getter, *Singer Internal Report Y257A834,* 1985 and Use of Silicone Gel Particle Getters for Microcircuits Packaging, *Proc. ISHM,* 1985.
34. Vaill, E. W.: Thermoset Molding, "Modern Plastics Encyclopedia," pp. 776–787, McGraw-Hill, New York, 1968.
35. Kookootsedes, G. J., and F. J. Lockhart: Silicone Molding Compounds for Semiconductor Devices, *One-hundred-fifty-third Meeting Am. Chem. Soc., Miami, Fla.,* April, 1967.
36. Zecher, R. F.: Designing Electrical Components for Transfer Molding, *Plastics Des. Processing,* December, 1968.
37. MacLeod, Ross W.: Final Protection of Printed Wiring Modules by Organic Materials, *Ann. Symp. Inst. Metal Finishing, Printed Circuit Group, Bristol, England,* September, 1968.
38. Schwartz, S. (ed.): "Integrated Circuit Technology," p. 144, McGraw-Hill, New York, 1967.
39. Wahlgren: Use of Epoxy Resins in High Reliability Transformers, *Electron. Equipt. Eng.,* January, 1956.
40. Hanson and Tuzinski: Strain Gage Evaluation of Flexible Epoxy Resins, *NEMA-AIEE Second Natl. Conf. Application Elect. Insulation, Washington, D.C., 1959.*
41. McKenna, E.: Mechanical and Environmental Tests: Encapsulated Transistors, *Harry Diamond Labs.* TR-1346, March, 1967.
42. Dallimore, G. R., F. Stukki, and D. Kasper: Measurement of Internal Stresses in Electronic Encapsulating Resins with a Small Solid State Transducer, *SPE J.,* **20:** (6): 544–546 (1964).
43. Johnson, L. I., and R. J. Ryan: Encapsulated Component Stress Testing, *Proc. Sixth Elect. Insulation Conf.,* September, 1965.
44. Isleifson, R. E., and F. D. Swanson: *AIEE Trans.,* **153:** 21.
45. Olyphant, M.: "Thermal Shock Tests for Casting Resins," Minnesota Mining and Manufacturing Co.
46. Hamill, A. T.: Westinghouse Goldilox Integrated Circuits, *Inform. Microelectron. Navy Equipt., U-Notes* 25, U.S. Naval Ammunition Depot, Crane, Ind., August, 1968.
47. Flood, J. L.: Reliability of Plastic Integrated Circuits, *Inform. Microelectron. Navy Equipt., U-Notes* 25, U.S. Naval Ammunition Depot, Crane, Ind., August, 1968.
48. Dalton, J. V.: Sodium Drift and Diffusion in Silicon Nitride Films, *Electrochem. Soc. Mtg.,* Cleveland, Ohio, May, 1966.
49. Snow, E. H., A. S. Grove, B. E. Deal, and C. T. Sah: Ion Transport Phenomena in Insulating Films, *J. Appl. Phys.,* **36:** 1664, 1965.
50. Yon, E., W. H. Ko, and A. B. Kuper: Sodium Distribution in Thermal Oxide on Silicon by Radiochemical and MOS Analysis, *IEEE Trans. Electron. Devices,* vol. ED-13 (1966).
51. Schneer, G. H., W. Van Gelder, V. E. Hauser, and P. F. Schmidt: A Metal Insulator-Silicon Junction Seal, *IEEE Trans. Electron. Devices,* vol. ED-15 (1968).
52. Johnson, G., and R. Benham: Performance of Epoxy Molding Compounds on DIP Integrated Circuits, *Soc. Matls. Process Engrs. Proc. Fifteenth Ann. Symp., Los Angeles,* April 30, 1969.
53. Kapfer, V. C., and A. L. Tamburino: Plastic Microcircuit Reliability, *RADC Status Rept., DOD-NASA-Ind. Meeting, Washington D.C.,* May 16, 1968.
54. Lee, S. M., J. J. Licari, and A. G. Valles: Properties of Plastic Materials and How They Relate to Device Failure Mechanisms, in M. E. Goldberg and J. Vaccaro (eds.), "Physics of Failure in Electronics," vol. 4. RADC Series on Reliability, Defense Documentation Center, Alexandria, Va., 1966.
55. Katsev, G. A., et al.: Effect of Absorption of Certain Amines on Semiconductor Surface on the Principal Parameters of Germanium Transistors, in A. N. Frumkin (ed.), "Surface Properties of Semiconductors," 1964.

56. Brandewie, J. V., P. H. Eisenberg, and R. A. Meyer: Investigation of Surface Failure Mechanisms in Semiconductor Devices by Envelope Ambient Studies, in M. E. Goldberg and J. Vaccaro (eds.), "Physics of Failure in Electronics," vol. 4, Rome Air Development Center, Defense Documentation Center, Alexandria, Va., 1966.
57. Licari, J. J., and G. V. Browning: Plastics for Packaging: Handle with Care, *Electronics,* April 17, 1967.
58. Kookootsedes, G. J., and F. J. Lockhart: Silicone Molding Compounds for Semiconductor Devices, *Mod. Plastics,* January, 1968.
59. Licari, J. J.: Selection and Control of Plastics for Semiconductor Packaging, *Inform. Microelectron. Navy Equipt. U-Notes,* 25, U.S. Naval Ammunition Depot, Crane, Ind., August, 1968.
60. Krasucki, Z., and H. F. Church: The Detection of Ionic Impurities in Insulation, *Inst. Elect. Engrs. Paper* 3819M, June, 1962.
61. Menzies, A. C.: *Anal. Chem.,* **32:** 898 (1960).
62. Alimarin, I. P., and M. N. Petrikova: "Inorganic Ultramicroanalysis," MacMillan, New York, 1964.
63. Sacharow, S.: The Permeability of Flexible Plastic Films, *Plastic Des. Processing,* September, 1965.
64. Private communication with Drs. S. Baird and Gordon Peattie of Texas Instruments, May, 1968.
65. Doctor and Franklin: Corrosion of Copper Wire by Resins, *Electron. Equipment,* July 1956.

10

Specifications

MATERIAL SPECIFICATIONS

The final choice of a coating, primer, or process is best implemented by means of written specifications. The material specification is a document used both for initial qualification of a material to specific engineering requirements and for receiving inspection of batches (lots) of an already qualified source to assure continued consistency and reliability. A material specification may also be used as a procurement document, and when called out on a purchase order, it legally binds the vendor to meet the quality-assurance provisions of the specification. Material specifications may be of two types. One type defines the minimum number of engineering requirements and the maximum allowable tolerances suitable for a specific use. This type of specification is usually written without concern for what materials are on the market and may be broad enough to be satisfied by a number of different coating types and formulations. Conversely, its provisions may be very tight, so that only one or even no source may be commercially available. An example of a specification of this type is given by MS001.

The second type of specification is written around a previously selected material. The actual formulation may be called out, or the physical, chemical, and electrical properties of the material may be narrowly defined so that only one formulation will be able to qualify. An example of a specification of this type is given by MS002. Both types of specifications are widely used in industry.

Note that in the second type of specification the material received must match an infrared spectograph as a condition of acceptability. The infrared spectra of organic materials differ widely, depending on their chemical structure, composition, and purity.[1,2]. Hence to ensure that the formulation has not been changed or the purity of the material altered, the infrared spectrum of the purchased material should be obtained and checked against the spectrum originally established as a standard. The infrared spectrum is fairly simple to obtain trom a sample of the material, and the test equipment, an infrared spectrophotometer, is available in many low-priced models.

For electronic parts and assemblies which are not coated in-house but are purchased already coated, an acceptance specification may be useful to assure quality control. In this case both nondestructive and destructive tests may be specified to assure that the part is adequately coated or that the coating fulfills certain requirements. In the case of low-cost small electronic components, destructive tests may be performed on a sampling basis from each large lot. Examples of tests that may be performed are adhesion, water soak, high-temperature exposure, radiation exposure, salt spray, and humidity with or without power. For larger or more expensive parts, nondestructive methods should be specified. Visual inspection under magnification, pencil hardness (followed by touchup), x-ray inspection, and ultraviolet-fluorescent inspection are examples of some nondestructive tests. A typical acceptance specification for aluminium parts coated with a high-electrical-insulation material is given by AS001.

Of course specifications should not be written where approved documents are already available and found to be applicable. Such documents may include military, federal, and ASTM specifications and Institute of Electrical and Electronics Engineers (IEEE), Society of Plastics Industries (SPI), National Electrical Manufacturer's Association (NEMA), Institute for Interconnecting and Packaging Electronic Circuits (IPC), and U.S. Underwriters' Laboratories standards. To familiarize the engineer with what is available, specification numbers and titles and brief descriptions are given in Tables 10-1 to 10-3. In addition, specification data for other materials and processes associated with coatings, such as solvents, resins, cleaning processes, and surface treatments, are given in Tables 10-4 and 10-5.

It should be emphasized that conformance of a material to military or federal specifications does not assure that it will be suitable for the intended application. Too often military specifications are not tight enough to provide the required reliability. In such cases the military specification should be supplemented with more specific provisions. An internal company specification may be written which calls out the military specification together with these other requirements.

PROCESS SPECIFICATIONS

The process specification is a written document which provides the step-by-step procedure to be used for manufacturing operation. An example of a typical process specification for application of a coating is given by PS001. The process specification should allow as much manufacturing freedom as possible within the confines of the engineering requirements. For example, instead of specifying a particular cure schedule, whenever possible a selection of cure schedules should be provided for. One practice that has proved convenient is to provide a time-temperature cure curve (Figure 10-1), so that manufacturing departments can select a cure schedule which will conform with oven temperatures which are already set. In all such cases, however, the equivalence of the cure schedules should be proved by assessing the properties of the cured coating.

If the method of application is not critical, the freedom to choose among spray, dip, or brush coating should also be allowed.

In summary, a rigid specification should be imposed only when absolutely essential; otherwise unnecessary cost is imparted to the final product.

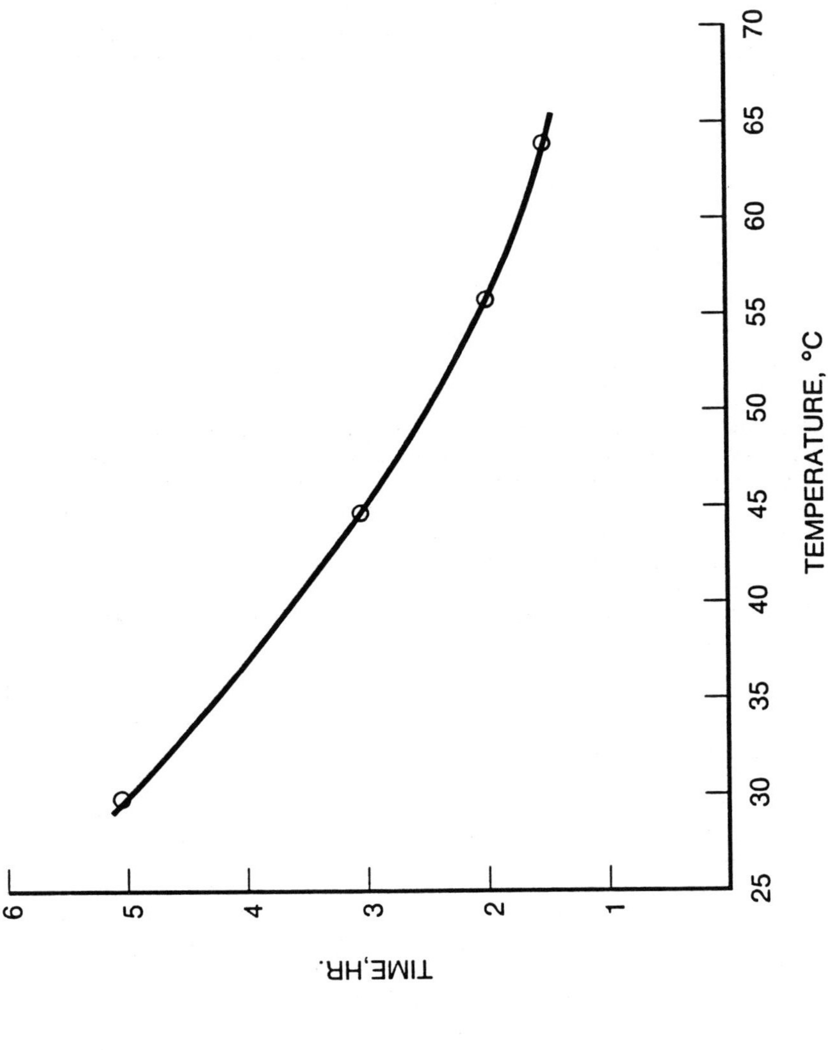

Figure 10-1: A time-temperature cure schedule for polyurethane conformal coating.

Number MS001	TITLE Coating, Polyurethane, Rapid Drying	PAGE 1

1. SCOPE

1.1 Description

This specification establishes the requirements for two types of rapid-drying low-temperature-curing polyurethane conformal coating systems. When properly mixed and cured, they provide electrical insulation and protection against environmental conditions.

1.2 Classification

1.2.1 The two types covered by this specification are:

Type I General Purpose
Type II High Adhesion

1.2.2 Type II shall be used unless otherwise specified.

2. APPLICABLE DOCUMENTS

The following documents, to the extent indicated, form a part of this specification. In the event of any conflict between the requirements of this specification and the listed documents, the requirements of this specification shall govern.

Federal Test Method Standard 141 Paint, Varnish, Lacquer, and Related Materials; Methods of Inspection, Sampling, and Testing

MIL-E-5272 Environmental Testing, Aeronautical and Associated Equipment, General Specification for

MIL-I-46058 Insulating Compound, Electrical (for Coating Printed Circuit Assemblies)

ASTM D 115 Methods of Test for Flash Point of Volatile Flammable Materials by Tag Open-cup Apparatus

ASTM E 168 Recommended Practices for General Techniques of Infrared Quantitative Analysis

3. REQUIREMENTS

3.1 Composition

The polyurethane coating system shall consist of a polyester resin base component, Part A, and an isocyanate catalyst component, Part B. They shall be furnished separately, and when mixed in equal parts by weight and applied by brush, spray, or dip methods, they shall produce a transparent coating conforming to the requirements of this specification.

3.2 Properties of the Mixed Uncured Coating Systems

The coating system shall meet the requirements of Table I for a minimum period of 4 hours after mixing equal weights of Parts A and B.

Number MS001	TITLE Coating, Polyurethane, Rapid Drying	PAGE 2

TABLE I

Property	Requirement	Test method
Viscosity, cps, maximum	500	4.4.2
Nonvolatile content, percent, minimum	35	4.4.1.1
Flash point, Tag open cup, °C, minimum	27	4.4.1.2
Work life, hours, minimum	4	4.4.3
Drying time	. . .	4.4.1.1
Tack free, minutes, maximum	60	
Dry hard, hours, maximum	2	

3.3 Properties of the Cured Coating

A film 0.001 to 0.002 inch thick of the coating prepared from a 50-50 mixture (by weight) by Parts A and B, and cured for 7 days minimum at room temperature shall meet the requirements of Table II.

TABLE II

Property	Requirement		Test method
	Type I	Type II	
Abrasion resistance, Taber wear index, CS-10 wheel, 1000 grams maximum	100	100	4.4.1.1
Water-vapor permeability, gram/hour/centimeter, maximum	1×10^{-6}	1×10^{-6}	4.4.1.4
Adhesion, percent, minimum			
Epoxy-glass laminate	100	. . .	4.4.4
Solder plate	. . .	80	
Anodized aluminum	. . .	80	
Gold plate	. . .	80	
Humidity resistance, ohms, minimum	. . .	. . .	4.4.1.3
Initial	1×10^{12}	1×10^{12}	
Post-humidity	1×10^{9}	1×10^{9}	
Salt-spray resistance, ohms, minimum	. . .	. . .	4.4.1.3
Initial	1×10^{12}	1×10^{12}	
Post-salt-spray	1×10^{9}	1×10^{9}	
Dielectric strength, volts/mil,	. . .	. . .	4.4.1.2
Condition I	2,500	2,500	
Condition II	1,200	1,200	

Number MS001	TITLE Coating, Polyurethane, Rapid Drying	PAGE 3

TABLE II (Continued)

Property	Requirement		Test method
	Type I	Type II	
Fungus resistance	Shall pass	Shall pass	4.4.1.3
Solderability and Repairability	Shall pass	Shall pass	4.4.5
Flexibility, ⅛ inch rod	Shall pass	Shall pass	4.4.1.1
Color			
Visible light	Transparent pink	Transparent pink	4.4.8
Ultraviolet light	Blue-white	Blue-white	

3.4 Shelf Life

The shelf life, as determined in paragraph 4.4.7, shall not be less than six months at 32°C

3.5 Infra-red Absorption

Materials qualified under this specification shall be identified by an infra-red analysis made in accordance with paragraph 4.4.6. Acceptance samples shall conform to this identification.

4. QUALITY ASSURANCE

4.1 Certification

The supplier shall certify that each shipment of material complies with all the requirements of this specification. The certification shall include the lot or batch number and the date of manufacture.

4.2 Qualification

The supplier shall submit a qualification sample that meets all requirements of this specification. The supplier shall certify that the sample submitted is representative of the material to be supplied under this specification.

4.3 Samples

4.3.1 Qualification Samples: Qualification samples shall be tested for, and shall meet, all requirements of this specification.

Number MS001	TITLE Coating, Polyurethane, Rapid Drying	PAGE 4

4.3.2 Acceptance Samples: Each batch or lot shall be inspected and an acceptance sample taken. Acceptance samples shall be tested for, and shall meet, the requirements of the tests specified in Table III and any other tests deemed necessary to ensure conformance to the requirements of this specification.

TABLE III

Viscosity
Nonvolatile content
Work life
Drying time, tack free
Drying time, dry hard
Adhesion, solder plate (Type II only)
Adhesion, gold plate (Type II only)

4.3.3 Preparation of Test Specimens:

4.3.3.1 Test specimens for those tests specified in Table IV shall be made of Grade GE-HT copper clad plastic laminate. They shall be etched with a test pattern in accordance with MIL-I-46058.

TABLE IV

Humidity resistance
Salt spray resistance
Solderability and repairability

4.3.3.2 Test panels shall be cleaned with a solvent composed of 50 ± 1 parts by volume of isopropyl alcohol and 50 ± 1 parts by volume of aliphatic naphtha. The solvent shall be applied by spraying or with a soft nylon bristle brush. Continue cleaning until all foreign particles are removed. Give the parts a final rinse by spraying or dipping using clean unused solvent.

4.3.3.3 After cleaning in accordance with 4.3.3.2, the parts shall be dried quickly by dry gas. The gas stream shall be applied at a minimum distance of 2 inches with the gas pressure not exceeding 30 psig.

4.3.3.4 Following solvent cleaning, the surfaces to be coated shall be thoroughly cleaned with an aqueous solution of 0.5 ± 0.2 ounce of Calgonite per gallon of water. The solution shall be at a temperature between 45°C and 70°C

Number MS001	TITLE Coating, Polyurethane, Rapid Drying	PAGE 5

4.3.3.4 The cleaning solution shall be applied by spraying or with a soft nylon brush. Continue until all foreign particles are removed and a continuous film of solution is formed over all surfaces. Flush the cleaning solution from all surfaces with distilled or deionized water. During the final rinse, which shall be a spray or stream of fresh distilled or deionized water, a continuous film of water shall form over the entire surface of the board. If the water drains and evaporates without breaking away at localized areas, the surface may be considered clean. Otherwise, repeat the operations specified in 4.3.3.2 through 4.3.3.5.

4.3.3.6 After cleaning, all parts shall be dried according to one of the following schedules in a vented, forced draft oven, or equivalent:

50-60°C, 1 hour, minimum
60-75°C, 30 minutes, minimum

4.3.3.7 Apply an 0.001 or 0.002 inch coating mixed in accordance with the manufacturer's instructions to the test specimen by spray, brush, or dip. Cure for 7 days minimum prior to testing.

4.4 Test Methods

4.4.1 Standard Methods:

4.4.1.1 Those tests listed in Table V shall be made in accordance with the specified method of Federal Test Method Standard 141.

TABLE V

Test	Method
Nonvolatile content	4041
Drying time	4061
Abrasion resistance	6192
Flexibility	6221

4.4.1.2 Those tests listed in Table VI shall be made in accordance with the specified ASTM Test Method.

TABLE VI

Test	Method
Flash point	D 1310
Dielectric strength	D 115

Number MS001	TITLE Coating, Polyurethane, Rapid Drying	PAGE 6

4.4.1.3 Those tests listed in Table VII shall be made in accordance with the specified procedure of MIL-E-5272.

TABLE VII

Test	Procedure
Fungus resistance	Procedure I
Humidity resistance	Procedure II
Salt spray resistance	Procedure III

4.4.1.4 Water vapor permeability shall be tested in accordance with the method specified in MIL-I-16923.

4.4.2 Viscosity shall be determined using a Model RVF Brookfield Viscometer at 25 ± 2°C, using a No. 1 spindle at 20 RPM.

4.4.3 Work life shall be determined using a sealed container approximately ¾ filled with mixed resin. Keep the container at 27 ± 5°C . The work life is elapsed period of time between mixing and the point where the viscosity exceeds 1000 cps.

4.4.4 Adhesion shall be evaluated on test surfaces after their exposure to humidity. Qualification samples shall be exposed to the humidity requirements of Procedure I of MIL-E-5272. The preparation of test specimens for acceptance samples may be accelerated by the use of a force cure for 4 hours at 70°C , followed by exposure to 100 percent relative humidity for 14–18 hours at 60–70°C. Inscribe a 0.5 to 1.0 inch cut in the coating. Apply a strip of 1 inch masking tape (3M No. 470 or equivalent) over the cut, using maximum finger pressure. Test for adhesion by stripping the tape with one quick pull at an angle approximately 180 degrees with respect to the panel surface. Examine for any separation of the coating from the surface and grade from 0 to 100 percent adhesion.

4.4.5 Solderability and Repairability: Prepare a test panel in accordance with paragraph 4.3.3. Wet the coated surface with Type W MIL-F-14256 rosin base soldering flux. Solder coat the tip of a 37½-watt soldering iron with QQ-S-571 flux-cored solder wire. Apply the tip of the soldering iron to the coated surface for 5 seconds. Failure is indicated by inability to solder through the film in one 5-second application or charring of the coating.

4.4.6 Infrared Analysis: An infra-red absorption curve shall be obtained on each part of the coating system. Samples shall be prepared as follows:

4.4.6.1 Place several drops of solution on an NaCl plate.

4.4.6.2 Flash off the solvents under a vacuum.

Number MS001	TITLE Coating, Polyurethane, Rapid Drying	PAGE 7

4.4.6.3 Cover the sample with an NaCl plate, taking care to minimize exposure to air.

4.4.6.4 Determine the absorption curve of the non-volatile component.

4.4.7 Shelf Life: Shelf life is defined as the minimum time, starting from the date of manufacture, that the material continues to meet the requirements of this specification when stored as packaged by the manufacturer at a temperature not exceeding 32°C.

4.4.8 Color shall be determined by visual inspection under the specified light sources. Visible light shall be normal room illumination. Ultraviolet light shall be any light source of less than 100 watts in the frequency between 2537 and 5661 angstroms.

5. PREPARATION FOR DELIVERY

5.1 Packaging

The material shall be packaged in accordance with the manufacturers' commercial practice to assure safe delivery by common carrier and prevent contamination by moisture, gases or other foreign material during normal handling.

5.2 Marking

Each container shall be permanently and legibly marked with the following:

Specification MS001
Type (I or II), Part (A or B) ________
Manufacturer's code ________
Manufacturer's name ________
Lot or batch number ________
Date of manufacture ________
Purchase order no. ________
Date of shipment ________
Mixing instructions ________
Storage temperature, not to exceed 90°F

NUMBER MS002	TITLE Coating, Polyurethane, Rapid Drying	PAGE 1

1. SCOPE

This specification establishes the requirements for a rapid drying, low temperature curing, two component polyurethane coating system. When properly mixed and cured it provides high electrical resistance and protection against environmental conditions.

2. APPLICABLE DOCUMENTS AND MATERIALS

2.1 Documents

MIL-C-27227 Coating, Polyurethane, Thermal Resistant, for Aircraft Application
Federal Test Method Standard No. 141 Paint, Varnish, Lacquer, and Related Materials, Methods of Inspection, Sampling, and Testing
ASTM D 115 Methods of Testing Varnishes Used for Electrical Insulation
ASTM D 257 Methods of Test for Electrical Resistance of Insulating Materials
ASTM D 1310 Methods of Test for Flash Point of Volatile Flammable Materials by Tag Open Cup Apparatus
ASTM E 168 General Techniques of Infrared Quantitative Analysis

2.2 Materials

Aroclor 1254 resin, Monsanto Chemical Co.: or equivalent
Butvar B-76-1 resin, Shawinigan Resin Corp.; or equivalent
Butyl acetate, normal, Spec. TT-B-838
Dibutyl tin dilaurate, Thermolite T-12, Metal and Thermit Corp.; or equivalent
Fluorescent dye, Rhodamine B, Eastman Organic Chemicals Dept.; or equivalent
Isocyanate resin, Mondur CB-60, Mobay Chemical Co.; or equivalent
Methyl isobutyl ketone, Spec. TT-M-268
Plastic laminate, copper clad, Grade GEB
Polyester resin, Type I, Multron R-4, Mobay Chemical Co.; or equivalent
Polyester resin, Type II, Multron R-12, Mobay Chemical Co.; or equivalent

3. REQUIREMENTS

3.1 Composition

The polyurethane coating system shall consist of a polyester resin base component, Part A, and an isocyanate catalyst component, Part B. They shall be furnished separately and, when mixed in the ratio specified in paragraph 3.1.3, they shall produce material conforming to the requirements of this specification.

NUMBER MS002	TITLE Coating, Polyurethane, Rapid Drying	PAGE 2

3.1.1 Composition of Part A: The polyester resin base, Part A, shall have the composition specified in Table I.

TABLE I

Ingredient	Parts by weight
Polyester resin, Type I	9.83 ± 0.05
Polyester resin, Type II	19.96 ± 0.10
Butvar B-76-1 solution*	1.50 ± 0.01
Aroclor resin, 1254	11.96 ± 0.06
Dibutyl tin dilaurate	0.120 ± 0.001
Fluorescent dye	0.020 ± 0.001
Methyl isobutyl ketone	56.61 ± 0.28

* 10 percent Butvar B-76-1 resin dissolved in butyl acetate.

3.1.2 Composition of Part B: The isocyanate catalyst component, Part B, shall be an adduct of tolylene diisocyanate and trimethylol propane dissolved in specific solvents. The composition shall be as specified in Table II.

TABLE II

Ingredient	Parts by weight
Isocyanate resin	49.0 ± 0.3
Methyl isobutyl ketone	51.0 ± 0.3

3.1.3 Composition of the mixed coating system. The composition of the mixed coating system shall be as specified in Table III.

TABLE III

Ingredient	Parts by weight or volume
Part A (paragraph 3.1.1)	50.0 ± 0.5
Part B (paragraph 3.1.2)	50.0 ± 0.5

3.2 Physical Properties

The components and the mixed coating system shall conform to the requirements of Table IV. The coating system shall be mixed according to paragraph 3.1.3, and shall be tested within 1 hour after mixing.

NUMBER	TITLE	PAGE
MS002	Coating, Polyurethane, Rapid Drying	3

TABLE IV

Property	Value			Test method
	Part A	Part B	Coating System	
Viscosity at 20 C with Brookfield #1 spindle, 60 rpm, cps	6–10	4–6	5–9	4.2
Solids content, percent	41–43	29–33	35–38	4.1
Flash point, °C, minimum	27	27	. . .	4.2
Specific gravity	0.94 ± 0.03	0.94 ± 0.03	. . .	4.1
Tack free time, minutes	. . .	. . .	45–60	4.1
Shelf life, minimum	3 months	3 months	. . .	4.6
Infrared curve	Curve 10-1	Curve 10-2	. . .	4.3

3.3 Properties of the Cured Coating System

A test specimen, prepared according to paragraph 4.4, shall conform to the requirements of Table V.

TABLE V

Property	Requirement	Test method
Abrasion resistance, wear index, maximum	50 milligrams	4.1
Pencil hardness, minimum	4 H	4.5
Insulation resistance, minimum	1×10^{12} ohms	4.2

4. QUALITY ASSURANCE

4.1 Those tests listed in Table VI shall be made in accordance with Federal Test Method Standard 141, using the specified method.

TABLE VI

Test	Method
Abrasion resistance	6192 (CS-10 wheel, 1000 grams, weight)
Solids content	4041 (Note 1)
Specific gravity	4183 (hydrometer method)
Tack free time	4061

NUMBER	TITLE	PAGE
MS002	Coating, Polyurethane, Rapid Drying	4

4.2 Those tests listed in Table VII shall be made in accordance with the specified ASTM method.

TABLE VII

Test	Method
Flash point	D 1310
Viscosity	D 115
Insulation resistance	D 257

4.3 Infrared Absorption

An infrared absorption curve, using a 0.026 millimeter liquid sample cell, shall be taken of the material as received in accordance with ASTM E 168. The recorded infrared spectrograms shall conform to Curves 10-1 and 10-2 and shall exhibit no major absorptions other than those indicated.

4.4 Cured Test Specimen

Test specimens shall be prepared from Grade GEB plastic laminate. A printed pattern shall be etched thereon. Test specimens shall not be given additional cleaning or preparatory treatments other than those given to production boards. Two coats of the coating solution, mixed in the ratio specified in Table III, shall be applied to the test specimen by dipping or spraying. The total dry film thickness shall be 2.0 ± 0.2 mils. After application of the first coat, the test specimen shall be air dried in a horizontal position for at least 30 minutes in a dust-controlled cabinet. The test specimen shall be cured for a minimum of 4 hours at 120 ± 10°F after application of the second coat prior to testing.

4.5 Pencil Hardness

The pencil hardness test shall be made in accordance with MIL-C-27227. The hardness of a flat surface of the coated object shall be tested, instead of the test panels specified in MIL-C-27227.

4.6 Shelf Life

Shelf life is defined as the minimum time, starting from the date of manufacture, that the material continues to meet the requirements of this specification when stored as packaged by the manufacturer at a temperature not exceeding 32°C.

5. PREPARATION FOR DELIVERY

5.1 Packaging

The material shall be packaged in accordance with the manufacturer's commercial practice to assure safe delivery by common carrier and prevent contamination by moisture, gases, or other foreign material during normal handling.

NUMBER	TITLE	PAGE
MS002	Coating, Polyurethane, Rapid Drying	5

5.2 Marking

Each container shall be permanently and legibly marked with the following:

Specification MS002
Part (A or B) ________________
Manufacturer's code ________________
Batch number ________________
Lot number ________________
Date of manufacture ________________
Purchase order no. ________________
Date of shipment ________________

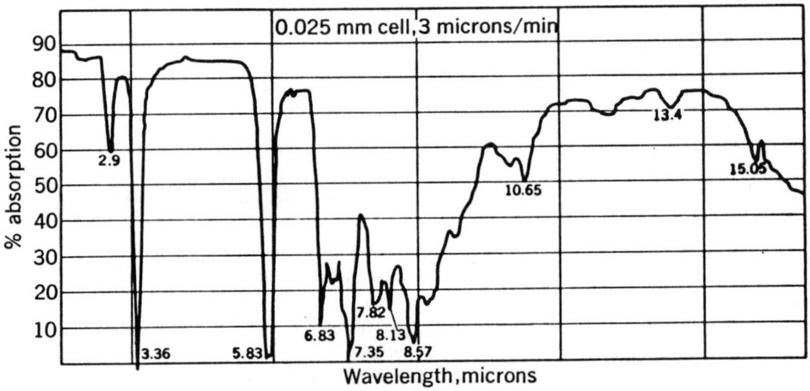

Curve 10-1 *Infrared spectrum of MS002, polyurethane part A.*

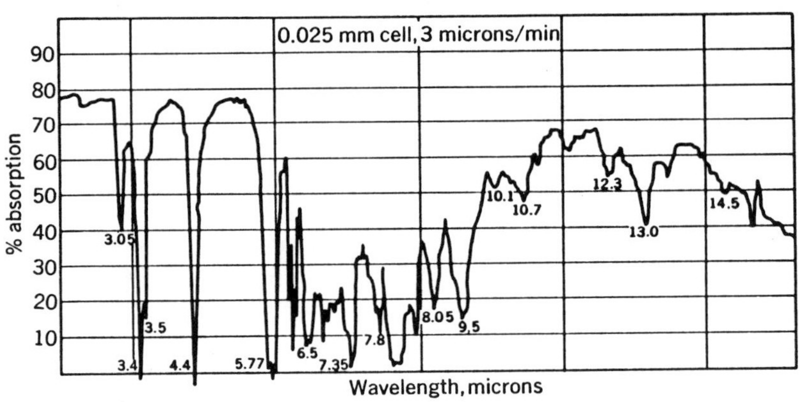

Curve 10-2 *Infrared spectrum of MS002, polyurethane part B.*

NUMBER AS001	TITLE Acceptance of Dielectric Coating on Metal Substrates for Printed Wiring Boards	PAGE 1

1. SCOPE

1.1 This specification establishes the requirements for an organic dielectric coating applicable to aluminum and copper heat sink substrates for printed wiring boards.

1.2 This coating may be used as a dielectric insulation on other heat sinks and parts, such as brackets, and electronic module housings.

1.3 This coating provides electrical insulation, thermal conductivity, corrosion resistance, nonhygroscopic properties, and a bondable surface.

2. APPLICABLE DOCUMENTS AND MATERIALS

The following documents, to the extent indicated, form a part of this specification. In the event of any conflict between the requirements of this specification and the listed documents, the requirements of this specification shall govern.

2.1 Specifications

PS062 Adhesive Bonding
PS012 Surface Preparation of Metals and Non-Metals for Adhesive Bonding
ASTM B 110 Method of Test for Dielectric Strength of Anodically Coated Aluminum
Federal Test Method Standard 141 Paint, Varnish, Lacquer, and Related Materials; Methods of Inspection, Sampling and Testing
MIL-E-5272 Environmental Testing, Aeronautical and Associated Equipment, General Specifications for
MIL-STD-202 Test Methods for Electronic and Electrical Component Parts

2.2 Materials

Glass Fabric, Impregnated, Epoxy Resin (Prepreg)

3. REQUIREMENTS

3.1 Appearance

The coating shall have no irregularities or scratches which would reduce the thickness of the coating below design requirements. There shall be no voids, flaking, peeling, or foreign inclusions which would affect the electrical insulation properties.

3.2 Coating Thickness

The dielectric coating thickness shall be 0.003 $\pm$ .001 inch.

NUMBER	TITLE	PAGE
AS001	Acceptance of Dielectric Coating on Metal Substrates for Printed Wiring Boards	2

3.3 Physical and Electrical Requirements

Applicable physical and electrical requirements are specified in Table I.

TABLE I

Physical and Electrical Requirements

Property	Requirement, Various Environments					Test Method
	After 4 hr @ 30-60% RH & 25±2°C (Test at 25±2°C)	After Humidity		After Thermal Cycling	At 120 ±5°C	
		24 hr @ 95-100% RH and 50°C	10 Cycles (240 hr) MIL-E-5272			
		4.11.2	4.11.1	4.12		
Dielectric withstanding[1] volts, minimum	1,000	1,000	1,000	1,000	...	4.5
Insulation resistance[1] megohms, minimum	50,000	50,000	50,000	50,000	...	4.6
Lap shear tensile strength psi, minimum	1,500	1,200	...	...	350	4.7
Abrasion resistance wear index, mg/1,000 cycles, max.	150	...	...	...	...	4.9
Hardness, shear-resistance units, minimum	2,000	...	...	...	...	4.10

[1] Based on a minimum coating thickness of .003 ± .001 inches, as required by specification.

3.4 Temperature-Pressure Mar Resistance

There shall be no evidence of depressions made in the coating by the conditions of the temperature-pressure mar test performed in accordance with 4.8.

3.5 Fungus Resistance: The coating shall not support fungus growth when tested in accordance with 4.13.

3.6 Salt Spray Resistance: There shall be no evidence of corrosion, peeling, blistering or pitting when tested in accordance with 4.14.

NUMBER AS001	TITLE Acceptance of Dielectric Coating on Metal Substrates for Printed Wiring Boards	PAGE 3

3.7 Chemical and Solvent Resistance: After testing in accordance with 4.15, the coating shall show no pronounced change in appearance, shall not be visibly softer on the immersed portion of the test specimen and shall meet the dielectric breakdown and insulation resistance requirements specified in Table I (room temperature test).

4. QUALITY ASSURANCE

4.1 Qualification Test

Standard 4 × 4 inch aluminum panels with centered ¼-inch diameter hole and 1 × 4 inch aluminum panels having 0.003 ± 0.001 inch thickness of dielectric coating on one side shall be tested to conform to the requirements of this specification. The coating manufacturer-applicator shall make no changes in the materials or processes without requalifying to this specification.

4.2 Production Acceptance Test

Visual inspection of workmanship in accordance with 4.3 shall be conducted on 100 percent of each production lot. Testing in accordance with 4.4, 4.5, and 4.6 shall be conducted on 10 percent of each production lot limited to a minimum of one piece. If the production part does not permit testing, because of its size or shape, suitable samples of the identical material, processed simultaneously and identically with the production parts, shall be used for test.

4.3 Appearance

The coating shall be examined visually without magnification for defects.

4.4 Coating Thickness

The thickness of the dielectric coating shall be measured by means of a Dermitron (Unit Processes Assemblies, Inc., New York, New York) or equivalent.

4.5 Dielectric Withstanding Voltage

The dielectric withstanding voltage shall be measured in accordance with ASTM B 110.

4.6 Insulation Resistance

4.6.1 Flat Surface: A circular copper electrode 3 inches in length with a contact area of 0.50 ± 0.02 square inches shall be used to measure the surface resistance of flat surfaces. The electrode shall be embedded in a circular "button" of polyester or acrylic plastic. The contact surface shall be ground and polished flat. The "button" diameter shall be 2.00 ± 0.002 inches and the weight of the entire assembly shall be 235 ± 5 grams. The electrode shall be placed on the flat surface being measured. A potential of 50 volts d.c. shall be applied between the electrode and the basis metal of the test part. Measure the resistance after 30 seconds at 50 volts. One reading shall be made for each five square inches of dielectric surface.

NUMBER AS001	TITLE Acceptance of Dielectric Coating on Metal Substrates for Printed Wiring Boards	PAGE 4

4.6.2 Hole Walls: The electric resistance of dielectric coatings in hole walls shall be measured according to 4.6.1, except that the electrode shall be a bare 6061-T4 aluminum rod. This electrode shall have a diameter not less than the diameter of the hole being measured less 0.05 inches. The electrode shall be inserted either through the hole or to its bottom during the measurement of the resistance. The test electrode shall be in contact with the hole wall during test.

4.7 Lap Shear Tensile Strength

The dielectric coated strip shall be bonded, in accordance with PS062, to an aluminum strip, using thin sheet prepreg, 0.002–0.003 inch thickness epoxy glass laminate, Type G-10. A minimum of three specimens shall be pulled and the results averaged.

4.8 Temperature Pressure Mar Resistance

Place 5 lengths of 18 ga. wire in contact with the coated specimen, parallel and equidistant. Preheat a 450 ± 25 gram rectangular metal block in a 175 ± 3°C oven for a minimum of 30 minutes. Place the preheated block on top of the wires and subject entire test assembly to 175 ± 3°C for 30 ± minutes. (See Fig. 10-2.) Remove the block and wires and visually examine coating for evidence of marring of the surface.

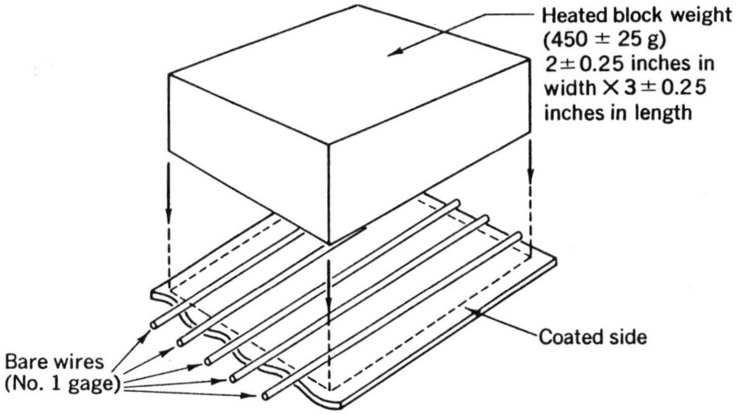

Fig. 10-2 Diagram of mar-resistance test showing relationship of weight and wires to test coating.

4.9 Abrasion Resistance

The resistance to abrasion shall be tested on standard 4 × 4 inch panels with centered ¼-inch diameter hole, having a 0.003 ± 0.001 inch thickness of dielectric coating. The wear index shall be determined in accordance with Federal Test Method Standard 141, Method 6192, using a 1000 gram load on CS-17

NUMBER AS001	TITLE Acceptance of Dielectric Coating on Metal Substrates for Printed Wiring Boards	PAGE 5

wheels, run for 100 cycles. Wheels shall be resurfaced by running against an S-11 abrasive disc for 25 cycles prior to the start of a test series and at least every 200 cycles. Report wear index in milligrams removed per 1000 cycles:

$$\text{Wear index} = \frac{(A - B)1000}{C}$$

where
A = Weight of test specimen before abrasion
B = Weight of test specimen after abrasion
C = Number of abrasion cycles recorded

4.10 Scratch Resistance (Hardness)

The hardness shall be determined by means of the Taber Shear/Scratch Tester in accordance with the instructions of the manufacturer (Taber Instrument Corp.). The test shall be performed on the same specimens as the abrasion resistance test, using an S-20 contour shear tool. Results shall be calculated in shear resistance units.

4.11 Humidity

4.11.1 Cycling Temperature and Humidity: Humidity cycling conditioning shall be in accordance with MIL-E-5272.

4.11.2 Non-cycling Humidity: Humidity conditioning shall be at 95 ± 5 percent R.H. at 50 ± 2°C,

4.12 Temperature Cycling

A temperature cycle shall consist of 15 ± 1 minute at 200 ± 5°C , followed by 15 ± 1 minute at -65 ± 5°C, with a transition period of 3 to 5 minutes.

4.13 Fungus Resistance: The fungus resistance of the coating shall be tested in accordance with MIL-E-5272, using Types I, II, III and IV.

4.14 Salt Spray Resistance: The salt spray resistance of the coating shall be tested according to MIL-STD-202, Method 101, Condition B, 5 percent salt solution.

4.15 Chemical and Solvent Resistance: Half-immerse a coated part or test panel in each of the following materials at 25 ± 2°C for 24 hours:

50/50 Isopropyl alcohol/naphtha
Perchloroethylene
Methyl ethyl ketone
Ferric chloride (25 percent solution)

Remove the coated part or test panel and dry for 10–20 minutes at 25 ± 2°C. Examine visually for obvious changes in appearance and check for softening. The coated part or test panel shall be tested in accordance with Dielectric Withstanding Voltage (4.5) and Insulation Resistance (4.6).

NUMBER PS001	TITLE Application of a Conformal Rigidizing Coating to Electronic Modules	PAGE 1

1. SCOPE

1.1 This specification establishes the procedure to be used for the application of a conformal coating to electronic modules.

1.2 This process is used to impart strength to the modules and to provide limited protection to strain sensitive components from forces induced by an embedment compound.

1.3 The coating applied is not intended for use on modules operating at temperatures above 120°C.

1.4 The coating shall not be applied to any part or surface where such application will interfere with the operation or performance of the equipment. Examples of such parts and surfaces are variable resistors or capacitors, contact areas of connectors, switches, sockets, edges which engage guide slots, test point terminals, and cold plate connections.

2. APPLICABLE DOCUMENTS AND MATERIALS

2.1 Documents

PS007 Preparation of Sealing and Masking Compound

2.2 Materials

Acetone, technical O-A-51
Dry gas Nitrogen or air with a dew point below +0°C and containing no particles larger than 100 microns in length and no more than 100 particles per cu. ft. in the 5–100 micron size range
Isopropyl alcohol, technical TT-I-735, Grade A
Maskant MS007
Naphtha, aliphatic TT-N-95, Type II
Polyurethane coating MS009
Tape, masking Number P-422, Permacel-LePage's Inc.; or Number 60 or 470, Minnesota Mining and Mfg. Co.
1,1,1-Trichloroethane 0-T-620

3. REQUIREMENTS

3.1 Safety, Storage, and Handling Precautions

3.1.1 All manufacturing operations described by this specification shall be individually evaluated for the selection of adequate ventilation and exhaust facilities by the Industrial Hygiene and Safety Office.

3.1.2 The Industrial Hygiene and Safety Office shall be consulted prior to the installation or modification of facilities necessary for the processing required by this specification.

NUMBER PS001	TITLE Application of a Conformal Rigidizing Coating to Electronic Modules	PAGE 2

3.1.3 Any resin or solvent solution splashed upon the facial area, or transferred there by contact with the hands or other means, shall be carefully removed with isopropyl alcohol, using care to protect the eyes. Eyes and eye areas affected by any of the processing materials shall be immediately flushed with large quantities of clean water. Persons experiencing contact with the processing materials in eyes and eye areas shall be referred to First Aid or the company hospital immediately after the flushing required previously.

3.2 Cleaning

3.2.1 All cleaning operations shall be performed as rapidly as possible to avoid the removal of identification and inspection marks.

3.2.2 The parts shall be cleaned with a solvent composed of 50 ± 1 parts by volume of isopropyl alcohol, (2.2), and 50 ± 1 parts by volume of naphtha (2.2). The solvent shall be applied by spraying or with a soft nylon bristle brush. Continue cleaning until all foreign particles are removed. Give the parts a final rinse by spraying with clean unused solvent.

3.2.3 After cleaning dry the parts quickly by dry gas (2.2). The stream of gas shall be sufficient to rapidly dry the cleaned part without dislodging or damaging attached components. The gas stream shall be applied at a minimum distance of 2 inches with the gas pressure not exceeding 30 psig.

3.2.4 After cleaning in accordance with 3.2.2 and drying in accordance with 3.2.3, place the part in a forced air oven for 25 ± 10 minutes at 75 ± 2°C.

3.2.5 Cleaned assemblies shall be protected from dust and recontamination at all times until the coating has been applied and cured.

3.3 Masking

3.3.1 Areas which will be exposed to the conformal coating material but which must be kept free of the coating, in accordance with 1.4, shall be masked. Either masking tape MS007 maskant (2.2) or a maskant prepared in accordance with PS007 may be used.

3.3.2 Masking tape shall not be stretched during its application. The tape shall be applied with sufficient pressure to effect a good bond, but the pressure shall not be sufficient to damage the part.

3.4 Preparation of the Coating Solution

3.4.1 All containers of resin mixed with hardener shall be labeled at the time of mixture with the expiration time of the mix.

NUMBER	TITLE	PAGE
PS001	Application of a Conformal Rigidizing Coating to Electronic Modules	3

3.4.2 The polyurethane coating shall be thoroughly mixed in the following ratio:

Material	Parts by weight
Resin	3
Hardener	1
Reducer	
For brush and dip application	1
For spray application	2

3.4.3 The mixed coating solution shall be degassed by placing it under a partial vacuum. Do not allow the solvent to boil.

3.4.4 The reactive components shall be weighed to an accuracy of ±2.0 percent.

3.4.5 Paper containers and wood spatulas shall not be used for mixing.

3.4.6 The mixed material shall be used within 4 hours.

3.5 Application of Coating

3.5.1 Dip Application:

3.5.1.1 Immerse the object to be coated in the coating solution prepared in accordance with paragraph 3.4. All surfaces to be coated shall be covered by the solution for at least 30 seconds, using agitation or gentle movement if necessary.

3.5.1.2 Withdraw the coated object and suspend the article in a nearly vertical position. Remove any excess resin beads or tears from the object with a soft nylon brush after the board ceases to drain.

3.5.1.3 After a minimum air dry time of 30 minutes, paragraph 3.5.1.1 and 3.5.1.2 shall be repeated until the required coating thickness is obtained.

3.5.2 Spray Application:

3.5.2.1 All surfaces to be coated shall receive a minimum of two spray applications of the coating solution prepared in accordance with section 3.4. Additional spray coats may be required to obtain the appearance and dry film thickness specified in section 4.

3.5.2.2 Spray the polyurethane coating using dry gas (2.2). The application shall be continued until the surface appears wet by reflected light, but shall not be thick enough to cause runs or sags. The part shall be allowed to air dry for a minimum of 30 minutes between the first and the second coat and for a minimum of 1 hour prior to the application, if necessary, of third or any subsequent application necessary to obtain the required coating thickness.

NUMBER PS001	TITLE Application of a Conformal Rigidizing Coating to Electronic Modules	PAGE 4

3.5.2.3 The three dimensional module shall be rotated 360 degrees in the X, Y, and Z axes to obtain complete coverage.

3.5.3 Brush Application: Apply the polyurethane coating solution, prepared in accordance with 3.4, to all surfaces. Brush out entrapped air bubbles and brush marks as carefully as possible with a soft nylon brush. Remove any excess resin beads or tears with the nylon brush. Repeat the application once after a minimum air dry time of 30 minutes.

3.6 Removal of Excess Resin

Excess resin accumulating on the surface or edges of parts shall be removed with a soft nylon brush before curing.

3.7 Coating Cure

3.7.1 After application to the circuitry assembly, the coating shall be air dried for 30 minutes minimum. It shall then be cured in an oven with forced air circulation for one of the following schedules:

(a) 5 days minimum air dry
(b) 6 hours minimum at 32°C
(c) 3 hours minimum at 45°C
(d) 2 hours minimum at 65°C
(e) 1 hour minimum at 75°C

3.7.2 Post cure 5 days at 70–85°F (not necessarily in a drying chamber) before exposure to chemical or environmental testing.

3.8 Clean-up

3.8.1 After the coating has been cured, per section 3.7, remove masking material, if any, and trim off flash material to desired contour. Any adhesive deposits left by masking tape shall be removed with acetone or 1,1,1-trichloroethane, using a wiping technique.

3.8.2 Every effort shall be made to clean up tools and equipment before the coating resin has an opportunity to set up. In the uncured state, the coating resin may be easily removed with a mixture of two parts acetone (technical grade) and one part isopropyl alcohol.

4. QUALITY ASSURANCE

4.1 Appearance

Without magnification, the cured coating shall present a smooth continuous surface free from thin areas, runs, or sags. Bubbles in the coating smaller in diameter than 0.020 inch shall not be cause for rejection. The coating shall be inspected for continuity under ultraviolet light.

4.2 Cure

The coating shall be tack-free to touch after curing, in accordance with 3.7.

ELECTRICAL-PERFORMANCE SPECIFICATIONS

Often the choice of coating is dictated, not by a coating specification, but by the properties it must have in order to meet an electronic-component or hardware specification. These specifications (some of which are given in Table 10-1.) define electrical tolerances, actual service conditions, and environmental conditions.

COMMERCIAL, MILITARY, AND FEDERAL SPECIFICATIONS

Hundreds of standard test methods, environmental and electrical requirements, and materials specifications have been generated by various government agencies and by national professional societies. This section describes most of the available specifications applicable to organic coatings. More complete listings may be found in *Plastec Note* 17[3] and in *Insulation,* Directory Encyclopedia Issue[4].

Many test conditions and acceptance criteria are of an accelerated nature, devised to allow extrapolation and prediction of long-term performance. However, results should be used with caution in predicting specific performance. Accelerated tests should be used for screening and selecting materials for further evaluation. Final tests should consist of either simulated tests or tests on prototypes. Simulated tests are not defined in the literature, since each application is different. A simulated test must be designed for each case. For example, a circuit-module coating should be evaluated by applying it to a conductor pattern that has been subjected to all the fluxing, soldering, and cleaning steps that it will be exposed to in its fabrication. Further simulated testing should include subjecting it to the environment it might be expected to see in operation. Ultimately, the unequivocal test is the performance of the coated part under actual operating conditions. Hence many design engineers will allow for prototype fabrication and testing of several units before full-scale production is begun.

These should be kept in mind in examining the lists of commonly used specifications in Tables 10-1 to 10-6.

TEST METHODS FOR ORGANIC COATINGS

The commonly used test methods and procedures for paints, primers, and insulation coatings are presented in Table 10-8. Many of the specifications are not coating specifications but are cited because the test procedures contained in them may be applicable to coatings. Often the only difference is the size and thickness of the sample used. Hence specifications for potting compounds, embedding compounds, wire, etc. will be listed whenever the method described may also be useful for the testing of coatings. A guide to abbreviations used in this section is given in Table 10-7.

Table 10-1: Electrical and Electronic Requirements

Document	Title	Description
ANA 400	Electronic Equipment: Aircraft and Guided Missiles, Applicable Documents	A comprehensive listing of specifications and other documents pertinent to aircraft and aerospace electronic systems
MIL-STD-454	Standard General Requirements for Electronic Equipment	The basic Department of Defense document standardizing the more common requirements for electronic equipment
MIL-STD-202	Test Methods for Electronic and Electrical Component Parts	Uniform standards for testing electronic and electric component parts, including environmental, physical, and electrical tests
MIL-E-5400	Electronic Equipment, Aircraft, General Specification for	General requirements for electronic gear, primarily in piloted aircraft, by classes defined as altitude and temperature range
MIL-I-983	Interior Communication Equipment, Naval Shipboard, Basic Design Requirements for	Basic design requirements, testing, and operating conditions for interior communication equipment on ships
MIL-E-4158	Electronic Equipment, Ground, General Requirements for	Requirements for design and fabrication of electronic ground equipment
MIL-E-8189	Electronic Equipment, Missiles, Boosters, and Allied Vehicles, Inactive Specification for	Design and general requirements for the design and manufacture of electronic systems for guided missiles
MIL-P-11268	Parts, Materials, and Processes Used in Electronic Equipment	Selection, application, and use of materials and processes in the construction of electronic communication equipment
MIL-E-11991	Electronic-Electrical and Electromechanical and Associated Weapons Systems Equipment, General Specification for	Requirements common to electronic, electrical, and electromechanical equipment for guided-missile weapon systems
MIL-E-16400	Electronic Interior Communication and Naval Equipment, Naval Ship & Shore: General Specifition	General requirements applicable to the design and construction of electronic equipment intended for naval ship-and-shore applications; ambient conditions within which equipment must operate
MIL-F-18870	Fire Control Equipment, Naval Ship and Shore: General Specification	General requirements for design and manufacture of surface, shore, antiaircraft, and underwater fire-control equipment
MIL-T-28800	Test Equipment for Use with Electrical and Electronic Systems, General Specification for	General requirements for design and construction of equipment used in testing electronic and fire-control equipment
MIL-STD-275	Printed Wiring for Electronic Equipment	Design principles governing the fabrication of printed wiring and the mounting of parts and assemblies
MIL-STD-750	Test Methods for Semiconductor Devices	Tests and environmental conditions for semiconductor devices
MIL-STD-883	Test Methods & Procedures for Microelectronics	Uniform methods and procedures for testing microelectronic devices, including monolithic, multichip, and film hybrid microcircuits, and microcircuit arrays for Military/NASA uses
NEMA MW 1000	Magnet Wire	Standards for magnet wire and wire coatings and acceptance tests
MIL-S-19500G	Semiconductor Devices, General Specification for	General requirements for semiconductor devices used in military equipment
MIL-S-19500G	Semiconductor Devices, General Specification for	General requirements for semiconductor devices used in military equipment
IPC-D-320	Printed Board Rigid, Single- and Double-Sided, and Product Specification	General requirements for printed circuit boards

Table 10-2: Finishing Systems

Document	Title	Description
MIL-HDBK-132	Military Handbook, Protective Finishes for Metal and Wood Surfaces	Details of inorganic and organic protective finish systems which have proved satisfactory at various ordnance corps installations
MIL-F-18264	Finishes, Organic, Weapons System, Application and Control of	Recommended materials and their characteristics, application equipment and procedures, and trouble-shooting in painting problems
MIL-F-7179	Finishes, Coatings, and Sealants, General Specification for the Protection of Aerospace Weapons Systems	Coating-system specifications, mainly for the protection of nonelectronic surfaces
MIL-F-14072	Finishes for Ground Signal Equipment	Design guide for selection of compatible materials and specifications for individual finishing system, by substrate and severity of exposure to prevent deterioration in the anticipated environments
MIL-STD-171	Finishing of Metal and Wood Surfaces	A comprehensive index of specifications and guide to the preparation and finishing of many types of surfaces
MIL-STD-186	Protective Finishing Systems for Painting and Finishing Rocket, Guided Missles, Support Equipment and Related Materials	Minimum requirements for procedures, materials, and systems for cleaning, plating, and painting of electronic components for rockets and guided missiles
MIL-STD-194	Painting and Finishing Fire Control Material	Minimum requirements for surface preparation, surface treatment, painting, and finishing of fire-control instruments
MIL-T-704	Treatment and Painting of Material	Materials and procedures for cleaning, treating, and painting surfaces and equipment to provide protection from environment and corrosion (including sources for color callout)
MIL-STD-808	Finish, Materials and Processes for Corrosion Prevention and Control in Ground Support Equipment	Procedures for cleaning, painting and plating of all ground and ground support equipment used for exposed equipment (Type I) and sheltered equipment (Type II)

Table 10-3: Coating Materials

Document	Title	Description
	Coatings, Insulation, Sealants, and Paints	
MIL-S-22266B	Insulating Compounds, Electrical (for Field Splicing Applications)	Room-temperature-curing epoxy-resin compounds for field-splicing and electronic insulation applications
MIL-S-23586	Sealing Compound, Electrical, Silicone Rubber, Accelerator Required	Requirements for RTV nonfuel-resistant silicone-rubber compounds for use from -80-400°F for electronic applications on military weapons systems
MIL-S-8516	Sealing Compound, Polysulfide Rubber, Electric Connectors and Electric Systems, Chemically Cured	Accelerated organic polysulfide liquid polymer for sealing and reinforcing electrical apparatus
MIL-P-22808	Paint, Epoxy, Hydraulic Fluid Resistant	An epoxy-polyamide coating system for protection of magnesium at temperatures to 500°F
TT-P-28	Paint, Aluminum, Heat Resisting (1200°F)	Paints with good salt-spray resistance, capable of useful service to 1200°F for many hours
MIL-I-46058	Insulating Compound, Electrical (for Coating Printed Circuit Assemblies)	Various types (epoxy, polyurethane, silicone, and polystyrene low dielectric) of conformal coatings for printed-circuit assemblies
MIL-I-24092	Insulating Varnish, Electrical, Impregnating, Solvent Containing	Characteristics of baking-type insulating varnishes, for electrical coils and structures
MIL-I-16923D	Insulating Compound, Electrical, Embedding	Requirements for several types of compounds for embedding or encasing electrical components
	Enamels	
MIL-E-15090	Enamel, Equipment, Light Gray (Formula No. 111)	Air-drying and baking enamel suitable for use on equipment, furniture, and switchboard installation
TT-E-527	Enamel, Alkyd, Lusterless	Alkyd type of combination air-drying and baking lusterless enamel as a finishing coat for primed metal
TT-E-529	Enamel, Alkyd, Semi-gloss	Alkyd semigloss enamel for exterior and interior wood and metal surfaces for military and special equipment
TT-E-489	Enamel, Alkyd, Gloss (for exterior and interior surfaces)	Covers alkyd and urea- or melamine-formaldehyde-modified alkyd enamels for use on primed exterior and interior wood and metal surfaces

(continued)

Table 10-3: (continued)

Document	Title	Description
MIL-E-7729	Alkyd Gloss Enamel	Baking enamel
TT-E-485	Enamel, Semi-gloss, Rust-inhibiting	Medium oil-alkyd and phenol-modified-alkyd enamels for use on metal, with good resistance to water, weathering, and hydrocarbon fluids
MIL-E-22118A	Enamel, Electrical Insulating	One grade and type of synthetic insulating enamel for electronic equipment and general use
TT-E-496	Enamel, Heat Resisting (400°F), Black	Air-drying and baking-type black heat-resistant enamel
	Primers and Surface Treatments	
MIL-C-8514	Coating Compound, Metal Pretreatment, Resin-Acid	Pretreatment wash-primer coating with adhesion-promoting and anticorrosive properties, for metals to be applied prior to a topcoating system; Air Force and Navy-Weps version
MIL-C-15328	Primer, Pretreatment (Formula No. 117 for Metals)	Wash primer similar to MIL-C-8514, Navships and DOD version
MIL-P-7962	Primer Coating, Cellulose-Nitrate Modified Alkyd Type, Corrosion-inhibiting, Fast-drying (for Spray Application over Pretreatment Coating)	Lacquer primer for use only over MIL-C-8514 and under nitrocellulose or acrylic-nitrocellulose lacquer
MIL-W-438B	Wire, Primer, Electrical	Requirements for electrical wire suitable for ignition or bridge wires
MIL-P-15930A	Primer Coating, Shipboard, Vinylzinc Chromate (Formula No. 120– for Hot Spray)	Coating for magnesium
MIL-P-23377	Primer Coating, Epoxy-Polyamide, Chemical and Solvent Resistant	Primer for aluminum alloys and for the separation of dissimilar metals
MIL-P-52192	Primer Coating, Epoxy	Coating for magnesium
TT-P-645	Primer Paint, Zinc Chromate, Alkyd Type	Corrosion inhibiting even under high-humidity conditions, for use on steel and aluminum surfaces
MIL-T-13867	Treatment, Moisture and Fungus Resistant, for Fire Control Electrical & Electronic Instruments and Equipment	Four types of treatments for increasing moisture and fungus resistance of fire-control electrical and electronic instruments and equipment
TT-P-605	Primer Coating, All Purpose Gray, Alkyd (in Pressurized Dispensers)	

(continued)

Table 10-3: (continued)

Document	Title	Description
TT-P-615	Primer Coating, Basic Lead Chromate, Ready Mixed	
TT-P-636	Primer Coating, Alkyd, Wood and Ferrous Metal	
TT-P-659	Primer and Surface Synthetic, Tints and White (for Metal and Wood Surfaces)	
TT-P-664	Primer Coating, Synthetic, Rust-Inhibiting, Lacquer-Resisting	
MIL-T-152B	Treatment, Moisture and Fungus Resistant, of Communications, Electronic, and Associated Electrical Equipment	Requirements for the application of moisture- and fungus-resistant varnish to materials not resistant to moisture and fungus
Varnishes and Lacquers		
MIL-V-173	Varnish, Moisture-and-Fungus-Resistant (for treatment of Communications, Electronic & Associated Electrical Equipment)	A commonly used moisture- and fungus-resistant varnish consisting of paraphenylphenol formaldehyde resin combined with tung oil and solvents
MIL-V-13497A	Varnish, Impregnating, Electrical Insulating (for Fire Control Instruments)	An electrical insulating varnish for impregnating electrical windings of motors
TT-V-109	Varnish, Spar, Alkyd Resin	A clear, air-drying alkyl resin varnish containing 30% phthalic anhydride and 45–55% oil acids dissolved in aliphatic and aromatic hydrocarbons
MIL-L-19537	Lacquer, Acrylic-nitrocellulose Gloss (for Aircraft Use)	General-purpose exterior coating for metals, resistant to diester lubricating oils
MIL-L-19538	Lacquer, Acrylic-nitrocellulose, Camouflage (for Air-Use)	Same as MIL-L-19537, except available in lusterless camouflage colors
MIL-L-14486	Lacquer, Vinyl Resin	For use as a finish coat
MIL-L-52043	Lacquer, Semigloss, Cellulose Nitrate	For use over magnesium
TT-L-32	Lacquer, Cellulose Nitrate, Gloss, for Aircraft Use	Durable cellulose-nitrate gloss lacquer
TT-L-50E	Lacquer, Nitrocellulose Acrylic and Acrylic Butyrate, Aerosol (in Pressured container)	Two types of air-dry lacquers in pressurized containers (aerosol), Type I, nitrocellulose, and Type II, acrylic

Table 10-4: Solvents, Resins, and Other Coating Ingredients*

Document	Title
TT-T-548	Toluol (Toluene)
TT-R-2660	Resin, Alkyd Solutions
MIL-R-21931	Resin, Epoxy
MIL-R-3043B	Resin Coating, Unpigmented, for Engine Components and Metal Parts
MIL-R-23461	Resin Compound, Thermosetting, Room Temperature Curing, for Metal Coating
MIL-R-15189B	Resin, Alkyd Phenolic
MIL-R-46068	Resin, Polyester, Bisphenol-A Type
MIL-R-82438	Resin, Polyester, Solutions
O-T-620	1,1,1-Trichloroethane
MIL-T-81533A	1,1,1-Trichloroethane
O-T-634	Trichloroethane

*This is only a short compilation of ingredients used in coating formulations. For a complete list see "Guide to United States Government Paint Specifications," National Paint, Varnish, and Lacquer Association, Washington, DC.

Table 10-5: Cleaning and Surface Treatments

Document	Title	Description
MIL-M-3171	Magnesium Alloy, Processes for Pretreatment and Prevention of Corrosion on	Dow 1, 7, 9, and 19 treatments
MIL-M-45202	Magnesium Alloys, Anodic Treatment of	Dow 17, HAE, and CR-22 anodic treatments
MIL-C-5541	Chemical Conversion Coatings on Aluminum Alloys	A chromate conversion coating for aluminum, covers Alodine treatment
MIL-T-12879	Treatments, Chemicals, Prepaint and Corrosion Inhibitive, for Zinc Surfaces	Prepaint and corrosion-inhibitive chemical treatments for plated, hot-dipped, and solid zinc surfaces
MIL-A-8625	Anodic Coatings for Aluminum & Aluminum Alloys	Chromic and sulfuric acid anodization processes
MIL-C-38334	Corrosion Removing Compound, Prepaint, for Aircraft Aluminum Surfaces	Nonflammable phosphoric acid-base liquid
MIL-S-5002	Surface Treatments and Inorganic Coatings for Metal Surfaces of Weapons Systems	Preparation, surface treatments, and metallic coatings for metal surfaces of weapons systems prior to application of organic finishes

Table 10-6: Test Procedures: General Specifications

Document	Title	Description
FED. STD. No. 141 (TTP-141)	Paint, Varnish, Lacquer, and Related Materials, Methods of Inspection, Sampling, and Testing	A compilation of 126 methods for determining physical and chemical properties of paint, varnish, and lacquer for conformance to federal and military specifications
ASTM D 115-55	Methods of Testing Varnishes Used for Electrical Insulation	Tests for electrical properties, percentage of solids, viscosity, drying time, and flexibility
MIL-STD-202	Test Methods for Electronic & Electrical Component Parts	Uniform methods for testing electronic and electrical component parts such as capacitors, resistors, switches, relays, transformers, and jacks, only for small parts such as transformers and inductors weighing up to 300 lb
FED. STD. No. 406	Plastics, Methods of Testing	A compilation of methods for testing plastics
MIL-I-16923	Insulating Compound, Electrical Embedding	Descriptions of and references to many electrical and other tests applicable to plastics and coatings
MIL-STD-810A	Military Standard Environmental Test Methods and Engineering Guidelines	A compilation of methods previously described in a variety of other specifications, such as MIL-E-5272, MIL-T-5422, and MIL-E-4970
MIL-E-5400	Electronic Equipment, Aerospace, General Specification for	General requirements for the design and manufacture of airborne electronic equipment for operation, primarily in piloted aircraft
MSC-PA-D-67-13	Apollo Spacecraft Nonmetallic Materials Requirements	Guidelines for selection and acceptance of nonmetallic materials and test requirements for control of flammability and toxicity of nonmetallic materials used in Apollo spacecraft
MIL-T-5422 (ASG)	Testing, Environmental, Aircraft Electronic Equipment	Testing procedures under simulated and accelerated environmental conditions, to meet requirements of MIL-E-5400
ASTM D 268	Methods of Sampling and Testing Volatile Solvents and Chemical Intermediates for use in Paint and Related Coatings and Materials	Specific methods for testing insulation, varnishes, and lacquers
ASTM D 411	Methods of Testing Shellac Used for Electrical Insulation	Specific methods for testing insulation, varnishes, and lacquers
ASTM D 902	Methods of Testing Flexible Resin Coated Glass Fabrics and Glass Fabric Tapes Used for Electrical Insulation	Specific methods for testing insulation, varnishes, and lacquers
ASTM D 1346	Methods of Testing Electrical Insulating Varnishes for 180°C and above	Specific methods for testing insulation, varnishes, and lacquers
ASTM D 2305	Methods of Testing Polymeric Films Used for Electrical Insulation	Specific methods for testing insulation, varnishes, and lacquers
IPC-A-25	Multipurpose Test Board Artwork Film	Printed circuit-board for testing purposes
IPC-B-25	Master Drawing, Multipurpose Test Board	Printed circuit-board for testing purposes

Table 10-7: Abbreviations and Sources of Documents

Abbreviation	Full Name	Source of Document
AATCC	American Association of Textile Chemists and Colorists	American Association of Textile Chemists and Colorists P.O. Box 12215, Research Triangle Park, NC
AMS	Aeronautical Material Specification	Society of Automotive Engineers 400 Commonwealth Drive Warrendale, PA
ANA	Air Force–Navy–Aeronautical	Superintendent of Documents U.S. Government Printing Ofc. Washington, DC
ANSI	American National Standards Institute	American National Standards Institute 1430 Broadway New York, NY
ASTM	American Society for Testing and Materials	American Society for Testing and Materials 1916 Race St. Philadelphia, PA
BM	Bureau of Mines	Superintendent of Documents, U.S. Government Printing Ofc. Washington, DC
CS	Commercial Standard	Superintendent of Documents, U.S. Government Printing Ofc. Washington, DC
ERF	Epoxy Resin Formulators	Society of Plastics Industry 355 Lexington Ave. New York, NY
FED. STD	Federal Standard	General Services Administration 26 Federal Plaza New York, NY 525 Market St. San Francisco, CA 7th and D Streets, S.W. Washington, DC 9th and Market St. Philadelphia, PA
IEC	International Electrotechnical Commission	American National Standards Association 1430 Broadway New York, NY
IPC	Institute for Interconnecting and Packaging Electronic Circuits	Institute for Interconnecting and Packaging Electronic Circuits 3451 Church St. Evanston, IL
Letter(s)-Letter-number	—	Same as for FED. STD
MIL or MIL-STD	Military or Military Standard	Naval Publications and Forms Center Tabor Ave. Philadelphia, PA
MSC or MSFC	Marshall Space Center or Marshall Space Flight Center	Scientific and Technical Information Division, NASA Washington, DC
NASA	National Aeronautics and Space Administration	Scientific and Technical Information Division, NASA Washington, DC
NEMA	National Electrical Manufacturers Association	National Electrical Manufacturers Association 2101 L Street, N.W. Washington, DC

Table 10-8: Specific Test Methods for Coatings

Type of test	ASTM	FED. STD. No. 141	FED. STD. No. 406	Others
Abrasion	D 1044, D 1242		1091	MIL-W-81044, 4.7.5.9
Abrasion resistance	D 968, D 1242, D 658, D 3359	6193 (jet), 6192 (Taber)		MIL-P-15143, 4.4.5; MIL-V-12276, 4.4.8; FED.STD.No. 601, 14111; IPC-TM-650
Adhesion	D 412, D 229 D 2197, D 1876 D 1000	6301.2 (wet tape), 6304.1 (knife test)	1111	MIL-V-12276, 4.6.3; MIL-C-22750, 4.6.3; MIL-C-6799, 4.6.9, 4.6.10, 4.6.11; MIL-C-3254, 4.5.11; MIL-C-16555, 4.5.18, 4.5.17; FED.STD. No. 601, 8031; IPC-TM-650 No. 601. 8031
Aging	D 756, D 1870 D 1046, D 3045			MIL-W-81044. 4.7.5.14; MIL-P-12420. 4.4.1; FED. STD. No. 601, 7001
Ash content	D 1951	5264		
Arc resistance	D 495			MIL-STD-202. Method 303
Bond strength	D 2519			
Brittleness	D 746			MIL-P-149, 4.6.5; FED. STD. No. 101, Methods 230, 231; MIL-C-13785
Capacitance	D 150		4021	MIL-STD-202. Method 305
Chemical and solvent resistance	D 1308, D 543			MIL-V-12276, 4.4.5
Color	D 1544	4252, 4251, 6123 4254.1, 5021, 4241.1		
Conditioning for testing	D 618, D 832			FED. STD. No. 601, Method 5111
Contaminants				MIL-P-23242, 4.6.10
Continuity of coating	D 1676			CS 274, 4.3.5
Corona	D 1868			
Corrosion	D 130, D 849 D 1616			MIL-P-23242, 4.6.17; FED. STD. No. 101, Method 299; VV-S-190, 4.5.4: MIL-STD-883, 2021.3 MIL-R-22628, 4.5.4
Corrosion resistance				MIL-P-149. 4.6.10; MIL-V-12276. 4.4.6
Corrosiveness	D 4350			MIL-C-3254. 4.5.9
Corrosivity index (water-extract conductance)			7071	
Crazing	F 484, F 791		6053	FED. STD. No. 101, Method 329; MIL-C-6799
Deterioration resistance in soil	D 2574, D 3273 D 3274			FED. STD. No. 101. Method 229; MIL-B-10930
Dielectric constant	D 150, D 229, D 669, D 1674, D 3032, D 2305			FED.STD. No. 101, Method 303; MIL-P-22035; MIL-STD-202, Method 301; MIL-1-16923, 4.6.6; MIL-I-46058
Dielectric proof voltage	D 1389			
Dielectric-strength breakdown voltage	D 149, D 1674 D 115, D 3394			MIL-I-2707, 4.7.4; MIL-I-16923, 4.6.4; FED. STD. No. 601, Method 13311; C59-48
Dissipation factor	D 150			MIL-I-16923, 4.6.6
Drying time	D 115, D 1640, D 1346	4061.2		MIL-C-27315, 4.7.6; MIL-R-15184, 4.5.9.3.2.2; MIL-R-15189, 4.5.9.3.2.2; MIL-C-6799, 4.6.5.1

Type of test	ASTM	FED. STD. No. 141	FED. STD. No. 406	Others
Electrical insulation resistance	D 257, D 229			MIL-STD-202, Method 302; (AS) C59.3; IEC 93; IEC 167; MIL-W-81044, 4.7.5.2; MIL-I-46058
Elongation and tensile strength of free films	D 2370			AATCC 76
Epoxy equivalent ...	D 1652			
Epoxy resin content .		7401, 7409		MIL-P-22808, 4.4.4.1
Expansion coefficient, thermal	D 696, D 229, D 1674			MIL-I-16923, 4.6.8; L-F-320, 4.3.6
Exposure (exterior of paints on metals)	D 1006, D 1014, D 4141, D 822, D 4364			
Extraction resistance (of films by chemicals)	D 1239	5151		
Film-forming characteristics ...				MIL-P-45021, 4.7.2; MIL-P-149, 4.6.6
Flame and glow resistance	D 568, D 635, D 747, D 777, D 1433			FED. STD. No. 101, Method 306; MIL-R-5001; MIL-R-16555; MIL-I-46058; MIL-R-17252; MIL-C-17435
Flammability	D 568, D 3713, D 3801, D 757	2021		J-C-98, Method 5211 (horizontal); J-C-98, Method 5221 (vertical); J-C-98, Method 5231 (spark); MIL-STD-202, Method 111 (external flame); MIL-C-16555, 4.5.6
Flash point	D 56, D 115, D 1310, D 92, D 93			FED. STD. No. 101, Method 307; MIL-P-149; MIL-C-13783; FED. STD. No. 810, Method 509
Flexibility	D 790, D 1676, D 522, D 4145, D 2794, D 115	6221, 6223.1		FED. STD. No. 601, Method 11041; MIL-P-15143, 4.6.2; MIL-C-18467, 4.3.3.9; MIL-C-18468, 4.6.7; MIL-P-23242, 4.6.3; MIL-I-46058
Flow properties (strippable coatings)				MIL-C-3254, 4.5.14
Fluorescent penetrant inspection				AMS 2645; MIL-I-6866
Friction				MIL-STD-601, Methods 8021, 8001, 8011
Fungus resistance ..	G 21			MIL-E-5272, 4.8; MIL-STD-810, Method 508.1; MIL-T-5422, 4.8; MIL-P-45021, 4.7.12; MIL-I-631, 4.6.16; MIL-I-46058
Gas transmission rate	D 1434			
Gelation				MIL-C-16555, 4.5.2; MIL-C-3254, 4.5.17
Gel temperature				MIL-P-45021, 4.7.10
Gel time	D 1955, D 2471, D 3056, D 4217			MIL-S-22266, 4.7.6; ERF-2; MIL-R-7575, 4.7.5
Hardness	D 676, D 1674, D 1474, D 2134, D 3363, D 4366, F 500, D 2582, D 2240			FED. STD. No. 101, Method 235; MIL-P-4745; MIL-G-12803; FED. STD. No. 601, Method 3041

Type of test	ASTM	FED. STD. No. 141	FED. STD. No. 406	Others
Heat resistance	D 115, D 2485 D 1932	6051		MIL-C-16555, 4.5.12; J-C-98, Method 4001; MIL-I-16923, 4.6.11; MIL-S-4383, 4.3.2.3
Homogeneity				MIL-P-149, 4.6.3; VV-S-190, 4.5.2
Humidity	D 2247, D 1674, G 21	6201		MIL-STD-202, Method 103; MIL-STD-810, Method 507.1; MIL-C-52210, 4.4.11; MIL-P-45021, 4.7.4; MIL-E-5272, 4.4; MIL-T-5422, 4.4
Hydroxyl number ..		7381		
Ignition properties ..	D 1929, D 3874			
Immersion resistance				MIL-C-18468, 4.6.6; MIL-C-18467, 4.3.3.11; TT-P-57, 4.3.2; MIL-STD-883, 1002
Impact resistance ...	D 3029, D 256			MIL-C-16555, 4.5.9; MIL-P-46056, 4.4.16
Knife test, primer coating	B 449	6304.1		MIL-P-46056, 4.4.11
Life, coating compound				MIL-P-23242, 4.6.2
Life cycle, wire insulation				MIL-W-81044, 4.7.5.13; MIL-W-81381, 4.7.5.12
Lifting properties ...		6251	6041, 6051	MIL-P-12420, 4.4.7; MIL-L-19537, 4.6.16; TT-L-20, 4.6.17
Magnetic particle inspection				AMS 2640; MIL-I-6868
Mar resistance	D 673			
Mildew resistance ..	G 21	6271.2		
Moisture absorption.	D 570			MIL-I-16923, 4.6.13; J-C-98, Method 8211
Moisture resistance..				MIL-STD-202, Method 106; MIL-STD-750, Method 1021.1; MIL-I-46058; MIL-STD-883, 1004.5
Moisture-vapor permeability	D 1434, D 1653 E 96			MIL-STD-202, Method 7032
Nonvolatile content .		4044, 4053.1, 4031		MIL-C-16555, 4.6.1; MIL-S-23586, 4.5.3; MIL-S-8784, 4.5.2; MIL-S-3254, 4.5.4
Oil resistance	D 115			MIL-C-22750, 4.6.5
Ozone resistance ...	D 574, D 2526, D 470			MIL-P-46076, 4.7.3.9
Peel strength	D 3807			MIL-R-23461, 4.6.2.2; MIL-C-27315, 4.7.7
Pencil hardness, primer coating ...	D 3363			MIL-P-27316, 4.5.2; IPC-TM-650
Pigment content	D 2698	4021.1		MIL-C-16555, 4.6.8
Pinholes and cracks in film				L-P-375, 4.3.6
Plasticity				MIL-P-149, 4.6.13
Pot life				ERF-13; MIL-S-23586, 4.5.20; MIL-S-24062, 4.7.3
Power factor and dielectric constant.				MIL-I-2707, 4.7.6; AMS 3780, Note 5
Q factor and dissipation factor				MIL-I-46058, 4.7.4

Type of test	ASTM	FED. STD. No. 141	FED. STD. No. 406	Others
Radiation resistance	D 1677, E 1026, E 1027			
Radiographic inspection				MIL-STD-202, Method 209; MIL-STD-453; AMS 2635; MIL-STD-883, 2012.5
Reflectance, strippable coating				MIL-C-3254, 4.5.18
Refraction, index of	D 542	4371		
Resin content				
Epoxy				MIL-P-46056, 4.4.4.1; MIL-C-46081, 4.4.4.1
Polyamide				MIL-P-46056, 4.4.5.1; MIL-C-46081, 4.4.5.2
Polyester				MIL-P-8013, 4.4.1.2
Vinyl				MIL-C-16555, 4.6.2
Roller coating		2112, 4335		
Sag resistance		4493, 4494.1		MIL-S-24062, 4.7.7, MIL-STD-810, Method 509.1
Salt fog	B 117, D 287			MIL-STD 883, 1009.6
Salt spray	B 117, G 43			MIL-STD-202, Method 1016; FED. STD. 101, Method 322; FED. STD. 151, Method 811.1; MIL-E-5272, 4.6; MIL-C-16555, 4.5.14; FED. STD. No. 810, Method 509
Sand and dust				MIL-E-5272, 4.11; MIL-STD-810, Method 510.1; MIL-T-5422, 4.7
Set time				MIL-R-3745, 4.5; MIL-R-15190, 4.5.10
Shrinkage	F 100, F 636			MIL-C-16555, 4.5.18.4
Softening	D 2134, D 1763, D 1525			
Solids content				MIL-C-18467, 4.3.3.4; MIL-R-3745, 4.4; MIL-S-4383, 4.4.1.1; MIL-V-12276, 4.4.4; FED. STD. No. 175, Method 4021
Solvent resistance	D 543, D 891			MIL-STD-202, Method 215; MIL-P-15143, 4.4.3; MIL-STD-883, 2015.5
Specific gravity	D 792, D 1346, D 115, D 2849 D 1475			MIL-I-16923, 4.6.2; FED. STD. No. 601, Method 14011
Sprayability		4331.1		FED. STD. No. 101, Method 122; MIL-C-3254, 4.5.3; MIL-P-46056, 4.4.10; MIL-C-52210, 4.4.8
Stability				
Material				MIL-C-16555, 4.5.3
Storage				MIL-P-46056, 4.4.8; MIL-C-6799, 4.6.13; MIL-P-52192, 4.4.8
Strippability				MIL-P-23242, 4.6.16; FED. STD. No. 101, Method 273; MIL-C-3254, 4.5.8; MIL-P-149, 4.6.9; MIL-P-45021, 4.7.6
Temperature, altitude				MIL-STD-810, Method 504.1; MIL-E-5272, 4.14; MIL-T-5422, 4.1

Type of test	ASTM	FED. STD. No. 141	FED. STD. No. 406	Others
Tensile strength and elongation				MIL-C-27315, 4.7.5; MIL-P-149, 4.6.12; MIL-P-45021, 4.7.11
Thermal stability (temperature resistance)	D 794, D 1932, D 2307, D 1830			MIL-P-15143, 4.4.7; MIL-C-52210, 4.4.9; MIL-C-16555, 4.5.4; MIL-STD-202, Method 108; MIL-I-2707, 4.7.5
Thermal conductivity	D 1674, C 177, C 518			MIL-I-16923, 4.6.9; AS Z98.1
Thermal shock				MIL-C-18467, 4.3.3.10; FED. STD. No. 101, Method 124, 276; MIL-STD-810, Method 503.1; MIL-STD-202, Method 107; MIL-STD-750, Method 1051.1; MIL-I-46058; MIL-E-5272, 4.3; MIL-STD-883, 1011.5 MIL-E-5272, 4.3
Thickness	D 1005, D 4414 D 1186			J-C-59; FED. STD. No. 151, Methods 520, 521.1
Tension, interfacial . .	D 1708			CS 274, 4.3.4
Toughness	D 256			MIL-P-4640, 4.5.6; MIL-R-15189, 4.5.9.3.2.4
Toxicity				MIL-P-15143, 4.4.9; BM-314
Transparency	D 1746			MIL-P-149, 4.6.7; FED. STD. No. 101, Method 282; MIL-F-22019; MIL-F-22191; MIL-P-45021, 4.7.3
Vibration				MIL-E-5272, 4.7; MIL-STD-810, Method 514.1; MIL-T-5422, 4.2; MIL-C-52210, 4.4.12; MIL-STD-883, 2005.2
Viscosity	D 1545, D 1824 D 1601, D 1243, D 1200, D 562 D 115, D 88 D 301, D 154, D 1084	4287		MIL-I-2707, 4.7.8; MIL-STD-101, Method 280; MIL-C-3254
Volatile matter	D 3960	4042.1		MIL-C-16555, 4.5.11; MIL-P-20689, 4.6.4
Volume resistivity . .	D 150			FED. STD. No. 601, Method 9111; MIL-I-22834, 4.5.3; MIL-I-16923, 4.6.5; MIL-R-21931, 4.3.3.11; MIL-P-13949, 4.6.2.10
Water				
In paints and varnishes	D 1364			
In solvents and thinners				
Water absorption . . .	D 1476			VV-S-190, 4.5.3
Water resistance	D 229, D 570			MIL-C-27315, 4.7.8; MIL-F-8799, 4.5.4.9; TT-L-20, 4.6.15; MIL-P-15184, MIL-R-15189
Water-vapor permeability	E-96			MIL-I-16923, 4.6.14; MIL-C-16555, 4.5.13; MIL-C-3254, 4.5.12
Weathering, accelerated	D 2565, G 27 D 4364	6132	6023	MIL-P-15143, 4.4.6; MIL-C-3254, 4.5.13; MIL-C-16555, 4.5.16

REFERENCES

1. Willard, H. H., L. L. Merritt, and J. A. Dean: "Instrumental Methods of Analysis," Van Nostrand, New York, 1952.
2. Bellamy, L. J.: "The Infra-red Spectra of Complex Molecules," Methuen, New York, 1954.
3. Beach, N. E.: Guide to Test Methods for Plastics and Related Materials, *Plastec Note* 17, Plastics Technical Evaluation Center, N.J., August, 1967.
4. *Insulation,* Directory Encyclopedia Issue, no. 7, sec. 4-4, June–July, 1968.

Appendix

The following tables have been compiled from data on physical, thermal, and electrical properties obtained from the literature and from numerous product bulletins. They should be used only as a general guide in comparing orders of magnitude of differences between various polymer types. Some values are averages of many independent values, whereas others are specific for a particular proprietary formulation. When possible, values from several sources have been given to show either reproducibility or range of results. For the most part, values are for polymer coatings, but where values were not available for coatings per se, those for unfilled casting resins or molding compounds of the same polymer type are given. In all cases, however, the source of information has been given, so that the reader may be able to trace back the exact nature of the material tested and the conditions under which it was tested.

Table A-1: Abbreviations and Symbols Used in Text

Å	Angstrom unit (1 Å = 10^{-8} cm)
AIEE	American Institute of Electrical Engineers
ASG	Aeronautical Service Group
ASTM	American Society for Testing and Materials
AWG	American Wire Gauge
BDMA	Benzyldimethylamine
CA	Chlorendic anhydride
CTFE	Chlorotrifluoroethylene
DEA	Diethylenetriamine
DDSA	Dodecenylsuccinic anhydride
DDS	Diaminodiphenyl sulfone

Table A-1: (continued)

DEAPA	Diethylaminopropylamine
DETA	Diethylenetriamine
DGEBA	Diglycidyl ether of bisphenol A
Dicy	Dicyandiamide
DTA	Differential thermal analysis
ETO	Ethylene oxide
eV	Electron volt
FEP	Fluorinated ethylene propylene
HHPA	Hexahydrophthalic anhydride
hz	Hertz (a unit of frequency equal to cycles per second)
khz	Kilohertz = 1,000 hertz
MDA	Methylene dianiline
MDI	*p, p'*-diphenylmethane diisocyanate
MEK	Methyl ethyl ketone
Mev	Million electron volts
Mhz	Megahertz (10^6 hz)
MIBK	Methyl isobutyl ketone
MPDA	*m*-phenylene diamine
NEMA	National Electrical Manufacturers Association
PA	Phthalic anhydride
PAPI	Polymethylene polyphenylisocyanate
pbw	Parts by weight
phr	Parts per hundred parts of resin
ppm	Parts per million
psi	Pounds per square inch
PVC	Polyvinyl chloride
RTV	Room-temperature vulcanizing
SPE	Society of Plastic Engineers
SPI	Society of the Plastics Industry
TCR	Temperature coefficient of resistance
TDI	Tolylene diisocyanate
TETA	Triethylene tetramine
TFE	Tetrafluoroethylene
TGA	Thermogravimetric analysis
TMA	Trimellitic anhydride or Thermomechanical analysis
torr	Millimeters of mercury pressure
ρ	Resistivity
	An aromatic ring structure such as benzene
	A saturated ring structure such as cyclohexane
⁻W⁻	Indicates a continuation of the polymer
—NCO	An isocyanate group
$\xrightarrow[\Delta]{}$	The application of heat

Table A-2: Thickness Units and Equivalents Commonly Used for Coatings

Mil	Inch	Centimeter	Millimeter	Angstrom	Micron	Microinch
1	0.001	0.00254	0.0254	254,000	25.4	1,000
1,000	1	2.54	25.4	2.54×10^{8}	25,400	1.0×10^{6}
394	0.394	1	10	1.0×10^{8}	1.0×10^{4}	3.94×10^{5}
39.4	0.0394	0.1	1	1.0×10^{7}	1.0×10^{3}	3.94×10^{4}
3.94×10^{-6}	3.94×10^{-9}	1.0×10^{-8}	1.0×10^{-7}	1	1.0×10^{-4}	3.94×10^{-3}
0.0394	3.94×10^{-5}	1.0×10^{-4}	0.001	10,000	1	39.4
0.001	1.0×10^{-6}	2.54×10^{-6}	2.54×10^{-5}	254	0.025	1

Table A-3: Thermal-Conductivity Units and Equivalents

Cal/sec cm °C	BTu/hr ft °F	WMK	W/cm °C
1	241.9	418.6	4.186
4.13×10^{-3}	1	1.73	0.0173
2.39×10^{-3}	0.578	1	0.01
0.239	57.8	100	1

Table A-4: Thermal-Conductivity Data

Material	Conductivity,* cal/(sec) (cm) (°C) x 10^{-4}	Source
Unfilled Plastics		
Acrylic	4–5	a
Alkyd	8.3	a
Depolymerized rubber	3.2	Hardman
Epoxy	3–6	b
Cured with diethylenetriamine	4.8	c
Electrostatic spray coating (Hysol DK-4)	6.6	Hysol
Electrostatic spray coating (No. 5133)	2.9	3M Co.
Epon 828 71.4%: DEA 10.7%	5.2	
Fluorocarbon, Teflon FEP	5.8	Du Pont
Teflon TFE	7.0	Du Pont
Phenolics	3–6	d
Polyamide, nylon 6/6	5.8	d
Nylon 6/10	5.2	d
Polyester	4–5	a
	2–4	d
Polyethylene	8	a
Polyimide, Pyre-M.L. enamel	3.5	e
Pyre-M.L. varnish	7.2	f
Polystyrene	1.73–2.76	g
	2.5–3.3	a
	2.4–3.3	d
Polyurethane, Magna Laminar X-500	2.3	Magna
Polyvinyl chloride	3–4	a
Polyvinyl formal	3.7	a
Polyvinylidene chloride	2.0	a
Polyvinylidene fluoride	3.6	h
Polyxylylene, parylene N	3	Union Carbide
Silicone, flexible casting resin	7	d
Gel coating	3.7	Dow Corning
	7 (150°C)	Dow Corning
RTV type	5–7.5	Dow Corning
Sylgard type	3.5–7.5	Dow Corning
Sylgard varnishes and coatings	3.5–3.6	Dow Corning
Filled Plastics		
Epon 828, cured with diethylene-triamine (A)	4	b
A + 50% aluminum	11	b
A + 50% beryllium oxide	12.5	b
A + 50% silica	10	b
A + 70% aluminum	13	b
A + 70% beryllium oxide	17.8	b
A + 70% silica	12	b
Epoxy, filled (2651 FT)	20.2	Emerson & Cuming
Flexibilized (B)	5.4	i
B + 64% by volume tabular alumina	50.0	i
B + 66% by weight tabular alumina	18.0	i

(continued)

Table A-4: (continued)

Material	Conductivity,* cal/(sec) (cm) (°C) x 10^{-4}	Source
. . . . Other Materials Used in Electronic Assemblies.		
Alumina (94%)	662	j
Alumina ceramic	256–442	a
Alumina (sapphire)	826	p
Aluminum	5664	q
Aluminum nitride	4060–4541	†
Beryllia (99%)	5163	p
Copper	9584	q
Glass, borosilicate	28	k
Potassium soda-lead	18	k
Silica (99.8%)	40	l
Gold	7600	q
Kovar**	395	m
	460	n
Mica	8.3–16.5	a
Nichrome***	320	o
Nickel	2173	q
Palladium	1716	q
Platinum	1711	q
Silicon	3561	p
Silicon carbide	2148	p
Silicon nitride	240–1000	r
Silver	10,253	q
Spinel ($MgO–Al_2O_3$)	281	q

*All values are at room temperature unless otherwise specified.

**Trade name of Western Electric Company. The composition is approximately 28% Ni, 18% Co, and 53% Fe.

***Trade name of Driver-Harris Co. The composition is approximately 80% Ni and 20% Cr.

a)*Materials Eng.,* Materials Selector Issue, Vol. 66, No. 5 (October, 1967).
b)Wolf, D.C., *Proc. Natl. Electron. Packaging Symp., New York,* June, 1964.
c)Lee, H., and K. Neville, "Epoxy Resins," McGraw-Hill, New York, 1957.
d)*Insulation,* Directory Encyclopedia Issue, No. 7 (June–July, 1968).
e)Pyre-M.L. Wire Enamel, *Du Pont Tech. Bull.* 19, August, 1967.
f)Pyre-M.L. Varnish RK-692, *Du Pont Tech. Bull.* 1, April, 1966.
g)Teach, W.C., and G.C. Kiessling: "Polystyrene," Reinhold, 1960.
h)Barnhart, W.S., R.A. Ferren, and H. Iserson: *Seventeenth Ann. Tech. Conf. SPE,* January, 1961.
i)Gershman, A.J., and J.R. Andreotti: *Insulation,* September, 1967.
j)ALSIMAG Ceramics- Chart No. 741, Technical Ceramic Products Div., 3M.
k)Shand, E.B.: "Glass Engineering Handbook," McGraw-Hill, New York, 1958.
l)Kingery, W.D.: "Oxides for High Temperature Applications," McGraw-Hill, New York, 1960.
m)Kohl, W.H.: "Handbook of Materials and Techniques for Vacuum Devices," Reinhold, New York, 1967.
n)Monack, A.J.: *Glass Ind.,* 23: 421 (1946).
o)Lyman, T. (ed.): "Metals Handbook," 8th ed., Vol. 1, American Society for Metals, Metals Park, Ohio, 1961.
p)*Merck Index,* 10th Edition.
q)"Handbook of Chemistry and Physics," 56th Edition, CRC Press, 1975–1976.
r)Werdecker, W. and Aldenger, F.: *IEEE Transactions on Components, Hybrids, and Manufacturing Technology,* Vol CHMT-7, (4), December 1984.

†W.R. Grace

Table A-5: Coefficients of Linear Thermal Expansion

Material	Expansion Coefficient (ppm/°F)*	Source
Unfilled Plastics		
Acrylic	30-50	a
Alkyd	8.2-31.7	a
Diallylphthalate	15-35	a
Epoxy	25-36	b
	28-50	c
Fluorocarbon, Teflon TFE	30	e
	55	a
Neoprene	340	a
Phenolic	33-66	a
	14-33	d
Phenoxy	30-35	a
Polyamide, nylon	55	a
Polyester	28-56	a
Polyethylene	83-167	a
Polyimide, Pyre-M.L. enamel	22	e
Pyre-M.L. varnish	22	c
Polypropylene	170	a
	61	d
Polystyrene	30-40	f
	33-44	a
Polyvinyl chloride	28-33	a
Polyvinyl formal	35.5-42.7	a
	35.6	d
Polyvinylidene chloride	87.8	a
Polyvinylidene fluoride	85	g
Polyxylylene, parylene C	20	h
Parylene N	38	h
Silicone	140-166	b
Other Materials Used in Electronic Assemblies		
Alumina ceramic	3.1-3.7	a
Alumina cermet	4.7-5.2	a
Aluminum	11.7-13.7	a
Aluminum nitride	1.5	l
Beryllia	9.4	i
Beryllium	6.4	a
Beryllium copper	9.3	a
Ceramic, electrical grade	2.0	a
Copper	9.8	a
Gallium arsenide	3.3	m

(continued)

Table A-5: (continued)

Material	Expansion Coefficient (ppm/°F)*	Source
Glass, borosilicate	2.6–2.8	Corning Glass
	1.8–2.5	a
	1.8	j
Lead silicate–soda lime	4.8–5.1	a
Silica	0.3–0.5	a
96% silica	0.44	i Corning Glass
Gold	7.9	a
Kovar	2.55	i
Nichrome	7.5	k
Nickel	6.8–9.2	a
Silicon	1.7	n
Silicon carbide	1.7	o
Silicon-nitride	1.4	a
	2.5	i
Silver	10.7	n
Steatite	3.3–4.0	a

*(ppm/°F)(1.8) = ppm/°C.

a)*Materials Eng.,* Materials Selector Issue, Vol. 66, No. 5, October, 1967-68.
b)Modern Plastics Encyclopedia, 1966.
c)Pyre-M.L. Varnish RK-692, *Du Pont Tech. Bull.* 1, April 1966.
d)*Insulation,* Directory-Encyclopedia Issue No. 7, June-July, 1968.
e)Pyre-M.L. Wire Enamel, *Du Pont Bull.* 19, August, 1967.
f)Teach, W.C., and G.C. Kiessling: "Polystyrene," Reinhold, NY, 1960.
g)Barnhart, W.S., R.A. Ferren, and H. Iserson, *Seventeenth Ann. Tech. Conf. SPE,* January, 1961.
h)Loeb, W.E., and C.E. White: *Proc. Natl. Electron. Packaging Symp.,* Long Beach, California, January 30, 1968.
i)Kohl, W.H.; "Handbook of Materials and Techniques for Vacuum Devices," Reinhold, NY, 1967.
j)MacAvoy, T.C., and E.A. Halaby: Substrates for Thin Film Circuitry, *IEEE Trans. Component Parts,* CP-11: 15-22 (March, 1964).
k)Kern, E.L., and L.A. Teichthesen: Passive and Process Materials for Semiconductor Device Fabrication, *Semiconductor Products Solid State Technol.,* October, 1965.
l)Dettmer, E.S., and Charles, H.K., Jr., *International Journal for Hybrid Microelectronics,* Vol. 10, No. 2, Second Quarter 1987, pp 9–18.
m)*Merck Index,* 10th Edition.
n)"Handbook of Chemistry and Physics," 56th Edition, CRC Press, 1975-76.
o)Werdecker, W. and Aldenger, F.: *IEEE Transactions on Components, Hybrids, and Manufacturing Technology,* Vol. CHMT-7, (4), Dec. 1984.

Table A-6: Water-Absorption Data

Material	Absorption in 24 Hr, %	Source
Acrylic	0.2–0.4 (40°C)	a
HumiSeal 1A14, 1F18, 1B12, 1B15	0–0.18	Columbia
Alkyd	0.4	b
Chlorinated polyether	0.01 (36°C)	a
Depolymerized rubber	0.5	Hardman
Diallylphthalate	0.09	c, d
Diallylisophthalate	0.10	d
Epoxy	0.04, 0.11–1.84	c, e
Cured with diethylenetriamine	0.11–0.12	c
HumiSeal 2A51, 2B13, 2B52, 2H55	0.1–0.31	Columbia
Fluorocarbon	0.01	a
Teflon FEP dispersion coating	<0.01	Du Pont
Teflon TFE dispersion coating	<0.1	Du Pont
Neoprene	0.05–0.35 (39°C)	a
Phenolic	0.15–0.6	a
	0.9	c
	0.1–0.2	d
Phenoxy	0.13	f
Polyamide	1.6–3.3 (38°C)	a
Polybutadiene	0.05–0.5 (40°C)	a
Polyester	0.15–2.5	d
Copolymerized with styrene	0.17–0.28	g
Polyethylene	0.01	c
	<0.01–0.02	d
Low density	<0.01 (37–39°C)	a
High density	<0.01 (37–39°C)	a
Polyimide	0.7	Union Carbide
Polyimide-amide	0.5–1.9 (in boiling water)	
Polypropylene	0.03	d
Polystyrene	0.03–0.3 (39°C)	a
	0.04	c
	0.03–0.05	d
Polysulfide, Thiokol	0.24–1.2 (40°C)	a
Polyurethane, HumiSeal 1A27	0.31	Columbia Technical
Polyvinyl alcohol	>30	a
Polyvinyl chloride	0.03 (38°C)	a
Plasticized	0.4	a
Polyvinyl fluoride, PT-207	0.04	Products Techniques
Polyvinyl formal, Formvar	1.0–1.3	a
Polyvinylidene chloride, Saran	<0.1 (39°C)	a
Polyvinylidene fluoride	0.04	h
Polyxylylene, parylene C	0.06	Union Carbide
Parylene N	0.01	Union Carbide
Silicone	0.15	c
	0.1–0.15	a
Flexible dielectric gel	0.10 (7 days)	Dow Corning
HumiSeal 1H34	0.14	Columbia Technical

* At room temperature, unless otherwise specified.

[a] Baer, E. (ed.): "Engineering Design for Plastics," Reinhold, New York, 1964.

[b] Volk, M. C., J. W. Lefforge, and R. Stetson: "Electrical Encapsulation," Reinhold, New York, 1962.

[c] Piser, J.: *Electronic Design*, vol. 17 (Aug. 16, 1967).

[d] *Insulation*, Directory Encyclopedia Issue, no. 7, June-July, 1968.

[e] *Celanese Tech. Bull.* 1164.

[f] Lee, H., D. Stoffey, and K. Neville: "New Linear Polymers," McGraw-Hill, New York, 1967.

[g] Parker, E. E. and E. W. Moffett: *Ind. Eng. Chem.*, **46;** 1615, 1954.

[h] Barnhart, W. S., R. A. Ferren, and H. Iserson: *Seventeenth Ann. Tech. Conf. SPE*, January, 1961.

Table A-7: Moisture-Vapor-Transmission Rates of Plastic Coatings and Films*

Coating or film	Moisture-vapor transmission rate, g/(mil)(100 in.^{2}) (24 hr)†	Source of data
Acrylic	35.5 (40°C)	a
Cellophane, type PVD uncoated film	134	Du Pont
Cellulose acetate, film	219	Du Pont
Epoxy, cured with anhydride	2.4	Autonetics
Cured with aromatic amine	1.8	Autonetics
Fluorocarbon, Teflon FEP	0.46 (40°C)	a
Teflon TFE	0.32 (40°C)	a
Teflon TFE dispersion coating	0.2	Du Pont
Mylar	1.9 (39°C)	a
Neoprene	15.5 (39°C)	a
Phenoxy	3.5	b
Polycarbonate	10	FMC
Polyethylene		
Low density	0.9–1.8 (37–39°C)	a
High density	2.6–0.43 (37–39°C)	a
Film	0.97	Dow Chemical
Polypropylene, Olefane	0.70	Avisun
Polystyrene	8.6 (39°C)	a
Polysulfide, Thiokol	43 (40°C)	a
Polyurethane		
Isocyanate–castor oil	4.3	Autonetics
Isocyanate polyester	8.7	Autonetics
Magna X-500	2.4	Autonetics
Polyvinyl fluoride, PT-207	0.7	Products Techniques
	2.97 (40°C)	a
Polyvinyl formal	7.1	a
Polyvinylidene chloride, Saran	0.097–0.45 (39°C)	a
Polyxylylene, parylene C	1	Union Carbide
Parylene N	14	Union Carbide
Silicone, gel coating	112.5	Autonetics
Methyl phenyl	38.3	Autonetics
RTV type	120.8	Autonetics
Semiconductor-junction coatings	14–94	Autonetics
Silicone alkyd	4.5–7.9	Autonetics

* At 25°C unless otherwise specified.

† g/(mil)(100 in.^{2})(24 hr) = g/(mil)(m^2)(24 hr) × 0.0645.

[a] Baer, E. (ed.): "Engineering Design for Plastics," Reinhold, New York, 1964.

[b] Lee, H., D. Stofey, and K. Neville: "New Linear Polymers," McGraw-Hill, New York, 1967.

Table A-8: Sward Hardness Values for Coatings

Coating	Sward hardness value (polished-glass standard = 100)	Source of data
Acrylic	24	[a]
	~20	[b]
Sherwin-Williams Kemacryl M49WC15	12	Autonetics
Alkyd	24	[a]
TT-E-489B, Class A*	13	Autonetics
TT-E-529, Class A*	7	Autonetics
TT-E-527*	3–4	Autonetics
MIL-E-7729, Type I*	8	Autonetics
MIL-E-7729, Type III†	8–9	Autonetics
MIL-E-15090, Type III, Class I*	5–7	Autonetics
Chlorinated rubber	24	[a]
	12	[c]
Epoxy	22	[c]
Epoxy amine	36	[a]
Epoxy ester	30	[a]
Epoxy furane	24	[a]
Epoxy melamine	36	[a]
Epoxy phenolic	44	[a]
Product Techniques PT-401*	18–22	Autonetics
Epoxy urea	34	[a]
Fluorocarbon, air dried	<10	[a]
Nitrocellulose, MIL-L-7178	11–12	Autonetics
Neoprene	<10	[a]
Phenolic	38	[a]
Polyester	30	[a]
Polyethylene, chlorosulfonated, Du Pont Hypalon	<10	[a]
Polyurethane	35–65	[a]
ASTM Type 1, oil-modified	9–58	[f]
ASTM Type 2, moisture-cured	12–67	[d, f]
ASTM Type 5	8–77	[f]
ASTM Type 4, catalyzed	12	[f]
Clear, MDI–castor oil cured	17	[e]
Clear, TDI–castor oil cured	10	[e]
Two-component, polyester-cured	68	[c]
Two-component, castor oil–cured	48	[c]
Polyvinyl	20	[a]
	6	[c]
Polyvinyl alkyd (1:1)	26	[a]
Silicone	16	[a]
Viton	<10	[a]

* Cured 7 days at room temperature.

† Cured 15 min at 160°C.

[a] *Materials Engineering*, Materials Selector Issue, vol. 66, no. 5, October, 1967.

[b] Teach, W. C., and G. C. Kiessling: "Polystyrene," Reinhold, New York, 1960.

[c] *Chem. Eng.*, Feb. 9, 1959, pp. 144–148.

[d] Dombrow, B. A.: "Polyurethanes," Reinhold, New York, 1965.

[e] Patton, T. C.: *J. Paint Technol.*, vol. 39, no. 512, September, 1967.

[f] *Spencer Kellogg Chem. Prod. Chart*, 1968.

Table A-9: Abrasion-Resistance Values for Organic Coatings

Coating	Taber wear index, mg	
	1,000 cycles	100 cycles
Alkyd	147	
Enamel cured 45 min at 120°C (TT-E-508)	51	
Enamel (cured 24 hr at room temperature, TT-E-508)	70	
Chlorinated rubber	200–220	
Epoxy, air-cured (PT-401)	208	
Amine-cured varnish	38	
Ester enamel	196	
Polyamide enamel	95	
Teflon-filled (PT-401)	122	
Teflon-filled, bake-type (PT-201)	136	
Epoxy polyamide, coating (1:1)	50	
Nitrocellulose lacquer, clear	96	
Phenolic, spar varnish	172	
Polyamide		29–31
Polyethylene		36
Polyurethane, ASTM Type 1	55–67	
ASTM Type 2, clear	8–24	
ASTM Type 2, pigmented	31–35	
ASTM Type 5	60	
Oil varnish	155	
Polyvinyl	85–106	
Polyvinylidene fluoride	17.6	
Polyxylylene, parylene C		4.4
Parylene D		30.5
Parylene N	9.7	1.6–2.6
Silicone, white enamel		113

Table A-10: Dielectric Strength*

Material*	Dielectric Strength (volts/mil)	Comments	Source
Polymer Coatings			
Acrylic	450–550	Short-time method	a
	350–400	Step-by-step method	a
	400–530	–	b
	1,700–2,500	2-mil-thick samples	Columbia
Alkyd	300–350	–	b
Diallylisophthalate	422	Step-by-step method	a
Diallylphthalate	275–450	–	b
	450	Step-by-step method	a
Epoxy	1,300	10-mil-thick dip coating	–
Cured with anhydride—castor oil adduct	650–730	–	Autonetics
Modified	1,200–2,000	2-mil-thick sample	Columbia
Fluorocarbon, Teflon FEP dispersion coating	4,000	1.5-mil-thick sample	Du Pont
Teflon TFE	400	60-mil-thick sample	c
	480	Short-time method	a
	430	Step-by-step method	a
Teflon TFE dispersion coating	3,000–4,500	1- to 4-mil thick sample	Du Pont
Neoprene	150–600	Short-time method	a
Phenolic	300–450	–	b
Polyamide	780	106-mil-thick sample	–
Polyamide imide	2,700	–	–
Polyester	250–400	Short-time method	a
	170	Step-by-step method	a
Polyether, chlorinated	400	Short-time method	a
Polyethylene	480	–	b
	300	60-mil-thick sample	c
	500	Short-time method	a
Chlorosulfonated	500	Short-time method	a
Polyimide	3,000	Pyre-M.L. 10-mil-thick sample	d
	4,500–5,000	Pyre-M.L. (RC-675)	e
	560	Short-time method, 80-mil-thick sample	a
Polypropylene	750–800	Short-time method	a
Polystyrene	500–700	Short-time method	a
	400–600	Step-by-step method	a
	450	60-mil-thick sample	c
Polysulfide	250–600	Short-time method	a
Polyurethane one-component	3,800	1-mil-thick sample	f
One-component (HumiSeal 1A27)	2,500	2-mil-thick sample	Columbia
Two-component, castor-oil cured	530–1,010	–	g
Two-component, 100% solids (PR-1538)	275	125-mil sample	Products Research
	750	25-mil sample	–

(continued)

Table A-10: (continued)

Material*	Dielectric Strength (volts/mil)	Comments	Source
Polyvinyl butyral	400	–	–
Polyvinyl chloride	300–1,000	Short-time method	
	275–900	Step-by-step method	–
Polyvinyl formal	860–1,000	–	b
Polyvinylidene fluoride	260	Short-time method, 500 volts/sec, 1/8 in sample	h
	1,280	Short-time method 500 volts/sec, 8-mil sample	h
	950	Step-by-step method (1-kvolt steps)	h
Polyxylylene, parylene C	3,700	Short-time method	Union Carbide
	1,200	Step-by-step method	Union Carbide
Parylene D	5,500	Short-time method	Union Carbide
	4,500	Step-by-step method	Union Carbide
Parylene N	6,000	Step-by-step method	Union Carbide
	6,500	Short-time method	Union Carbide
Rubber, depolymerized	360–380	–	Hardman
Silicone, flexible dielectric gel	800	–	Dow Corning
HumiSeal 1H34	1,500	2-mil-thick sample	Columbia
RTV types	550–650	–	G.E., Stauffer
Sylgard 182	500	–	Dow Corning
Other Materials Used in Electronic Assemblies			
Alumina ceramics	200–300	–	b
Alumina (sapphire)	1,200	–	i
Aluminum nitride	127–356	–	i
Beryllia	240	–	i
Boron nitride	900–1,400	–	–
Electrical ceramics	55–300	–	b
Forsterite	250	–	b
Glass, borosilicate	4,500	40-mil-thick sample	c
Silicon carbide	18	–	i
Steatite	145–280	–	b

*Samples are standard 125 mils thick unless otherwise specified.

a) *Insulation, Directory,* Encyclopedia Issue, 7, June–July, 1968.
b) *Materials Eng.,* Materials Selector Issue, Vol. 66, No. 5 (October, 1967).
c) Kohl, W.H.: "Handbook of Materials and Techniques for Vacuum Devices," Reinhold, New York, 1967.
d) Learn, J.R., and M.P. Seegers: Teflon-Pyre-M.L. Wire Insulation System, *Thirteenth Symp. Tech. Progress Com. Wire Cables, Atlantic City, NJ,* Dec. 2–4, 1964.
e) Milek, J.T.: Polyimide Plastics: A State of the Art Report, *Hughes Aircraft Rept.* S-8, October, 1965.
f) *Hughson Chem. Bull.* 7030A.
g) *Spencer-Kellogg Bull.* TS-6593.
h) Kynar Vinylidene Fluoride Resin, *Pennsalt Chem. Corp. Prod. Sheet* KI-66a, 1967.
i) Dettmer, E.S. and Charles, H.K., Jr., *International Journal for Hybrid Microelectronics,* Vol. 10, No. 2, Second Quarter 1987, pp 9–18.

Table A-11: Arc Resistance

Material	Arc resistance, secs	Source
Acrylic	35–55	a
Diallylisophthalate	123–128	b
Diallylphthalate	118	b
Epoxy	45–120	b
Epocast 202/D40 (A)	64	d
A + 100% alumina	78	d
A + 100% gypsum	102	d
A + 75% mica	79	d
A + 100% silica	124	d
Modified	45–55	a
Epoxy polyamide, 50% epoxy, 50% Versamide 115	76	c
60% epoxy, 40% Versamide 125	82	c
Fluorocarbon, Teflon FEP	>300	e
Teflon TFE	>300	e
Phenolic	Tracks	b
Polyamide, nylon 6/6	140	b
Nylon 6/10	130	b
Polyimide	183	f
Polystyrene	60–80	b
Polyurethane, one-component, air-dried	40	a
Two-component, castor oil-cured	88–140	g
Polyvinylidene fluoride	50–70	h
Silicone	120–200	Dow Corning
RTV types	90–130	Dow Corning

[a] HumiSeal Protective Coatings, *Columbia Tech. Corp. Tech. Data Sheet.*
[b] *Insulation*, Directory-Encyclopedia Issue, no. 7, June–July, 1968.
[c] Lee, H., and K. Neville: "Handbook of Epoxy Resins," McGraw-Hill, New York, 1967, page 10-8.
[d] Delmonte, J.: *Plastics Technol.*, March, 1958.
[e] Teflon for Electrical and Electronic Systems, *Du Pont Tech. Bull.*, 1964.
[f] Milek, J. T.: Polyimide Plastics: A State of the Art Report, *Hughes Aircraft Rept.* S-8, October, 1965.
[g] *Spencer Kellogg Bull.* TS-6593.
[h] Kynar, *Pennsalt. Chem. Corp. Prod. Sheet* KI-66a.

Table A-12: Volume Resistivities

Material	Volume resistivity, ohm-cm at 25°C	Source
Acrylic	10^{14}–10^{15}	a
	$>10^{14}$	b
	7.6×10^{14}–1.0×10^{15}	Columbia
Alkyd	10^{14}	a
Diallylphthalate	10^{8}–2.5×10^{10}	a
Epoxy, cured with diethylenetriamine	2×10^{16}	c, d
Epoxy polyamide	1.1–1.5×10^{14}	
Fluorocarbon, Teflon FEP	$>2 \times 10^{18}$	b
Teflon TFE	$>10^{18}$	b
Phenolic	6×10^{12}–10^{13}	a
Polyamide	10^{13}	
Polyamide imide	7.7×10^{16}	e
Polyether, chlorinated	10^{15}	b
Polyethylene	$>10^{16}$	
Chlorosulfonated	10^{14}	b
Polyimide	10^{16}–10^{18}	f
Polypropylene	10^{10}–$>10^{16}$	a
Polystyrene	$>10^{16}$	b
Polysulfide	2.4×10^{11}	g
Polyurethane, one-component	5.5×10^{12}	h
	2.0×10^{12}	h
	4×10^{13}	Columbia
Two-component (PR-1538)	1×10^{13}	Products Research
	5×10^{9} (150°C)	Products Research
Polyvinyl chloride	10^{11}–10^{15}	b
Polyvinylidene chloride	10^{14}–10^{16}	a
Polyvinylidene fluoride	2×10^{14}	h
Polyxylylene, parylene	10^{16}–10^{17}	Union Carbide
Rubber, depolymerized	1.3×10^{13}	Hardman
Silicone, flexible clear	2×10^{15}	Dow Corning
Flexible dielectric gel	1×10^{15}	Dow Corning
HumiSeal 1H34	3.3×10^{14}	Columbia
RTV types	6×10^{14}–3×10^{15}	Stauffer

[a] *Materials Eng.*, Materials Selector Issue, vol. 66, no. 5, October, 1967.
[b] *Insulation,* Directory-Encyclopedia Issue, no 7, June–July, 1968.
[c] Lee, H., and K. Neville: "Epoxy Resins," McGraw-Hill, New York, 1957.
[d] Tucker, Cooperman, and Franklin: Dielectric Properties of Casting Resins, *Electronics Equipment*, July, 1956.
[e] Freeman, J. H.: A New Concept in Flat Cable Systems, *Fifth Ann. Symp. Advanced Techniques Aircraft Elec. Syst., Washington, D.C.*, October, 1964.
[f] Milek, J. T.: Polyimide Plastics: A State of Art Report, *Hughes Aircraft, Rept.* S-8, October, 1965.
[g] Hockenberger, L.: *Chemie-Ing. Technol.*, vol. 36, 1964.
[h] *Hughson Chem. Tech. Bull.* 7030A; Kynar, *Pennsalt Chem. Prod. Sheet* KI66a, 1967.

Table A-13: Dielectric Constants of Polymer Coatings (at 25°C)

Material	60 to 100 Hz	10^6 Hz	$>10^6$ Hz	Source
Acrylic		2.7–3.2		a
Alkyd			3.8 (10^{10} hz)	b
Asphalt and tars			3.5 (10^{10} hz)	b
Cellulose acetate butyrate		3.2–6.2		a
Cellulose nitrate		6.4		a
Chlorinated polyether	3.1	2.92		a
Chlorosulfonated polyethylene, Hypalon	6.19	~5		Du Pont
	7–10 (10^3 hz)			c
Diallylisophthalate	3.5	3.2	3 (10^8 hz)	a,c
Diallylphthalate	3–3.6	3.3–4.5		a,c
Epoxy, anhydride–castor oil adduct	3.4	3.1	2.9 (10^7 hz)	Autonetics
Cured with diethylenetriamine	4.1	4.2	4.1	e
Cured with dodecenylsuccinic anhydride (100:132 pbw)	2.82			d
Cured with metaphenylenediamine	4.6	3.8	3.25 (10^{10} hz)	e
Cured with methylnadic anhydride (100:84 pbw)	3.31			d
One-component	3.8	3.7		Conap
One-component	3.8	3.5		Conap
Two-component	3.7			Conap
Two-component	3.3	3.1		Conap
Epoxy polyamide, 40% Versamid 125, 60% epoxy	3.37	3.08		e
50% Versamid 125, 50% epoxy	3.20	3.01		e
Fluorocarbon, Teflon FEP dispersion coating	2.1 (10^3 hz)			Du Pont
TFE	2.0–2.08	2.0–2.08		Du Pont
Teflon TFE dispersion coating	2.0–2.2 (10^3 hz)			Du Pont
Phenolic		4–11		a
	5–6.5	4.5–5.0		c
Polyamide	2.8–3.9	2.7–2.96		
Polyamide-imide	3.09	3.07		
Polyester	3.3–8.1	3.2–5.9		c
Polyethylene		2.3		a
	2.3	2.3		c
Polyimide, Pyre-M.L. enamel	3.8	3.8		f
Du Pont RC-B-24951	3.0 (10^3 hz)			g
Du Pont RC-5060	2.8 (10^3 hz)			g
Du Pont RK-692 varnish	3.8			g
Polypropylene		2.1		a
	2.22–2.28	2.22–2.28		c
Polystyrene	2.45–2.65	2.4–2.65	2.5 (10^{10} hz)	b
Polysulfides	6.9			h
Polyurethane, one-component	4.10	3.8		Conap
Two-component (PR-1538)	6.8 (10^3 hz)	4.4		Products Research
Two-component, castor oil–cured		2.98–3.28		i
Polyvinyl butyral	3.6	3.33		a

Table A-13: (continued)

Material	60 to 100 Hz	10^6 Hz	$>10^6$ Hz	Source
Polyvinyl chloride	3.3–6.7	2.3–3.5	. . . .	[a]
Polyvinyl chloride–vinyl acetate copolymer	3–10	. . . .	. . . .	[a]
Polyvinyl formal	3.7	3.0	. . . .	[a]
Polyvinylidene chloride	. . . .	3–5	. . . .	[a]
Polyvinylidene fluoride	8.1	6.6	. . . .	[j]
	8.4	6.43	2.98 (10^9 hz)	[k]
Polyxylylene, parylene C	3.10	2.90	. . . .	Union Carbide
Parylene D	2.84	2.80	. . . .	Union Carbide
Parylene N	2.65	2.65	. . . .	Union Carbide
Rubber, depolymerized	4.1–4.2	3.9–4.0	. . . .	Hardman
Shellac, natural, dewaxed	3.6	3.3	2.75 (10^9 hz)	[l]
Silicone, flexible dielectric gel.	3.0	. . . .		Dow Corning
RTV types	3.3–4.2	3.1–4.0	. . . .	G.E., Stauffer
Sylgard type	2.88	2.88	. . . .	Dow Corning
Wax, paraffinic	2.25	2.25	2.22 (10^{10} hz)	[l]

[a] *Materials Eng.*, Materials Selector Issue, vol. 66, no. 5, October, 1967.
[b] Volk, M. C., J. W. Lefforge, and R. Stetson: "Electrical Encapsulation," Reinhold, New York, 1962.
[c] *Insulation,* Directory Encyclopedia Issue, no. 7, June–July, 1968.
[d] Coombs, C. F.: "Handbook of Printed Circuits," McGraw-Hill, New York, 1967.
[e] Lee, H., and K. Neville: "Handbook of Epoxy Resins," McGraw-Hill, New York, 1967.
[f] Learn, J. R., and M. P. Seegers: Teflon-Pyre-M.L. Wire Insulation System, *Thirteenth Symp. Tech. Progr. Com. Wire Cable, Atlantic City, N.J.*, Dec. 2–4, 1964.
[g] Experimental Polyimide Insulating Varnishes, RC-B-24951 and RC-5060, *Du Pont Bull.* H65-4, January, 1965.
[h] Hockenberger, L.: *Chemie-Ing. Technol.*, vol. 36, 1964.
[i] *Spencer Kellogg Bull.* TS-6593.
[j] Barnhart, W. S., R. A. Ferren, and H. Iserson: *Seventeenth Ann. Tech. Conf. SPE,* January, 1961.
[k] Kynar, *Pennsalt Chem. Corp. Prod. Sheet* K166a, 1967.
[l] Von Hippel, A. (ed.): "Dielectric Materials and Applications," M.I.T., Cambridge, Mass., 1961.

Table A-14: Dielectric Constants of Materials Other Than Polymers*

Material	Condition	Dielectric constant
Air, gas	760mm, 20°C	1.000590
Ammonia, liquid	760mm, -78°C	25
Benzene	760mm, 20°C	2.284
Carbon tetrachloride	760mm, 20°C	2.238
Methanol	760mm, 20°C	33.62
Acetone	760mm, 25°C	20.7
Isopropanol	760mm, 25°C	18.3
Hexane	760mm, 20°C	1.890
Toluene	760mm, 25°C	2.379
Water	760mm, 25°C	79.45

Table A-15: Dielectric Constants of Inorganic Insulators

Material	Dielectric Constant (at 1 MHz)	Source
Alumina		
ceramic	4.5-8.4	a
sapphire	11.4	*b*
Aluminum nitride	8.8-10	*b*
Beryllia	6.5	*b*
Boron nitride	4.2	c
Corderite	4.5-5.4	a
Forsterite	6.2	a
Gallium arsenide	11	*d*
Glass		
Aluminum silicate	6.3	c
Borosilicate	4.0-5.1	c
Silica (quartz, fused)	3.75-4.1	a
Silicon carbide	40	*b*
Silicon	13	*d*
Steatite	5.5-7.5	a

a "Handbook of Chemistry and Physics," 56th Edition, CRC Press, 1975-1976.
b Dettmer, E. S. and Charles, Jr., H. K.: *International Journal for Hybrid Microelectronics*, Vol 10 No 2, Second Quarter 1987, pp9-18.
c Materials Selector Issue, *Materials Eng., vol. 66, no. 5, October 1967.*
d "Merck Index", 10th Edition.

Table A-16: Dissipation Factors (at 25°C)

Material	60 to 100 hz	10^6 hz	>10^6 hz	Source
Acrylics	0.04–0.06	0.02–0.03	–	a
Alkyds	0.003–0.06	–	–	a
Alumina, 99.5%	–	0.0001	–	h
Aluminum nitride	0.0007–0.002	–	–	i
Beryllia, 99.5%	–	0.0003	–	h
Diallylisophthalate	0.008	0.009	0.014 (10^{10} hz)	b
Diallylphthalate	0.010	0.011	0.011	b
Epoxy, cured with anhydride—castor oil adduct	0.0084	0.0165	0.0240	Autonetics
One-component	0.011	0.004	–	Conap
	0.008	0.006	–	Conap
Two-component dip coating	0.027	0.018	–	Conap
Epoxy polyamide, 40% Versamid 125, 60% epoxy	0.0085	0.0213	–	c
50% Versamid 115, 50% epoxy	0.009	0.0170	–	c
Fluorocarbon, Teflon FEP	<0.0003	<0.0003	–	a
Teflon FEP dispersion coating	0.0002–0.0007	–	–	Du Pont
Teflon TFE	<0.0003	–	–	a
	0.00012	0.00005	–	Union Carbide
Glass, borosilicate	–	0.013–0.016	–	Corning Glass
Silica	0.0006	0.0001	0.00017 (10^{10} hz)	d
96% silica	–	0.0015–0.0019	–	Corning Glass
Phenolic	0.005–0.5	0.022	–	a
Polyamide	0.015	0.022–0.097	–	–
Polyester	0.008–0.041	–	–	a
Polyether, chlorinated	0.01	0.01	–	a
Polyethylene, chlorosulfonated	0.03	0.07 (10^3 hz)	–	b
Linear	0.00015	0.00015	0.0004 (10^{10} hz)	d
Polymethylmethacrylate	0.06	0.02	0.009 (10^{10} hz)	d
Polystyrene	0.0001–0.0005	0.0001–0.0004	–	–
Polyurethane, One-component	0.038–0.039	0.068–0.74	–	Conap
	0.02	–	–	Conap
Two-component, castor oil-cured	–	0.016–0.036	–	e
Polyvinyl butyral	0.007	0.0065	–	–
Polyvinyl chloride	0.08–0.15	0.04–0.14	–	a
Plasticized	0.10	0.15	0.01 (10^{10} hz)	d
Polyvinyl chloride-vinyl acetate copolymer	0.6–0.10	–	–	–
Polyvinyl formal	0.007	0.02	–	–
Polyvinylidene fluoride	0.049	0.17	–	f
	0.049	0.159	0.110	g
Polyxylylenes, parylene C	0.02	0.0128	–	Union Carbide
Parylene D	0.004	0.0020	–	Union Carbide
Parylene N	0.0002	0.0006	–	Union Carbide
Rubber, depolymerized	0.007–0.013	0.0073–0.016	–	Hardman
Silicon carbide	0.050	–	–	i
Silicone, flexible clear	0.001	–	–	Dow Corning
Flexible dielectric gel	0.0005	–	–	Dow Corning
RTV types	0.011–0.02	0.003–0.006	–	G.E.
Sylgard type	0.001	0.001	–	Dow Corning

a)*Machine Design,* Plastics Reference Issue, Vol. 38, No. 14 (1966).
b)*Insulation,* Directory-Encyclopedia Issue, No. 7, June-July, 1967.
c)Lee, H., and K. Neville: "Handbook of Epoxy Resins," McGraw-Hill, New York, 1967.
d)Mathes, K.: *Elec. Insulation Conf.,* 1967.
e)*Spencer Kellogg, Div. Textron Corp. Tech. Bull.* TS-6593.
f)Barnhart, W.S., R.A. Ferren, and H. Iserson: *Seventeenth Ann. Tech. Conf. SPE,* January, 1961.
g)Kynar, *Pennsalt Chem. Corp. Prod. Sheet* K166a, 1967.
h)*Machine Design,* Design Guide, September 28, 1967.
i)Dettmer, E.S., and Charles, H.K., Jr., *International Journal for Hybrid Microelectronics,* Vol. 10, No. 2, Second Quarter 1987, pp 9–18.

Index

Ablative coatings
epoxies - 36
silicones - 42
Abrasion resistance - 135–136
alkyds - 2
epoxies - 2, 31, 97
fluorocarbons - 70
parylenes - 78, 80
phenolics - 2, 260
polyamides - 97
polyethylenes - 97
polyurethanes - 1, 2, 14, 15, 97
polyvinyl chloride - 92
polyvinyl fluoride - 92
polyvinylformal - 260
table of values - 373
tests for - 135
Acrylics - 90, 91, 94, 95, 133, 167, 182
electron beam curing - 188
ultraviolet light curing - 183, 186
Adducts
urethane - 3–5, 8, 14
Adhesion - 318–319
acrylics - 94
anodize - 133
epoxies - 31, 32, 45
wire coatings - 265
Adhesives
die attachment - 111, 112
electrically conductive - 111, 112
Alkyd - 1, 2, 29, 30
corrosion protection - 88
polyesters - 86, 87
radiation effects - 244
thermal stability - 33
Alpha particle barriers
polyimides - 59, 288
silicones - 288
Aluminum - 131, 133, 134
corrosion - 305, 306, 314
Amide-imide - 63–65, 254–258
Amides - 5
Amine curing agents
for epoxies - 21–25
for polyurethanes - 11
for silicones - 39
Anhydride
curing agents for epoxies - 26, 27
halogenated - 27
Anodize coatings - 130–133
Application methods - 162–176
automatic dispensing - 294, 298
brush - 36, 163, 175

cast - 36, 298, 299
dip - 36, 163, 166, 176, 215, 264, 298
flow - 175
fluidized bed - 163, 168–173, 300
impregnation - 163, 172, 173
roller - 163, 175
screen-printing - 164, 173, 174, 210
spin - 164, 173, 174, 184, 215, 298
spray - 36, 162–166, 175, 215
transfer molding - 295–297, 299
vapor deposition - 164, 184–186
Arc resistance - 124
epoxies - 33, 124, 202
phenolics - 100
polystyrene - 93
silicone - 50
table of values - 376

Beryllium - 131
Biocides - 143
Bisphenol A - 16–18, 21, 26, 31, 32, 37
halogenated - 18
Boron trifluoride - 21–23
Breakdown voltage (see Dielectric strength)

Capacitance - 113–115
Capacitor insulation
fluorocarbons - 67
polyvinyl - 91
thin films - 182, 189
Castor oil - 10, 13, 29
Chloride ion contaminants - 314
Chlorinated rubber - 2
Circuit board coatings (see also Printed circuit boards) - 36 42–44, 125, 162–164, 177, 182, 196–210
Cleaning - 150–161, 318
methods - 156–161
plasma - 157, 159
solvents and solutions - 154–156
tests - 161
ultrasonic - 157–159
vapor degreasing - 157
Clean rooms - 154, 156
Cobalt naphthenate - 11–13
Coil - 250–276
polyurethane insulation - 8
silicone impregnant - 42, 43
Contaminants - 150–154
ammonia - 307
catalysts - 316
corrosion effects - 125, 305, 314–318
effect on semiconductor devices - 287, 304, 305
effects on arc resistance - 124
effects on electrical properties - 106, 109, 110
ionic impurities - 106, 110, 151, 154, 203, 305, 308, 310
outgassing - 307
particulates - 154
prevention - 154
removal of - 154–161
in spacecraft - 227–229
tests for - 308
types - 151
Copper - 131, 316, 317
Corrosion
of aluminum - 305
of copper - 316, 317
due to ionic contaminants - 150–154, 305
due to metal particles - 153
due to microorganisms - 140–142
due to moisture - 305, 314–318
due to outgassing - 242
of thin-film resistors - 305
Corrosion resistance/protection - 129–135
acrylics - 94, 133
alkyds - 88
epoxies - 31, 36, 133, 134
epoxy polyamides - 97, 133
fluorocarbons - 67, 69, 70
fluxes - 133

metals - 130–135
polyamides - 97
polyimides - 59, 64
polyurethanes - 9, 133
polyvinyl chloride - 97
polyvinyls - 91
silicones - 42
Cure
completeness of - 268, 269, 308
curing agents - 21
mechanisms - 21–28
schedules - 325, 326
Curing methods
electron bombardment - 188
gamma irradiation - 190, 191
heat curing - 177
infrared - 178
photocuring - 178–187
plasma polymerization - 189–190
step curing - 178

DGEBA - 16, 17
Diallylphthalate - 95–96
Dicyandiamide - 28
Dielectric constant
acrylics - 94, 95, 116
cure conditions - 118–121
definition - 114, 115
diallylphthalate - 115
epoxies - 32, 76, 115, 118–121
fluorocarbons - 67, 68, 72, 73, 115, 116, 238–240
frequency, effect on - 114, 115
parylenes - 76
phenolics - 100, 115
polyamides - 116
polyethylene - 115, 116, 185
polyimides - 58, 61, 65, 66
polystyrenes - 93, 115, 116
polysulfides - 101
polyurethanes - 76
polyvinylchloride - 116
polyvinyl formal - 91
radiation, effect on - 238–240
silicones - 45–47, 49–52, 76, 115
table of values - 378–380
temperature, effect on - 118, 119
Dielectric strength - 122–124
acrylics - 95
epoxies - 32, 76
fluorocarbons - 68, 72, 73, 122, 123
parylenes - 76, 77, 83
phenolics - 100
polyimides - 61, 65
polystyrene - 93
polyurethanes - 76
polyvinyl formal - 91
silicones - 45, 50–52, 76, 123
table of values - 374, 375
temperature, effect on - 123
wire coatings - 266, 267
Diglycidyl ether of Bisphenol A - 16, 17
Dip coating - 166–168, 176, 215
Dissipation factor - 117, 121
acrylics - 94, 95
cure conditions, effect on - 119–121
definition - 117, 118
epoxies - 32, 76, 118–121
fluorocarbons - 67, 68, 72, 238–240
frequency, effect on - 119, 120
parylenes - 76
phenolics - 100
polyimides - 61, 66
polystyrene - 93
polysulfides - 101
polyurethanes - 76
radiation, effect on - 238–240
silicones - 45–47, 49–52, 76
table of values - 381
temperature, effect on - 118, 119
Dye-penetrant test - 314, 315

Electrical properties - 103–124
arc resistance - 124, 376
capacitance - 113, 115
dielectric constant - 114–121, 378–380
dielectric strength - 122–124, 374, 375

dissipation factor - 117–121, 381
impurity, effects on - 105, 106, 109
insulation resistance - 104, 105
moisture, effects on 110–112
surface resistivity - 104, 105, 109–111
volume resistivity - 104–109, 377
Electromagnetic interference (EMI) - 111, 112, 139, 140
Electron-bombardment polymerization - 188
Electrostatic coating - 165–167, 172
Electrostatic discharge (ESD) - 111, 139, 141
Electrostatic spray - 96
Epichlorohydrin - 16, 17, 19
Epoxies - 15–36, 100
abrasion resistance - 31
adhesion - 30, 31, 319, 320
amine-curing - 22–25, 109
anhydride-curing - 22, 26, 27, 109–111
Bisphenol A - 16, 21, 26
catalysts for - 21, 23
chemistry of - 15–30
classifications - 29
curing agents - 21–29, 110, 111
density - 21
electrical properties - 31–33, 76, 105–111, 118, 119
equivalents - 17, 18
esterified - 29, 36
fillers - 33, 145
flame retardant - 18, 27
flexibilizers - 36, 105, 106
fluidized bed - 171
fungus resistance - 141, 142
halogenated - 18
latent curing agents - 28
Lewis catalysts - 23
moisture resistance - 31, 109, 110
plastic packaging - 280–283
polyamide - 27, 28, 31, 36, 105, 106
polysulfide - 101, 116, 117, 317
primers - 135
properties - 30–37, 167
purity - 309
radiation resistance - 242, 244
reworkability - 209, 210
salt spray resistance - 31
screen-printable - 173
thermal conductance - 145
thermal stability - 19, 21–23, 27, 31, 33–35
vibration damping - 205
Epoxy polyamide - 2, 96, 97, 105, 106, 126, 205
Epoxy resins
Bisphenol A types - 16–18, 21
list of and suppliers - 17, 18
novolac types - 18, 19
peracetic acid types - 19–21
polyolefin types - 20
ESD (see Electrostatic discharge)
Expansion coefficient (see Thermal expansion)

Failure, causes
ionic contaminants - 305, 308, 309
moisture - 109, 110, 116, 200–203
shrinkage stresses - 206
thermal expansion mismatch - 206
vibration - 205
FEP Teflon (see also Fluorocarbons) - 67–70, 72, 171
Fillers
arc resistance, effect on - 124
conductive - 111, 112, 140
EMI gaskets - 140
in epoxies - 33
moisture permeability, effect on - 127
outgassing, effect on - 240
in phenolics - 99
in phenoxies - 37
in polyvinyls - 92
radiation effects - 242
thermal conductance, effect on - 144, 145

Flame retardancy -
epoxies - 18, 27
polyimides - 61
polyvinyls - 92
silicones - 42
Flammability - 231
Flexibilizers - 228, 229
Fluidized bed - 96, 168–172, 300
Fluorocarbons - 65–75, 190
abrasion resistance - 69, 70
chemistry of - 66
corrosion resistance - 69–71
dispersion coatings - 72–74, 155–157
electrical properties - 67, 68, 72, 116, 122, 123
Freon solvents - 155–157
friction coefficient - 69
mechanical properties - 68, 72
properties of - 67, 73
solvent resistance - 69–71
Teflon FEP - 66, 69, 70, 72, 171
Teflon TFE - 65, 66, 69, 70, 72, 109, 122, 171, 188, 190, 272
thermal stability - 67, 68
Flux - 133, 155, 159, 268
Formvar - 90, 91, 240 (see also Polyvinyl formal)
Freons (see Fluorocarbon solvents)
Friction, coefficient of - 136, 137
nylon - 136, 264
parylene - 78, 80
Teflon - 68, 136, 264
various coatings - 137
Fungus resistance - 141, 142, 144

Gamma radiation curing - 190, 191
Gel coatings - 42, 43
Gloh-top coating - 271, 279, 288, 289, 294, 300
Glow-discharge polymerization - 189, 190
Glyptals - 87, 88

Hardness - 136, 137, 139 (see also Sward hardness)
HAST - 314
Hermeticity - 320
High-speed multichip circuits - 221–224
Humidity (see Moisture or Water)
Hybrid microcircuits - 154, 159, 210
cleaning - 159, 161
particle immobilizing - 82, 221, 293, 294
Hybrid wafer scale integration - 221, 288

Impregnants - 14, 28, 36, 42, 43, 252, 262
Infrared curing - 178
Infrared spectrograms - 324
Inspection - 207, 209
Insulation
for capacitors - 67
classes - 45
for magnet wire - 7, 8, 42, 58, 64, 65, 67, 250–252
Insulation resistance - 104–111, 203, 205–207
acrylics - 95, 206
fluorocarbons - 206
parylenes - 78
silicones - 206
Integrated circuit coatings - 42
Interlayer dielectrics
polyimide - 59, 173, 223
Isocyanates - 3–5
blocked types - 7
moisture sensitivity - 9

Junction coatings - 286, 287, 289, 298 (see also Barrier coatings)

Latent curing agents - 28
Lewis catalysts - 21, 23
Linoleic acid - 28, 96
Linseed oil - 29

Magnesium - 130–132, 134
Magnesium-lithium - 133
Magnet-wire (see Wire)
Microbial protection - 140–144

Microelectronic circuits - 188
Moisture
  electrical properties, effects on - 109, 110, 116, 207
  failure mechanisms - 203
  printed circuit boards, effect on - 200, 201, 207
Moisture curing - 13–16
  polyurethanes - 7, 135
Moisture permeability - 125–129
  factors affecting - 129
  filler, effects on - 127
  fluorocarbons - 73
  parylenes - 78
  table of values - 371
  tests for - 313, 314
Moisture resistance - 125–129, 310–319
  epoxies - 30, 32
  fluorocarbons - 73
  parylenes - 78, 82
  polyesters - 86
  polyimides - 65
  polyurethanes - 200, 201
  silicones - 42
Moisture-vapor transmission (see Moisture permeability)
Multilayer circuits - 210, 211, 218

Naphthenates - 11–13, 29, 39
Nickel - 131, 140
Novolac - 18, 19, 32, 312
Nylon - 96, 97, 99, 253–255, 258, 259, 264, 268

Oleoresinous coatings - 253
Outgassing
  alkyds - 33, 34, 36
  epoxies - 33, 34, 36
  fluorocarbons - 68, 69
  phenolics - 33, 307
  polyesters - 33
  polyimides - 61, 62
  radiation-caused - 240, 241
  silicones - 229, 230
  in space environment - 226–231
  thick-film resistors, effect on - 220, 221
  toxicity - 231, 232

Particle getters - 288, 293
Particle immobilizers - 293–294
  perylene - 82, 221, 293
  silicones - 293
Particle impact noise detection - (see PIND test)
Parylene (see also Polyxylylenes) - 74–83, 185–186, 287, 293
Passivation coatings - 291–292
  inorganics - 305, 320
  parylenes - 82
  plasma polymerized - 189, 190
  polyimides - 59, 287, 288
  silicones - 286, 287
Permeability (see Moisture permeability)
Phenol - 7, 97
Phenolics - 2, 30, 31, 61, 63, 96–100
  chemistry - 97–99
  electrical properties - 100
  purity - 307, 309
  radiation effects - 241, 244
Phenoxies - 37
Photocuring - 178–182, 186, 187
  acrylics - 183
  epoxies - 183, 184
  polyimides - 184, 291, 292
  silicones - 184
Photoresists - 173, 178, 184, 211–215
Photosensitive coatings - 61, 211–215, 290–292 (see also Photocuring and Photo-resists)
PIND test - 154
Planarization coatings
  polyimides - 59, 290
Plasma etching - 210
Plasma polymerization - 189, 190
Plasticizers - 141, 142
  radiation effects on - 242
Plastic packaging - 280–285, 300–320
Polyacetylene - 112, 113

Polyamides (see also Nylon) - 22, 27, 28, 31, 96–97, 105, 106, 116, 241, 254
Polybenzimidazole - 64
Polycarbonate - 64
Polyesters - 5, 32, 85–88, 95
  chemistry - 85–87
  corrosion - 86
  hydrolytic stability - 86
  outgassing - 241
  properties - 88, 167, 262
  thermal stability - 261
  wire insulation - 251, 254–262
Polyethylene - 89, 90, 116, 171, 185, 238
Polyimides - 55–65
  barrier coatings - 287
  chemistry - 55–61
  electrical properties - 65, 66, 261, 273
  flame resistance - 61
  high-speed circuits - 223
  interlayer dielectrics - 173, 223, 288, 290–292
  moisture resistance - 57, 65
  outgassing - 61, 62
  patterning - 56, 57, 290
  photosensitive - 61, 290, 291
  properties - 61
  radiation resistance - 240, 261, 272, 273
  solubility - 58
  solvent resistance - 59
  thermal stability - 55, 56, 58, 61–63, 261, 270
  wire insulation - 251, 254, 255, 259, 261
Polymer thick films - 218–220
Polymethylmethacrylate - 89, 116, 241 (see also Acrylics)
Polyols - 6–8
Polyphenylene - 64
Polyphenylether - 64
Polyphenylsulfide - 64
Polypropylene - 89, 90, 171
Polystyrene - 88, 91, 93, 182, 188, 238, 241
  radiation resistance - 243
Polysulfides - 100, 101, 116, 117, 317, 318
Polyurethanes - 1–16, 30
  abrasion resistance - 1, 2, 260
  blocked types - 7, 8
  catalysts for - 10, 11
  chemistry - 2
  circuit board coatings - 196, 198, 199
  classifications - 5
  electrical properties - 10
  fungus resistance - 141
  hardness - 138, 139
  list and suppliers - 16
  material specifications - 327–338
  moisture cured - 7, 13–16, 135
  moisture protection - 200, 201
  oil modified - 12
  plastic packaging - 283, 284
  process specification - 344–347
  reworkability - 209, 210
  salt spray protection - 200, 201
  thermal stability - 33, 234, 235, 270
  toughness - 2
  toxicity - 3
  uralkyds - 12, 13
  vibration damping - 205
  water absorption - 126
  wire insulation - 251, 254, 255, 259, 260, 268, 270
Polyvinyl - 2, 85, 88–93, 171, 182
  acetates - 91, 92, 259
  alcohols - 91, 92
  butyral - 90, 91, 130, 134, 253, 259
  chloride - 90–92, 242
  fluoride - 90, 92
  formal - 90, 91, 240, 241, 251–260, 268, 269, 274
  fungus resistance - 142
  wire insulation - 251–258
Polyvinylidene chloride - 91, 127
Polyvinylidene fluoride - 188
Polyxylylenes - 73–83, 185–186 (see also Parylene)
  abrasion resistance - 78, 80
  applications - 80, 82, 223

chemistry - 73–75
coefficient of friction - 78, 80
coefficient of thermal expansion - 80
corrosion protection - 80
electrical properties - 76–78
moisture permeability - 78
particle immobilizing - 82, 221, 293
permeability to gases - 80
properties - 75–80
reworkability - 79, 210
solvent resistance - 79
thermal conductivity - 80
thermal stability - 79–82
water absorption - 80
Power factor (see Dissipation factor)
Prepreg, epoxy - 28
Pressure-cooker test - 314, 320
Primers - 14, 37, 45, 50, 90, 130, 133–135, 164, 171, 318, 352
Printed circuit boards
acrylic coated - 94, 183, 197
cleaning - 159, 160
coatings - 125, 162–164, 177, 182, 196–210
epoxy coated - 197, 198
fabrication of - 211, 212
multilayer - 210, 211
parylene coated - 80, 82, 199
polyurethane coated - 196, 198–201
reliability - 200–203
silicone coated - 184, 199, 200

Radiation
coatings, effects on - 237–247
nuclear - 237
space environment - 236
wire coatings, effects on - 271–275
Radio frequency interference (RFI) - 139, 140
Reliability - 320
Resistivity (see Volume resistance, insulation resistance)
Resistors
polymer thick film - 218–220
thick film - 133, 219, 221
thin film - 133
Resists
photoresists - 211–215
screen-printed - 215, 216
solder maskants - 216, 217
Rework - 48, 58, 79, 209, 210
RTV (see Silicones)

Salt spray resistance
epoxies - 30, 31, 135
magnesium-lithium - 134
metal-filled epoxies - 112
plastic packaged devices - 316
polyurethanes - 200–202
wash primers - 134
Screen-printing - 173, 210
Semiconductors - 104, 105, 112, 125
barrier coatings for - 182, 279, 286, 287
Silicones - 30, 37–52
adhesion - 319, 320
alkyd types - 41
catalysts for - 48
chemistry - 38–42
circuit board coatings - 199, 200
electrical properties - 45–47, 49, 109, 247
EMI gaskets - 140
fluorosilicones - 43
fungus resistance - 141
gels - 288, 289, 293
halogenated - 42
microcircuit packaging - 284–287, 308
migration - 45
outgassing - 229, 230
particle immobilization - 293
plasma polymerization - 189
properties - 42–52
purity - 307–309
reworkability - 209
RTV types - 39, 48–52, 285, 286

solvent resistance - 233, 234
Sylgards - 41, 46, 47, 50, 109, 247
thermal stability - 33, 41, 42, 44, 49
Silver
corrosion of - 131
EMI coatings - 139, 140
epoxy filler - 112
polymer thick films - 218, 219
Solderability - 268
Solder flux (see Flux)
Solder maskants - 216, 217
Solvent resistance
epoxies - 30
fluorocarbons - 71
parylenes - 79
polyimides - 59, 64
polysulfides - 234
silicones - 233, 234
wire coatings - 267, 268
Solvents
cleaning - 154-156
fluorocarbon - 155-158
particle contaminants in - 154, 155
specifications - 354
Space environment, effects - 226-231, 236-247, 271-275
Specifications - 324-361
acceptance - 339-343
cleaning - 354
coatings - 351
electrical - 348, 349
enamels - 351
finishes - 350
material - 324, 325, 327-338
primers - 352
process - 325-326, 344-347
resins - 354
solvents - 354
sources - 356
surface treatments - 354
testing - 355
test methods - 357-361
varnishes - 353
Spin coating - 164, 173-174, 184, 215
Spray coating - 162-166
Sterilization - 232-235
Stoichiometry - 308
Stresses
resin shrinkage - 301-304
tests - 304
thermal expansion - 301-304
Styrene (see Polystyrene)
Surface resistivity - 109-111
Surface treatment
radiation grafting - 190
specifications - 354
Sward hardness - 137, 138
table of values - 372
Sylgards (see Silicones)

Taber abrasion resistance - 2, 15, 135, 136
Teflon (see TFE Teflon and Fluorocarbons)
Test methods - 204
abrasion resistance - 135, 136
adhesion - 319
continuity of wire coatings - 267
corrosion - 316-318
differential scanning calorimetry - 269
dye penetrant - 314, 315
electrical overload - 268
flammability - 231
humidity - 128, 203, 204, 207, 208
impurities, in coatings - 308-310
impurities, in solvents - 155
moisture transmission - 313
outgassed products - 231
particle impact noise detection - 154
pressure cooker - 314
salt spray - 204
solderability - 268
solvent resistance - 267, 268
specifications - 348, 349, 357-361
surface cleanliness - 161
Sward hardness - 137, 138
thermal - 268, 269, 271

Tetrafluoroethylene (see TFE Teflon and Fluorocarbons)
TFE Teflon (see also Fluorocarbons) - 67-71, 171, 264
  radiation effects on - 238-240
Thermal conductivity - 142-147
  coefficient of - 143
  fluorocarbons - 69
  parylene - 80
  phenolics - 100
  polyimides - 61-63
  silicones - 50-52
  table of values - 366, 367
  test methods - 145-147
  units - 365
Thermal expansion
  fluorocarbons - 69
  measurement - 271
  phenolics - 100
  polyimide - 61, 62
  table of values - 368, 369
Thermal stability
  acrylics - 94
  coatings - 233, 234
  epoxies - 19, 21-23, 27, 31, 33-35, 234
  magnet wire coatings - 250-252
  parylenes - 79, 81, 82
  phenoxies - 37
  polyimides - 55, 56, 58, 61, 62, 234, 270
  polystyrenes - 234, 235
  polyurethanes - 234, 235, 270
  polyvinylformal - 269
  silicones - 41, 42, 44, 45
  wire coatings - 269-271
Thermogravimetric analysis - 271
Thermomechanical analysis - 271
Thermoplastic flow - 266
Thermoplastics - 37, 96, 165, 170, 171, 182, 218, 250, 300
Thick-film circuits - 173, 217, 218
  capacitors - 173, 217
  cermets - 218
  coatings for - 220, 221
  conductors - 173, 217
  polymer thick films - 218-220
  resistors - 173, 217
Thin-film circuits
  coatings for - 217
  deposition methods - 182-191
  failure mechanisms - 305
  properties - 191-192
Tin-lead solder - 133, 268
Tolylene diisocyanate - 3, 260
Toxicity
  of epoxy catalysts - 21-23
  of outgassing products - 231, 232
Transfer molding - 295-299
Transformer coatings - 163, 172, 262
  epoxy - 36, 262
  fluorocarbons - 67
  polyamides - 256
  polyesters - 254-256
  polyimides - 254-256
  polysulfides - 101
  polyurethanes - 254, 256
  polyvinylformal - 254, 255
  silicones - 42, 44

Uralkyds - 12, 13, 15, 16
Ureas - 5
Urethane adducts - 3, 5
UV curing (see Photocuring)

Vacuum impregnation - 172, 173
Vapor deposition - 184-191
  polyethylene - 185
  polyxylylene - 185-186
  Teflon - 185
Vibration damping - 300, 301
Vinyl (see also Polyvinyl) - 2, 85, 88-93, 95, 142
Volume resistivity - 104-109
  epoxies - 31-33, 76, 105-109
  fluorocarbons - 72, 73, 104, 105, 109
  parylenes - 76, 109
  phenolics - 100
  polyethylene - 109
  polyimides - 61
  polystyrenes - 93, 104, 105
  polysulfides - 101
  polyurethanes - 76

polyvinyl chloride - 109
silicones - 45-47, 49, 50, 76, 109
table of values - 377

Wash primers (see Primers)
Water absorption - 125, 126, 311, 312
acrylics - 95
epoxies - 31, 126, 127, 311, 312
epoxy-polyamides - 126
fluorocarbons - 73
nylon - 99
phenolics - 100, 311
polyamides - 97, 99
polyurethanes - 9, 126
polyvinyl acetates - 92
polyvinyl alcohols - 92
silicones - 311, 312
table of values - 370
Water extract resistivity - 308, 309
Wire - 250-276
coatings/insulation - 7, 8, 42, 58, 64, 65, 67, 240
enamesl/varnishes - 90-92, 94, 97, 234, 260-263
magnet wire coatings - 7, 8, 42, 58, 64, 65, 67, 250-252, 274, 275
test methods - 265-271
Wire bonds - 161
corrosion - 314
failure of - 303